T5-AGK-955

Members of The Vincent Memorial Laboratory Staff

RUTH M. GRAHAM

DELORIS C. MacKINNEY

MAUD H. RHEAULT

MARJORIE H. SOULE

KATHERINE A. RUDOLF

EILEEN GRAY

AUGUSTA BURKE

MARY S. BRADFORD

The Cytologic Diagnosis of Cancer

BY THE STAFF OF THE

VINCENT MEMORIAL LABORATORY

OF THE VINCENT MEMORIAL HOSPITAL

A GYNECOLOGIC SERVICE AFFILIATED WITH

THE MASSACHUSETTS GENERAL HOSPITAL

BOSTON, MASSACHUSETTS

∴

THE DEPARTMENT OF GYNECOLOGY

HARVARD MEDICAL SCHOOL

∴

PUBLISHED UNDER THE SPONSORSHIP OF

THE AMERICAN CANCER SOCIETY

W. B. Saunders Company

PHILADELPHIA · LONDON

Reprinted October, 1954, and February, 1958

TO

George N. Papanicolaou, M.D., Ph.D.

WHOSE OUTSTANDING CONTRIBUTIONS IN THE FIELD

OF EXFOLIATIVE CYTOLOGY HAVE OPENED A NEW

AND PRODUCTIVE APPROACH TO THE EARLY

DIAGNOSIS OF MALIGNANT DISEASE

Foreword

This book is the result of six years' experience in establishing the value of cytologic technics in the diagnosis of cancer of the cervix, endometrium, and other organs. In 1942 the Vincent Memorial Hospital Laboratory was opened at the Massachusetts General Hospital. As the Vincent Memorial Hospital was not to be built until the end of World War II, our laboratory facilities were limited. It was necessary to concentrate upon a problem that could be carried out in limited quarters. At this time the Laboratory Staff was interested in the work on vaginal smears begun by Drs. Papanicolaou and Shorr. Shortly before commencing our work, Dr. Papanicolaou, in 1941, published his first clinical paper on the diagnosis of cancer by means of the vaginal smear. In as much as the Vincent Memorial Hospital was a gynecologic unit and cared for many genital cancers, it was natural that our investigations were concentrated upon this problem. For the next few years increasing numbers of patients and slides were studied in our laboratory and various articles published in medical journals. At the cessation of the war, interest was expanded to include the study and the diagnosis of cancers wherever satisfactory secretions could be obtained. The material, therefore, included the urinary tract, the lungs and bronchi, and the stomach. The various investigations produced ever-increasing accuracy in the diagnosis and greater satisfaction with the method. The credit for the research must be given to Mrs. Ruth M. Graham, who has, with her staff, been responsible for all the cytologic diagnoses. Her group of technicians have developed into highly trained cytologists, and their diagnoses have been accepted by the whole hospital staff. The mistakes have been few, and the aid obtained from the cytologic reports has made possible the cure of many very early and otherwise unrecognizable cancers. The high esteem in which Mrs. Graham is held and the obvious value of her cytologic diagnosis of cancer have made the Vincent Memorial Hospital Laboratory a very important part of our hospital group.

This book is a report of Mrs. Graham's stewardship of the laboratory, and she has written the text and selected the photomicrographs. She has been ably helped by her entire staff in the format of the book and in the assembling of the material. Mrs. Deloris MacKinney has contributed the many illustrative drawings which help clarify each photomicrograph. The attempt has been made to exclude doubtful cells and questionable material, and it has been the endeavor of the Laboratory group to present as clearly as possible the known facts and the known cells and to avoid controversial cytologic considerations. It is hoped "The Cytologic Diagnosis of Cancer" will be helpful to those interested in this method of diagnosis of malignant tumors.

Elsewhere in the book credit has been given to all the workers—both doctors and technicians—who have contributed to its making.

Joe V. Meigs, M.D.

Maurice Fremont-Smith, M.D.

Preface

In the past few years there has been increasing interest in the cytologic diagnosis of malignancy. Since the publication of the monograph by Papanicolaou and Traut in 1943, many workers have learned the method and have applied it successfully, especially in the diagnosis of uterine carcinoma. Abroad, the interest has focused more on examination of sputum specimens for possible carcinoma of the lung. Gradually, the usefulness of the method has broadened to include the study of many other types of body secretions such as gastric secretion and urine.

The Vincent Memorial Laboratory has used the cytologic method since 1942. For some time our attention was confined to the application of the vaginal smear in uterine carcinoma. In 1946 our interests broadened to include the study of other body secretions. During these years we have studied approximately 7700 cases by vaginal smears, 450 by smears of sputum or bronchial aspirations, 400 by smears of urine sediment, 400 cases by smears of gastric secretion, and 250 cases by examination of sediment of serous fluid. These figures indicate cases, not smears. If individual smears were considered, the figures would be tripled, since three slides are done per specimen with the exception of the vaginal smears. In these the total numbers of smears is at least double the number of cases, since, in many instances, more than one slide per patient has been done.

This book is an attempt to present this material, not from a clinical viewpoint, but as conclusions drawn in the cytologic identification of cells as malignant. It is intended as a guide for those interested in the actual interpretation of smears. We have not included any statistics on the accuracy of the method since those are available in the numerous papers in the literature.

The format of the book is as follows: Chapters begin with a histologic section of the tissue under discussion as a point of orientation. This is followed by a black and white photomicrograph and colored drawing of a field of classical desquamated cells derived from that epithelium. These are typical cells of their type and illustrate the essential characteristics and points of differentiation. Unfortunately all cells present are not of this typical type, but show varying degrees of deviation from the classical form. These variations are given on pages following. Finally, difficulties in interpretation are discussed and general criteria for identification listed. There is, of necessity, repetition, since benign cells are compared to malignant in chapters dealing with normal epithelium and malignant to benign cells in chapters dealing with the carcinoma arising from that epithelium.

In the choice of fields to illustrate the numerous forms which one type of cell may assume, we have attempted to include all cells which have led to difficulties in interpretation, in our own laboratory, in errors made by students in the laboratory, and slides sent to us for consultation. We have only included illustrations of cells which we now interpret as fulfilling definite criteria for individual cell types. To include borderline cells of questionable origin did not seem appropriate.

With increasing experience in the cytologic method, a rough classification of cell types gradually developed. We feel that even if a strict classification cannot be employed in all instances, some attempt at classification is helpful for teaching purposes. It is easier to learn the criteria for individual cells and the points of differentiation from other cells if even general rules are used. We have presented here the classification which has proved useful in our own interpretation of slides and which has also facilitated our teaching of the method.

Each field in the book has three separate illustrations, a low power photomicrograph, a high power photomicrograph and a drawing. Since smears are screened under low power, we felt it essential that an illustration of their appearance at that magnification be given. The high power photomicrograph represents the field as seen under higher magnification at one focal level.

Because photomicrographs show only one plane, and, in many instances, groups of cells overlap, the drawing of the same field has been included to represent greater depth and to present additional detail. All the plates in the book are at the uniform magnification. The low power is comparable to a field seen with a 10× ocular and 4× objective (×40 magnification). The high power represents a field seen with a 10× ocular and 43× objective (×430 magnification).

The technic of preparing the different secretions is given in a separate chapter. An appendix is included, giving the histologic diagnosis for all illustrations of carcinoma cells. The bibliography is comprehensive and includes all the references we have encountered on exfoliative cytology. Evidence of the increasing interest in this field is shown by the fact that 78 per cent of the publications are after 1943.

In the discussion of vaginal smears, there is one distinct omission. There is no consideration of the smears of pregnancy. Since the Massachusetts General Hospital has no obstetrical service, we have had little experience with this type of material, and thus did not feel qualified to include these in our discussion. In addition to the diagnostic application of vaginal smears, our interest at the Vincent Laboratory has been focused on the effect of radiation on the cells of the cervix and vagina in treated cases of carcinoma of the cervix. We have included a detailed discussion of these effects, not only as an aid in diagnosis, but also because of its possible application in prognosis. Though we have used the vaginal smear almost exclusively in the diagnosis of uterine cancer, the cellular criteria and points of differentiation are also applicable to other types of smears such as cervical, endocervical and endometrial.

It should be pointed out that the use of both smears of gastric secretion and of urine sediment is still in the research stage. Much work remains to be done in both the technical details and identification of cells. The criteria in sputum and vaginal secretion are better defined, but much experience with the cytologic method is required before all the cellular variations can be catalogued with reasonable accuracy. After any experience with the method, the student will be impressed with the great degree of variation from one smear to another. This variation makes the study of exfoliated cells much more interesting than it would be if the cellular pattern were more regular, but also accounts for the difficulties of the method. This book is presented with the hope that it will be regarded as only an introduction to the study of exfoliated cells in the diagnosis of malignancy.

Preface

This laboratory manual has been prepared by the staff of the Vincent Memorial Laboratory. The drawings were done by Deloris C. MacKinney. The photomicrographs were the responsibility of Maud H. Rheault. Marjorie H. Soule assisted in the writing and in the organization of the material. Other members of the staff assisted with the many technical details of preparation.

The staff would like to express its appreciation to the American Cancer Society and to their representative, Dr. Charles C. Lund, for the sponsorship of this book and for the financial support which made possible the publication of the colored plates.

We are indebted to Dr. Robert Fennell, of the Department of Pathology, for help in the choice of the histologic sections. The Ortho Research Corporation has generously made available large quantities of the counterstains EA-50 and OG-6 for our use. We would like to thank the Photographic Department of the Massachusetts General Hospital for allowing us to use their facilities for the developing and printing of the photographs, and for their helpful advice and criticism. We are indebted to Dr. John B. Graham for his critical reading of the text in manuscript. Dr. Howard Ulfelder is responsible for the technical improvements described for gastric aspirations. Our publishers, the W. B. Saunders Company, have been extremely helpful in advice and especially lenient in allowing us to include additional figures. We would like to express our appreciation for their helpful and considerate cooperation. Finally, we are greatly indebted to Dr. Joe V. Meigs and to Dr. Maurice Fremont-Smith whose help and encouragement have made possible the publication of this book.

STAFF OF THE VINCENT
MEMORIAL LABORATORY

Contents

CHAPTER I

NORMAL CELLS FROM CERVICAL AND VAGINAL SQUAMOUS EPITHELIUM

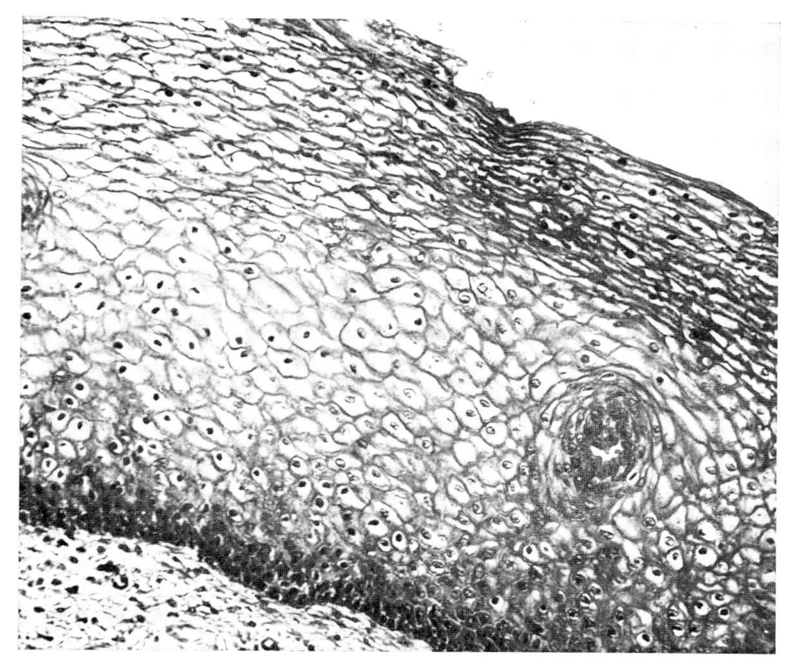

FIG. 1

HISTOLOGIC SECTION: CERVICAL SQUAMOUS EPITHELIUM

The squamous epithelium of the female genital tract is composed of three layers. Beginning at the basement membrane they are: germinal, transitional or basal, and superficial layers. The cells from the germinal layer are seldom encountered in the vaginal smear. The cells of the transitional zone are commonly referred to in exfoliative cytology as basal cells. They have been divided into two groups according to their size, shape, denseness and amount of cytoplasm.

The inner layer basal cell originates in the deeper portions of the transitional zone and classically is a small round cell four to five times the size of a leukocyte. The cytoplasm is a deep basophilia. The rim of cytoplasm is usually equal to the width of the nucleus. The outer layer basal cells come from the upper portions of the transitional zone. The cells retain their round form though the cytoplasm has increased, being equal to at least twice the width of the nucleus. The cytoplasm is less dense and stains a lighter blue. The nucleus in both inner and outer layer basals has finely vesicular chromatin and is oval or round in shape with a good nuclear border. Cellular borders are usually quite sharp and individual cells appear as definite units.

DESCRIPTION OF BASAL EPITHELIAL CELLS

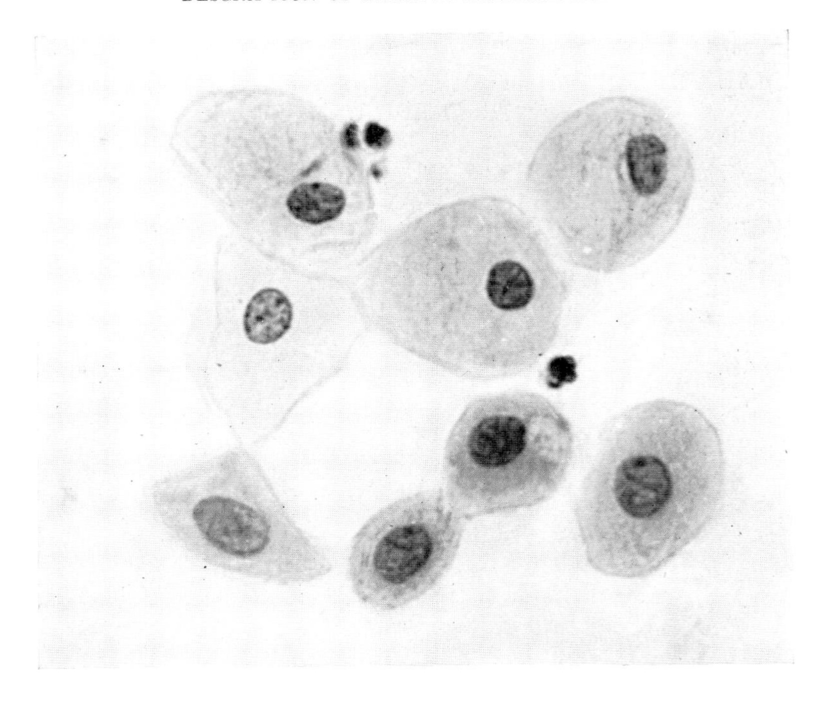

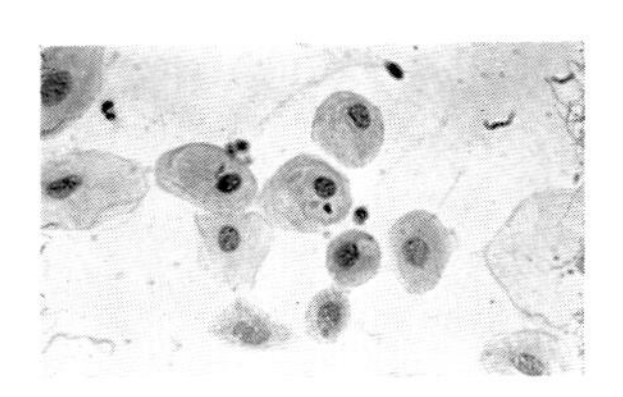

Low Power: Discrete epithelial cells, both nuclear and cellular variation is slight.

High Power:

A. Characteristics of Nucleus: 1. Clear, regular nuclear borders. 2. Slight variation in chromatin content. The chromatin is finely granular. (See cells 2 and 4.) Individual particles are very small and appear occasionally in clumps. 3. The shape of the nuclei is round or oval. Compare cell 2 with 5. Notice the nuclei vary only slightly in size, cell 3 having a smaller nucleus than the rest, owing to degeneration which has shriveled the nucleus, leaving a perinuclear vacuole.

B. Characteristics of Cytoplasm: 1. Definite cellular borders. 2. Cells 1, 2, 3 and 8 have a large amount of transparent cytoplasm around their nuclei and are called outer layer basal cells. The early precornified cell, 4, which is square in shape, also has thin transparent cytoplasm. Cells 6 and 7 have less cytoplasm, which stains deeper and is more dense, and are known as inner layer basal cells. Cell 5 is slightly aberrant because of the elongated shape of the cytoplasm. 3. Cells 6 and 7 show vacuoles in the cytoplasm, an early sign of degeneration. 4. Staining reaction—basophilic in most instances, however, degenerate basals usually stain acidophilic.

C. General Characteristics of Group: Individual cells each with a vesicular nucleus surrounded by an adequate amount of cytoplasm. Good nuclear-cytoplasmic ratio.

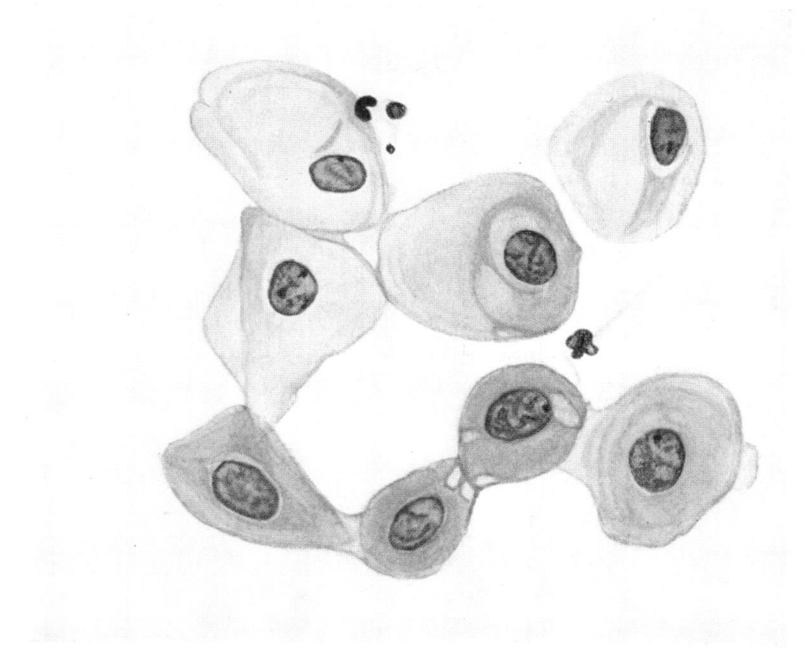

PLATE 1

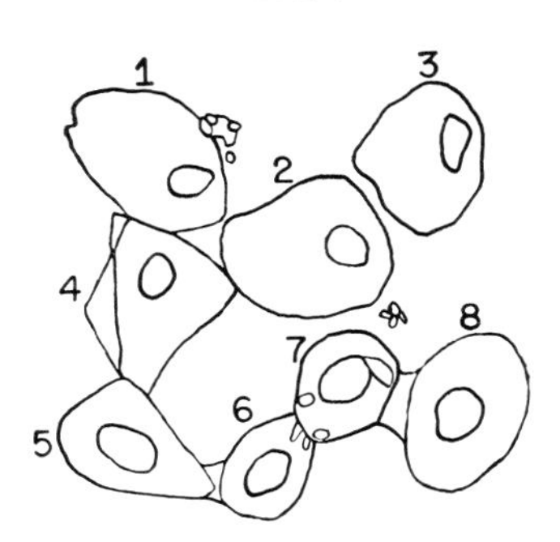

KEY TO BASAL CELL PLATE

1. Outer layer basal, vesicular nucleus, thin transparent cytoplasm.
2. Outer layer basal, finely granular nucleus, unevenness in density of cytoplasm.
3. Outer layer basal, degenerate, partially pyknotic nucleus, perinuclear vacuole.
4. Early precornified cell, showing square shape and folding of transparent cytoplasm.
5. Outer layer basal, oval nucleus, cellular form slightly elongated.

6 and 7. Inner layer basals, dense cytoplasm showing beginning of vacuolization, round vesicular nuclei.

8. Outer layer basal, with central nucleus containing finely divided chromatin.

FIG. 2

BASAL CELLS: CYTOPLASMIC CHANGES—LOSS OF CELL BORDERS

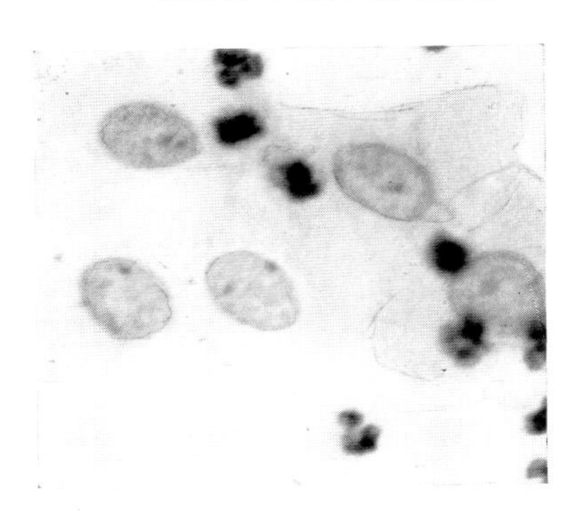

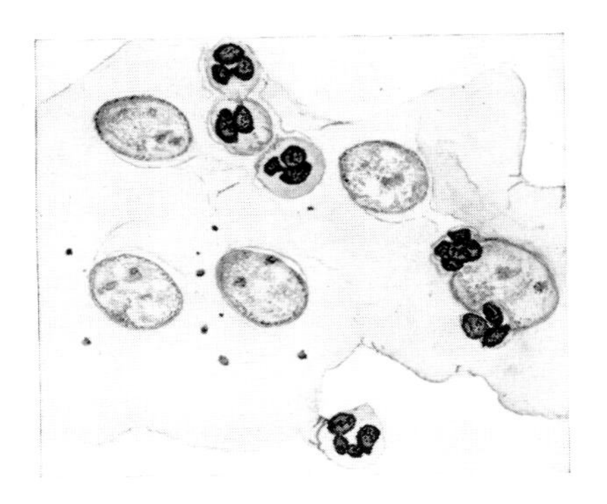

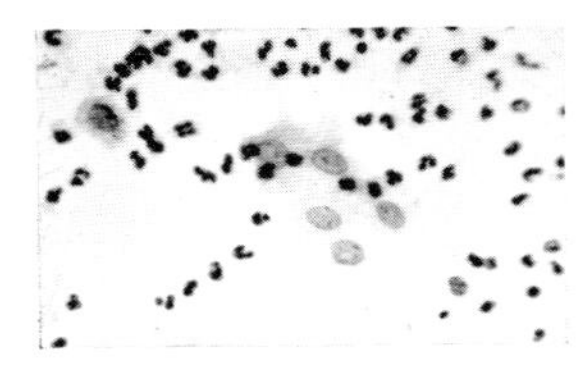

Low Power: Group of nuclei with indistinct cytoplasm.
High Power: The nuclei are the size of outer layer basals. (See Plate 1.) They are bland with faint clumps of chromatin and do not vary in shape. Because of degeneration the cytoplasm is indistinct and appears only as a background to the nuclei. No individual cellular borders are visible. This type of degeneration often accompanies trichomonas infestation.

FIG. 3

BASAL CELLS: CYTOPLASMIC CHANGES—GLYCOGEN DEPOSITS

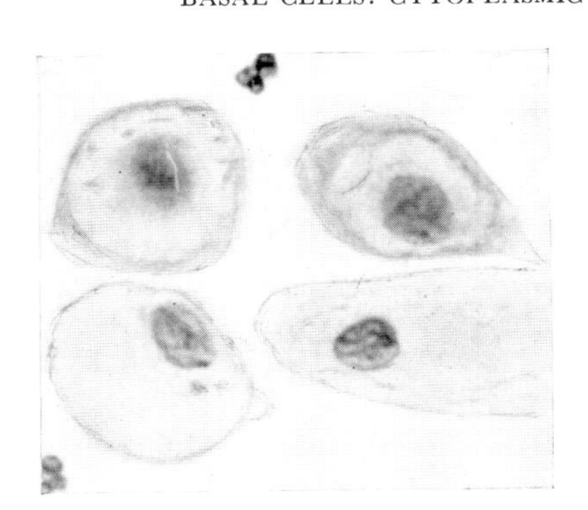

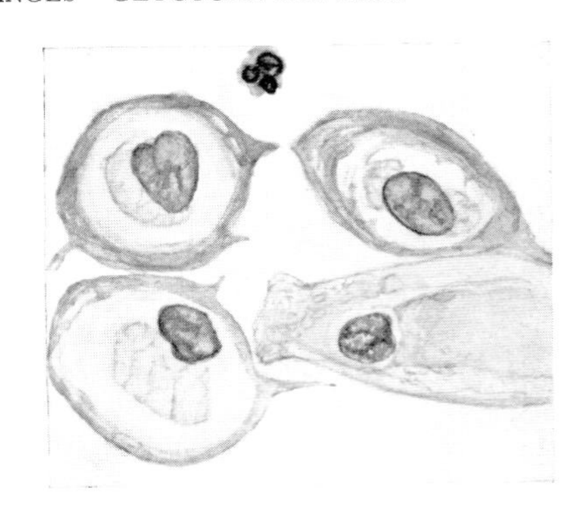

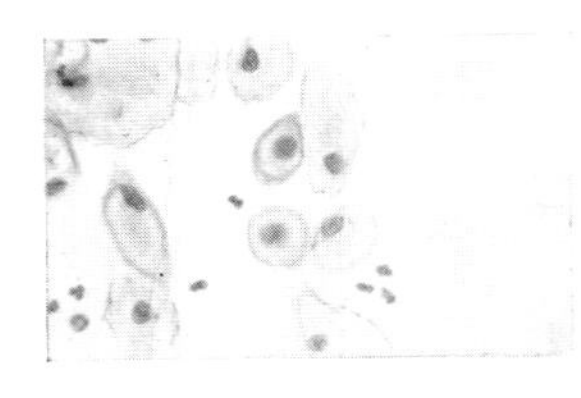

Low Power: Group of inner and outer layer basal cells with thick cellular borders.
High Power: These cells are distinctive in that their cellular borders are much thicker than usual. The border stains a much deeper blue than the remaining cytoplasm. The central deposits around the nuclei are irregular and take a deep yellow stain. This peculiarity has been interpreted as glycogen deposits.

Fig. 4

BASAL CELLS: CYTOPLASMIC CHANGES—PERINUCLEAR VACUOLE

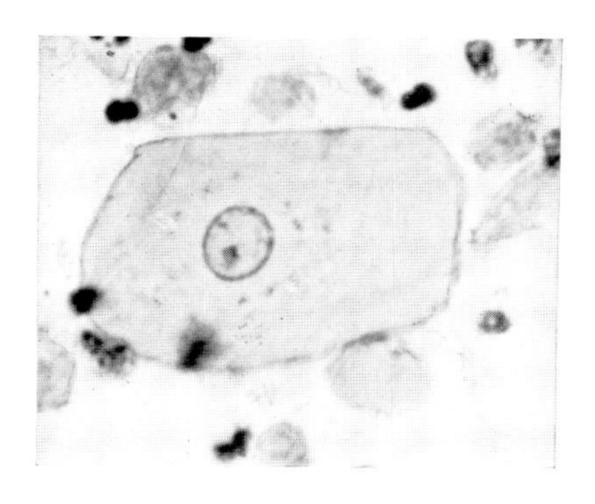

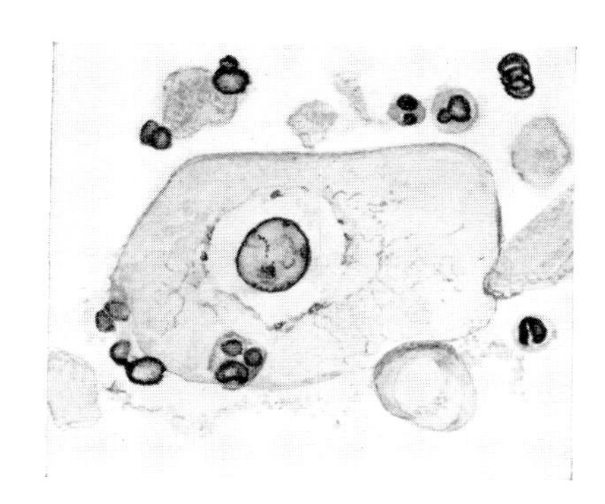

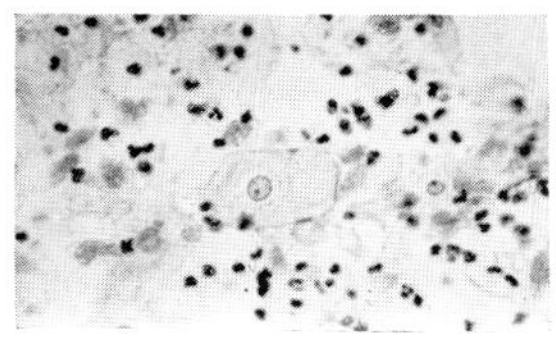

Low Power: Trichomonas, leukocytes and one preserved outer layer basal cell.
High Power: The small irregular cells are trichomonas and appear as brownish-gray, nondescript blobs. Careful examination sometimes reveals a faint line indicating a nucleus. Trichomonas cause changes in precornified and outer layer basal cells often producing active nuclei, loss of cell borders and perinuclear vacuoles similar to the vacuole pictured above.

Fig. 5

BASAL CELLS: CYTOPLASMIC CHANGES—ABERRANT SHAPES

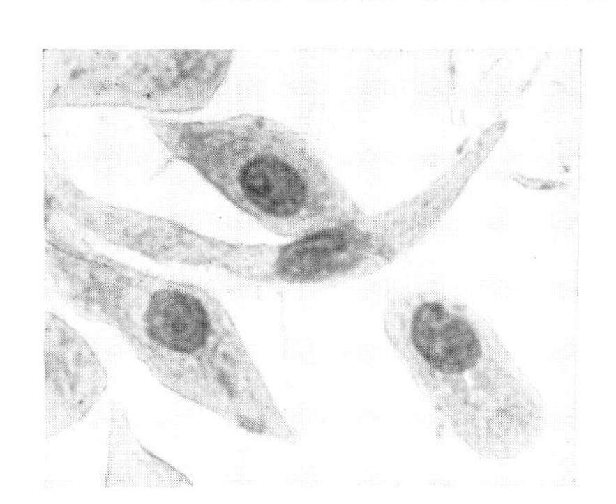

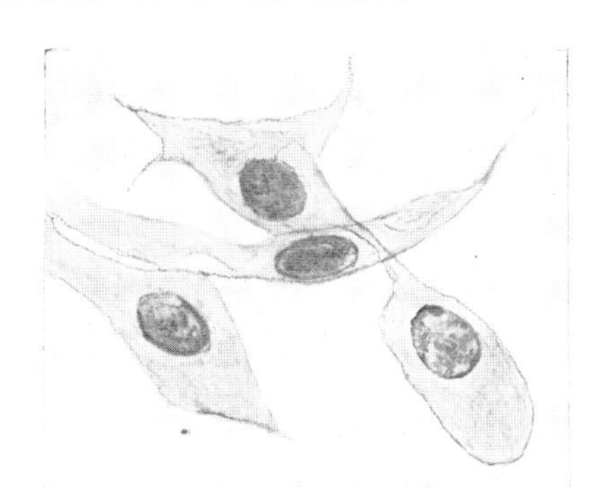

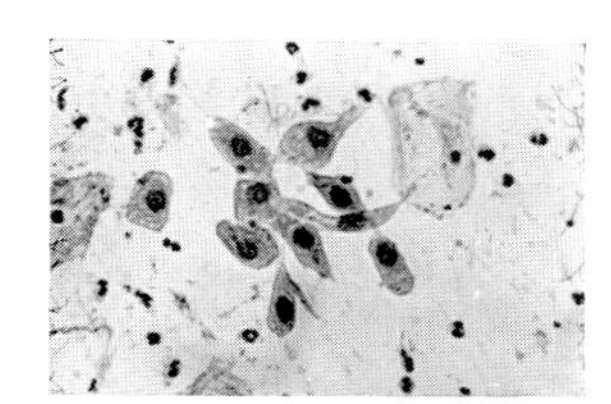

Low Power: Group of epithelial cells with abnormally shaped cytoplasm.
High Power: The nuclei are round and of the same structure as the basal nuclei shown in Plate 1. These are called aberrant basal cells because of the abnormal elongated shape of their cytoplasm which usually stains acidophilic. Three of the four cells show tail projections coming from the cytoplasm.

FIG. 6

BASAL CELLS: NUCLEAR CHANGES—SWELLING

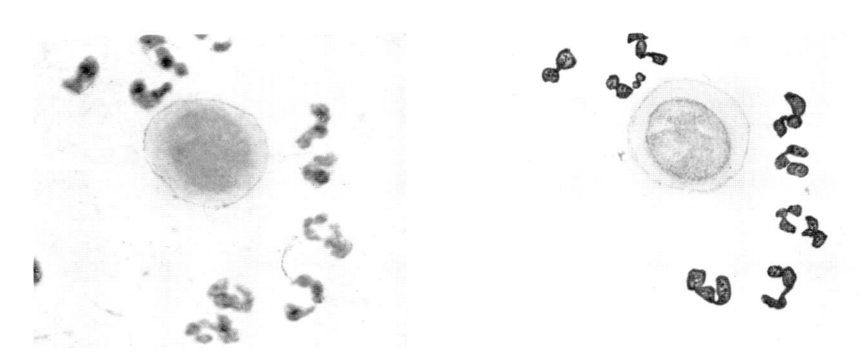

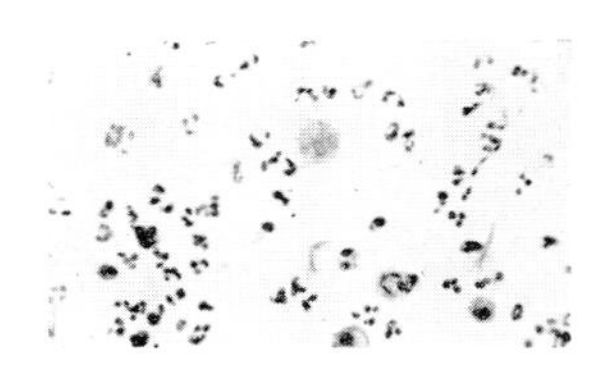

Low Power: Cell appears only as a lightly staining nucleus. No cytoplasm is visible.

High Power: The nucleus is larger than normal with less cytoplasm than usual, but both nuclear and cellular borders can be identified. The chromatin content of the nucleus is more finely divided than in a well preserved cell. Individual chromatin particles are difficult to distinguish and the nucleus has a swollen and smooth appearance.

FIG. 7

BASAL CELLS: NUCLEAR CHANGES—ACTIVE NUCLEUS

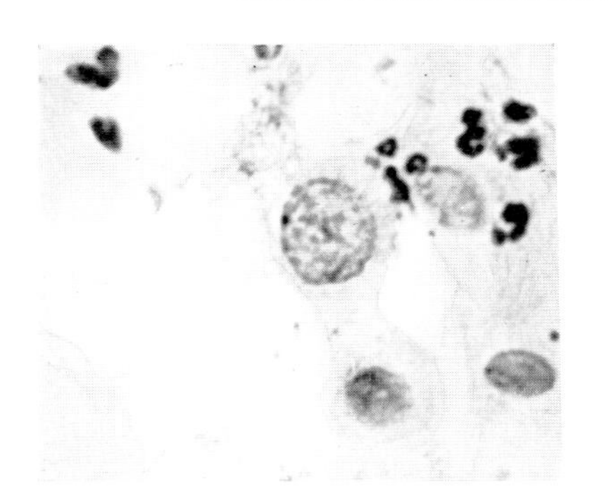

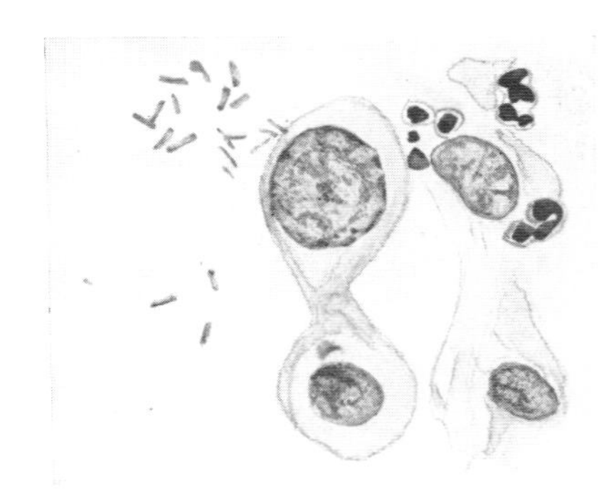

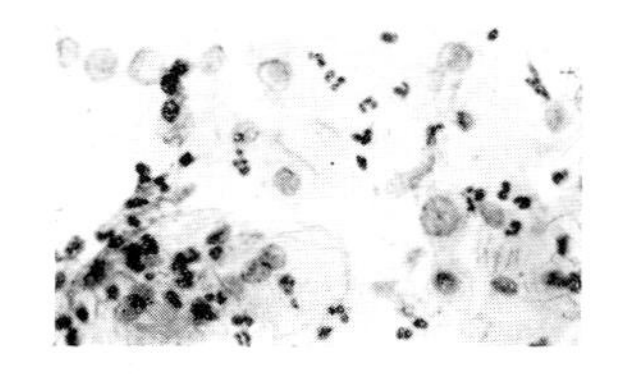

Low Power: Field of epithelial cells, one being larger than the rest.

High Power: The nucleus is large and round, clumps of chromatin being more prominent than usual, particularly at the nuclear border, giving it a darker look. We identify this cell as a basal because the clumps of chromatin are evenly distributed, the nuclear border is not irregular and a clear cytoplasmic border is visible.

Fig. 8

BASAL CELLS: NUCLEAR CHANGES—PYKNOSIS

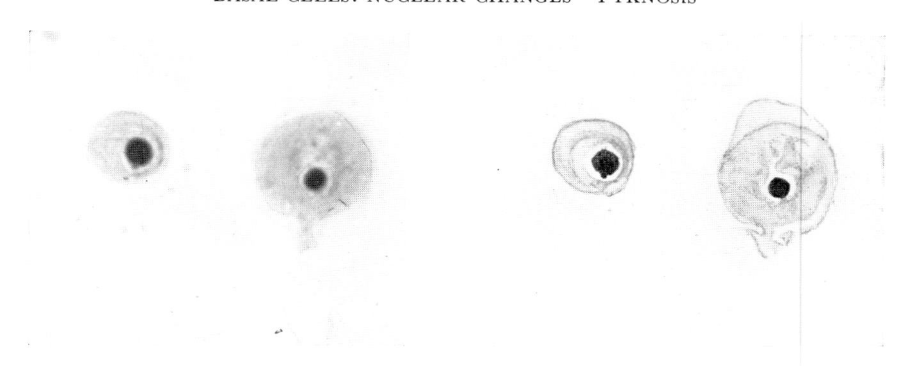

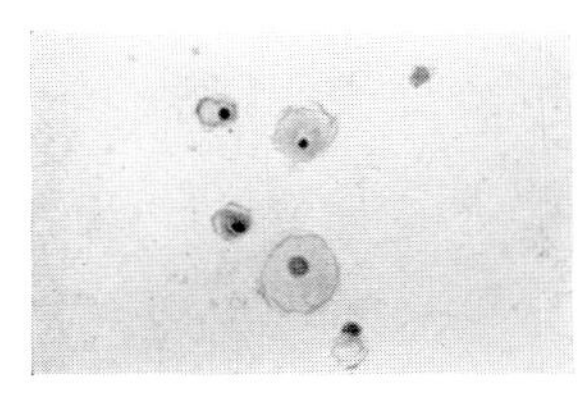

Low Power: Several epithelial cells with small dark nuclei.

High Power: The chromatin of these nuclei has condensed until no structure can be observed. The condensation leaves a perinuclear vacuole formerly occupied by the whole nucleus. We identify these cells as basal, instead of cornified, because of their small size and round appearance. Cytoplasm of cell on the right presents an irregular border, another sign of degeneration.

Fig. 9

BASAL CELLS: NUCLEAR CHANGES—KARYORRHEXIS

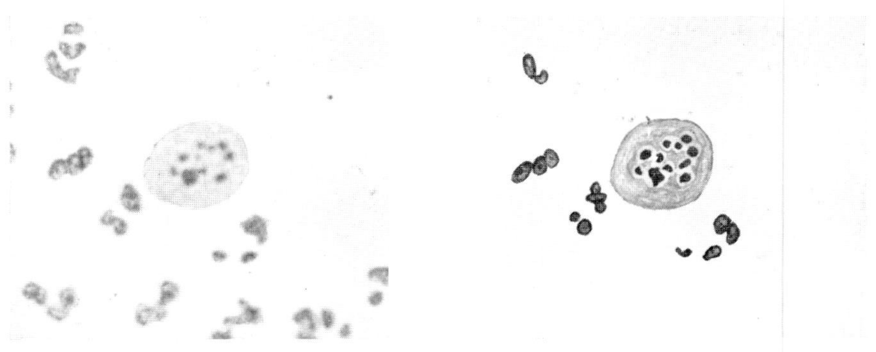

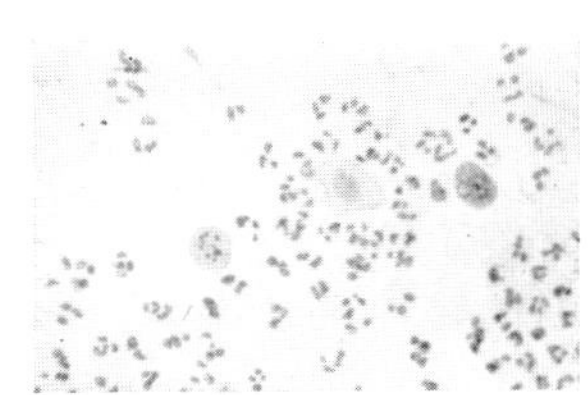

Low Power: Inner layer basal cells with indefinite nuclear patterns.

High Power: The chromatin of the nucleus has separated into many individual clumps which appear as dense black particles. The nuclear border has disappeared. This particular pattern of the nucleus is seen as the last stage of disintegration. It should be distinguished from the prophase of mitosis where chromatin particles are much finer.

A smear which shows only basal cells may be interpreted as showing atrophy of the female genital tract. If the ovaries are producing no estrin, the squamous epithelium is extremely low so that only the transitional layer remains and the vaginal secretion contains only basal cells. Such smears are seen in postmenopausal women. However, it should not be assumed because the estrin cycle has ceased that the vaginal smear will show only basal cells. We have found that the majority of postmenopausal smears still contain some superficial cells of the squamous epithelium.

Difficulties in Interpretation: The completely preserved basal cell is easily recognized as a benign epithelial cell as the adequate amount of cytoplasm and small regular nucleus easily identify it. It is only when there is degeneration of the cells or unusual activity in the basal layer that identification of this type of cell becomes difficult.

Occasionally when the genital tract is extremely atrophic, cells from deep in the mucosa may be aspirated. These appear as sheets of regular nuclei without cellular borders. The nuclei may be quite active, *i.e.*, the chromatin in small clumps. However, the nuclei are even in size and shape and show little variation from one to another. These cells usually are present when the vaginal smear has been obtained with some difficulty. If the vagina is completely dry it is perhaps better to omit the smear, since continued suction will pull off these sheets of cells which occasionally make interpretation difficult.

It has been our experience that the type of degeneration which is most confusing is that caused by trichomonas. The nuclei of basal cells in such smears are often extremely active (see Fig. 7) and commonly the cellular borders are lost and only a background of cytoplasm remains. (See Fig. 2.) There are two points which identify such cells as atypically benign rather than malignant. The nuclei are evenly spaced in the background of cytoplasm and, although the nuclei may appear somewhat hyperchromatic because of the clumping of the chromatin, they are still even in size, vary little in shape and have smooth regular borders.

General Criteria for Identification of Basal Cells: Inner layer basal cells: Small round cell with round or oval nucleus, usually centrally located. Cytoplasm is adequate for size of cell. Occur usually as single cells, only occasionally are tightly clumped. If in a group, focusing up and down will establish cellular borders for individual cells.

Outer layer basals: round or oval cells, usually twice the size of inner layer basals. Increase in size is due to increase in cytoplasm which is more transparent than in inner layer basals. Nucleus is of the same type and size as inner layer basals.

The superficial cells of the squamous epithelium of the genital tract are designated as precornified and cornified cells according to the nomenclature used in exfoliative cytology. The precornified cells are from the zone of epithelial cells directly above the basal or transitional layer. (See Fig. 1.) They are large, thin, transparent cells with an abundance of cytoplasm which usually stains basophilic. Their nuclei are vesicular. The cornified cells are the surface layer of cells, and appear in smears as large cells with thin transparent cytoplasm, usually acidophilic with deep staining pyknotic nuclei.

It is important to remember that as the cells of the squamous epithelium mature, the cytoplasm increases in amount and the nucleus appears smaller.

The outer layer basal has more cytoplasm than the inner layer basal cell, the precornified more than the outer layer basal cell. The cornified cell appears to have still a slight increase, since the nucleus has condensed and become pyknotic. It is obvious that the identification of certain cells as inner layer basals or outer layer basals becomes a matter of personal interpretation. A good many cells are borderline. It is equally evident that whether a cell is an outer layer basal or a precornified is again a matter of interpretation. We depend on the nucleus entirely for determining whether a cell is a cornified or precornified cell. If a superficial cell has a pyknotic nucleus it is considered cornified whatever its cytoplasmic stain, and if the nucleus is vesicular it is considered to be precornified. This we realize is an arbitrary classification.

In the development of epithelial cells, the nucleus and cytoplasm do not mature at the same rate. Classically, a precornified cell should have a vesicular nucleus and a basophilic cytoplasm. The cornified cell should have a pyknotic nucleus and an acidophilic cytoplasm. The majority of the superficial cells do fulfill the above criteria, but a substantial number show the reverse, *i.e.*, a vesicular nucleus with acidophilic cytoplasm; pyknotic nucleus with basophilic cytoplasm. Since in the best of hands any staining procedure varies, we have considered that a classification would be more accurate based on nuclear changes alone. If superficial cells with pyknotic nuclei are considered as cornified and those with vesicular nuclei as precornified, the classification may be carried over to cells stained by other methods than the Papanicolaou method.

It is essential to point out why identification of these various cell types is important. They cause the least difficulty in determining whether a cell is malignant or benign. However, since the cornified cells are seldom present if the ovaries are not functioning and their numbers are related qualitatively to the amount of circulatory estrin if the vaginal smear is to be used as an indication of the activity of the ovaries, these cells are extremely important. In the normal cyclic woman the cornified and precornified cells are in abundance. In the menopausal group, the basal cells predominate. Between these two extremes are many cases where estrogen activity is of real clinical importance. We have not found in our laboratory that accurate quantitative data regarding estrin production can be demonstrated by the use of the vaginal smear. We have used a rough qualitative measure for reporting the amount of circulatory estrin as mirrored in the changes of the genital squamous epithelium. Our criteria for measuring estrin effect is as follows: no estrin effect, nothing but basal cells; very slight estrin effect, basal cells with majority of precornified cells; slight estrin effect, only precornified cells; moderate estrin effect, majority of cells precornified, some cornified cells; marked estrin effect, majority of cells cornified, some precornified cells. This type of data has proved helpful clinically.

DESCRIPTION OF CORNIFIED AND PRECORNIFIED EPITHELIAL CELLS

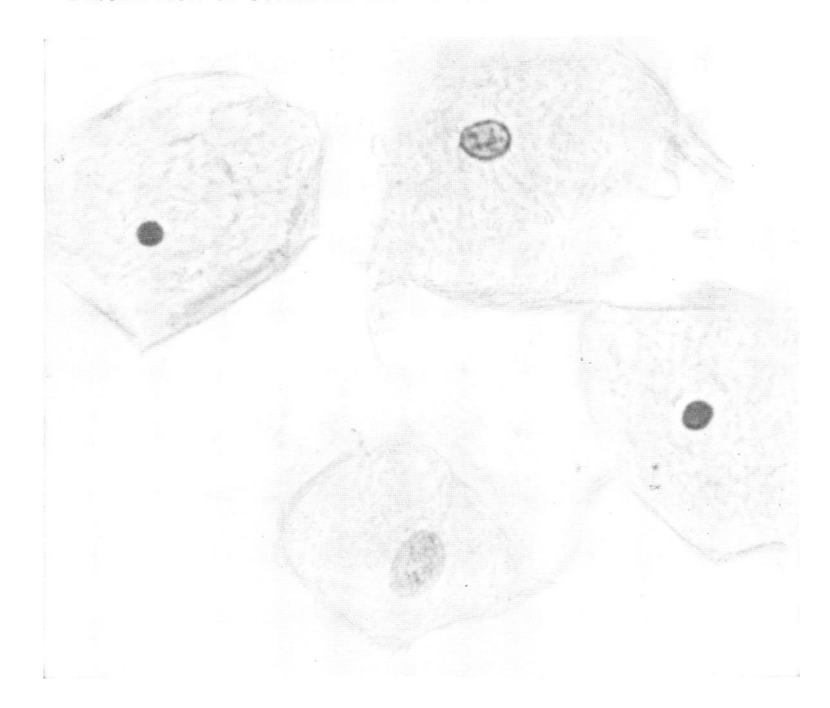

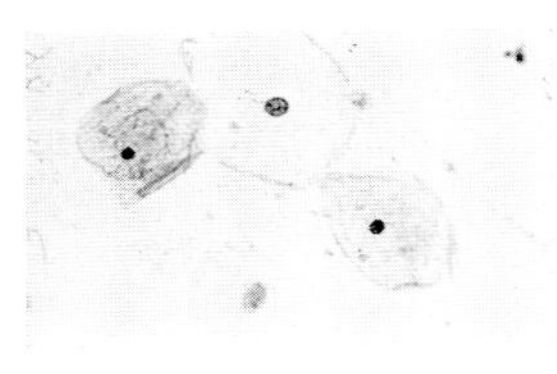

Low Power: Individual epithelial cells with small nuclei and large amount of cytoplasm.

High Power:

A. Characteristics of Nucleus: 1. Even, well defined nuclear borders. 2. Difference in chromatin content. Cells 1 and 4 have small pyknotic nuclei which show no individual chromatin particles, since the nucleus is condensed. These nuclei are characteristic of cornified cells. Precornified cells have larger and more vesicular nuclei. Cell 2 shows more clumping of the chromatin particles than cell 3, which is smoothly granular. 3. The shape of the nuclei in cornified and precornified cells is round or oval and size variation is slight. The nuclei of the cornified cells show progressive condensation. In cell 4 the nucleus has condensed, leaving a surrounding space previously occupied by the nucleus.

B. Characteristics of Cytoplasm: 1. Irregular cellular borders. 2. The general shape of the cytoplasm is square, but in many instances becomes folded or wrinkled. (See cells 1 and 3.) Both cornified and precornified cells are thin and transparent in appearance and the cytoplasm is smooth. 3. Staining reaction: acidophilic in the cornified cells and basophilic in the precornified cells in most instances. However, staining reaction is not specific.

C. General Characteristics of Group: Flat, large, transparent cells varying in staining reaction of cytoplasm from basophilic to acidophilic. Nuclei vary from vesicular to pyknotic.

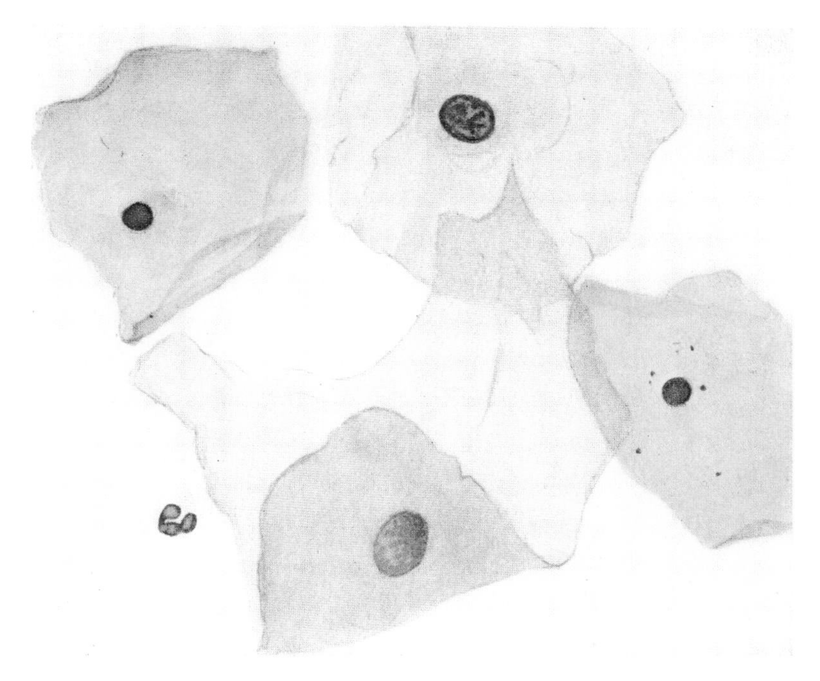

PLATE 2

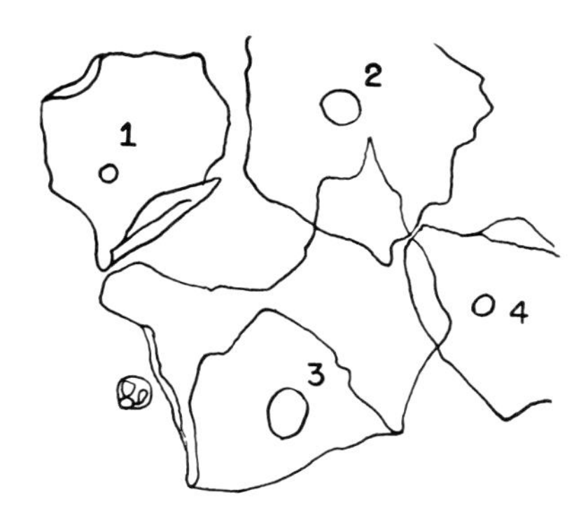

KEY TO CORNIFIED AND PRECORNIFIED CELL PLATE

1. Flat cornified cell with pyknotic nucleus and fairly even cellular border.
2. Precornified cell showing wrinkled, transparent cytoplasm and a vesicular nucleus.
3. Precornified cell with large amount of folded transparent cytoplasm and bland nucleus.
4. Cornified cell with small dense nucleus, irregular cytoplasm with granules and evidence of a perinuclear vacuole.

Fig. 10

SUPERFICIAL CELLS: CYTOPLASMIC CHANGES—CYTOLYSIS

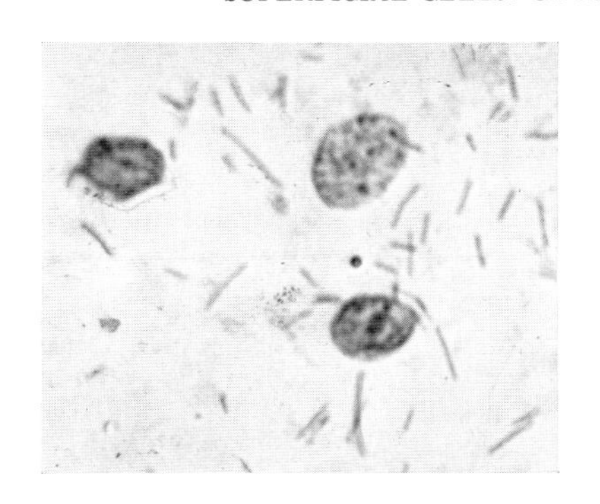

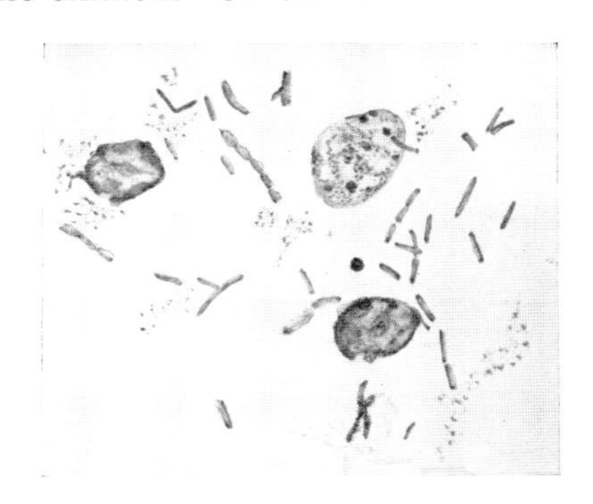

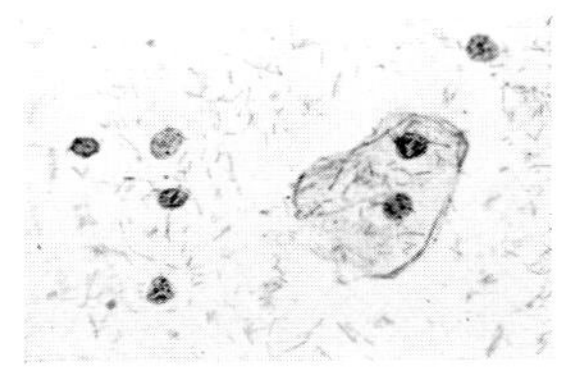

Low Power: Extruded nuclei, two cells with indistinct cellular borders.
High Power: Light staining nuclei with evenly distributed chromatin particles and no sharp nuclear borders. Careful examination shows that there is no cytoplasm surrounding the nuclei. The Döderlein bacilli cause degeneration of the cytoplasm of the precornified cells, leaving only their nuclei. It is important to notice that these nuclei vary only slightly in size.

Fig. 11

SUPERFICIAL CELLS: CYTOPLASMIC CHANGES—KERATINIZATION

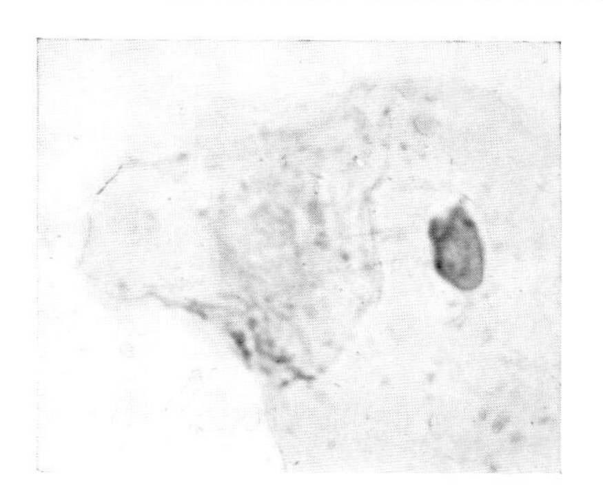

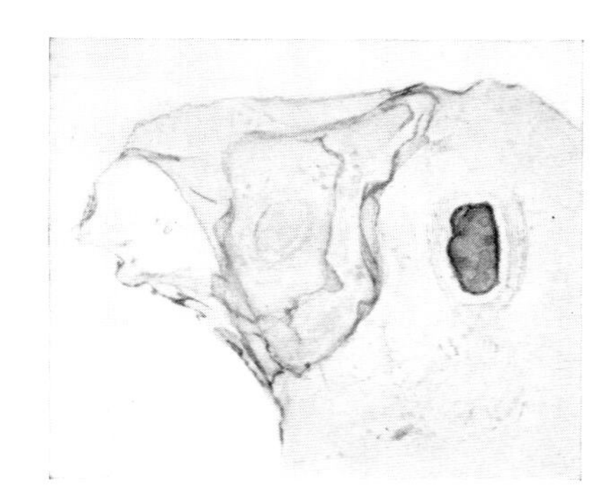

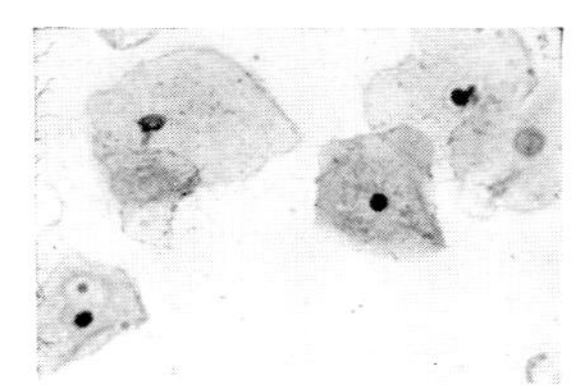

Low Power: Precornified cell with degenerated cell superimposed.
High Power: Precornified cell with a vesicular nucleus and adequate cytoplasm. The keratinized cell has matured to the stage where the nucleus has completely disappeared and the cellular border has become wrinkled and irregular. Keratinized cells usually stain orange rather than acidophilic or basophilic and are considered as evidence of a high estrin effect.

Fig. 12

SUPERFICIAL CELLS: NUCLEAR CHANGES—ACTIVE CHROMATIN

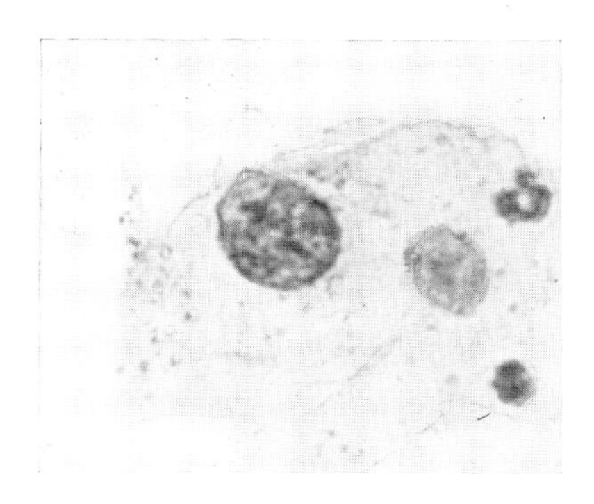

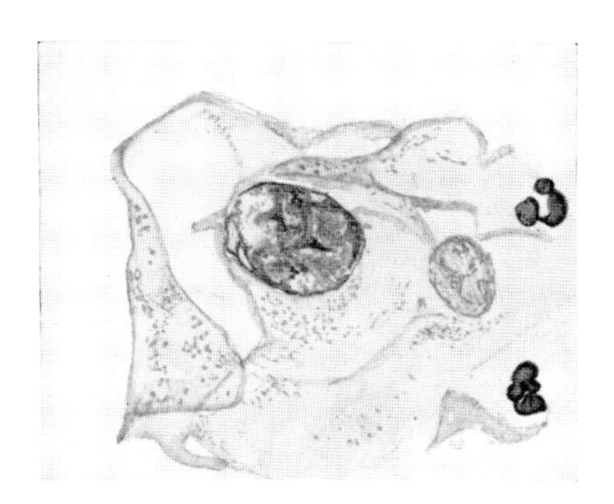

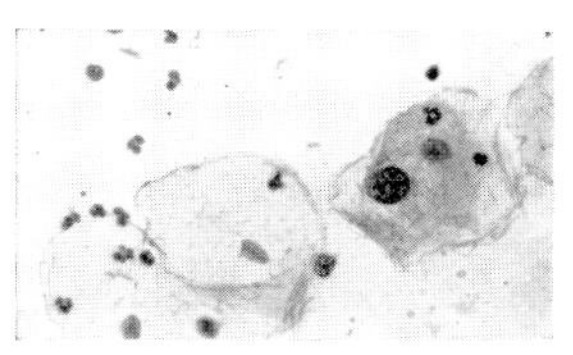

Low Power: Superficial cells, one having an enlarged nucleus.

High Power: Two precornified cells superimposed. The cellular borders are uneven and the cytoplasm is folded. One nucleus is twice as large and active as the adjacent vesicular one, but note the chromatin content is evenly spread with an occasional clumping. We interpret this nucleus as benign, since it has ample cytoplasm surrounding it.

Fig. 13

SUPERFICIAL CELLS: NUCLEAR CHANGES—KARYORRHEXIS

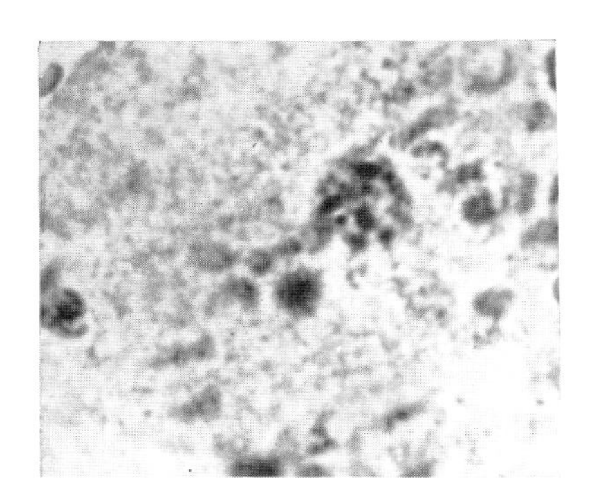

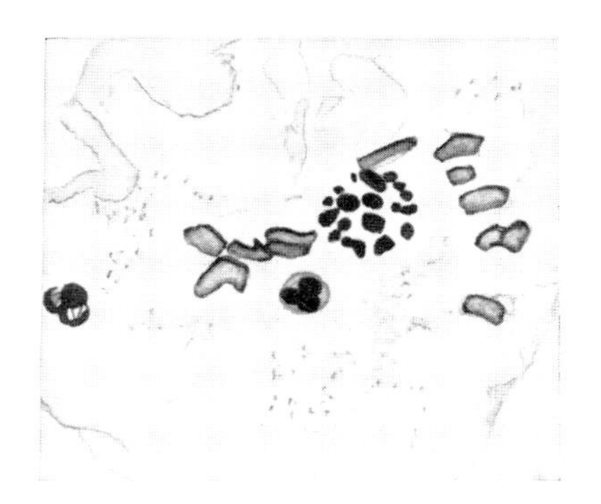

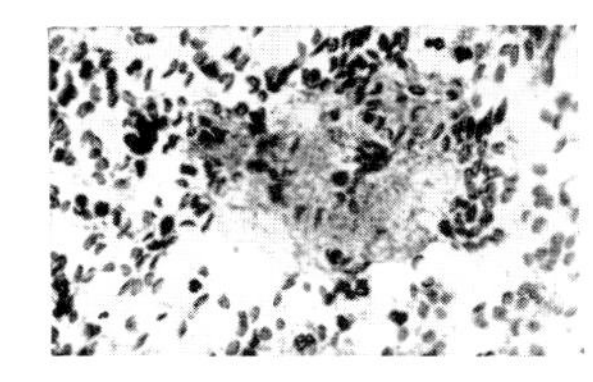

Low Power: Precornified cell with degenerate nucleus.

High Power: Breaking up of the nucleus, or karyorrhexis, has taken place, owing to degeneration, and the nuclear border has been lost. Coarse clumps of chromatin are distributed irregularly or in a peripheral manner. This is to be distinguished from mitosis, in which the chromatin material is finely clumped. (See Fig. 32.)

FIG. 14

SUPERFICIAL CELLS: CYTOPLASMIC CHANGES—CURLING

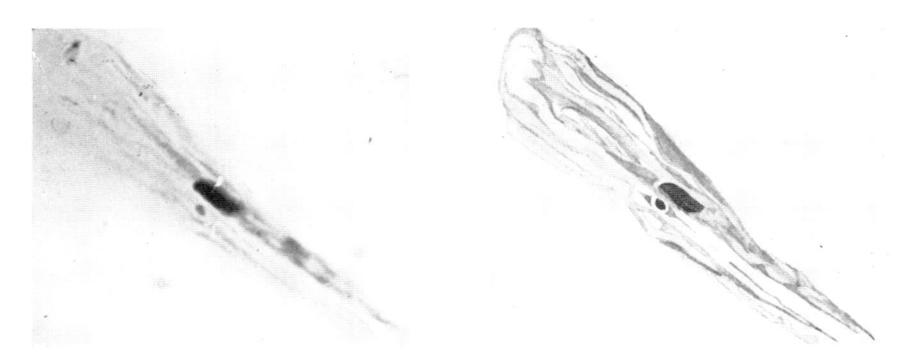

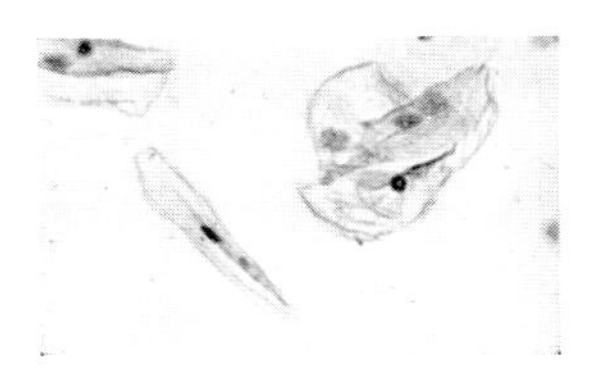

Low Power: Elongated cornified cell with pyknotic nucleus.

High Power: The nucleus is dense and individual chromatin particles cannot be identified. The cytoplasm is curled, making the cell appear elongated instead of flat and square. This cell may be confused with a malignant fiber cell but careful examination reveals a small pyknotic nucleus and adequate curled cytoplasm which identify it as benign.

FIG. 15

SUPERFICIAL CELLS: CYTOPLASMIC CHANGES—PEARL FORMATION

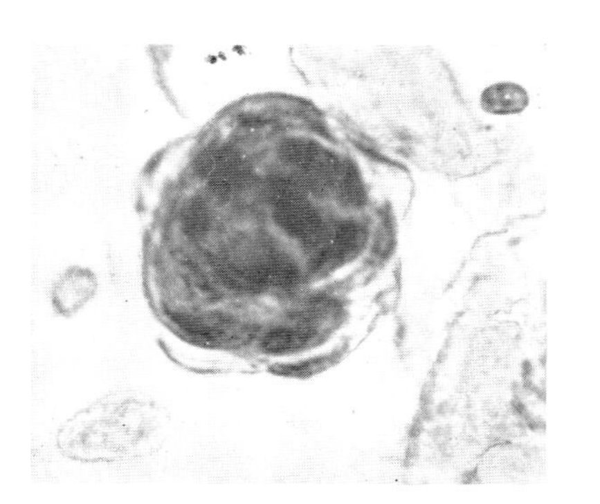

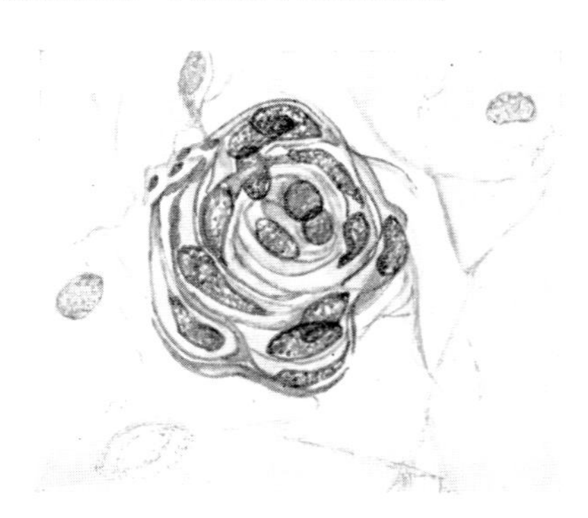

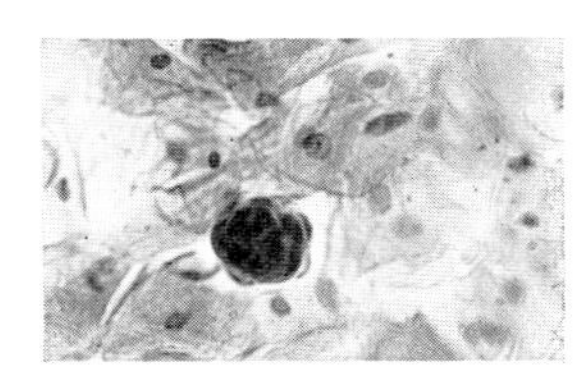

Low Power: The epithelial pearl appears as a small cluster of dark nuclei.

High Power: Group of precornified cells in a spiral pattern. The nuclei are aberrant in shape and vary slightly in size, but the chromatin is evenly distributed and nuclear borders are not definite. The cytoplasm is also aberrant and cellular borders are indistinct. These pearls occur in normal vaginal smears.

Superficial squamous cells are the most frequent of any type of normal epithelial cell seen in the vaginal secretion. They are present in the normal cyclic woman, their relative numbers varying according to the period of the cycle. Because the hormones secreted by the ovary have been shown to have such a pronounced effect on the mucosa of the entire female genital tract, many attempts have been made to determine the time of ovulation by examination of the vaginal secretion. A critical evaluation of this method of timing ovulation leaves much to be desired. Many factors such as infection or trauma confuse the picture. We recognize that by examination of the vaginal secretion the time of ovulation in an occasional case may be decided, but definitely feel that the variations seen due to other factors than hormone secretion exclude the use of the vaginal smear as a routine procedure for this purpose.

Difficulties in Interpretation: In general, the recognition of the superficial squamous cells is less difficult than that of any other cells encountered in the vaginal secretion. We have not made a "false positive" report on the basis of marked atypicality of these cells. There are only three instances in which any doubt as to the origin of the cells would occur. The first is that of cytolysis, *i.e.*, specific and entire degeneration of the cytoplasm of precornified cells, leaving free nuclei. (See Fig. 10.) This should cause no difficulty, since the nuclei have smoothly granular chromatin and do not vary in size. They can be easily identified as precornified by comparison with the nuclei of preserved precornified cells in the smear.

The second group which perhaps might cause occasional difficulty are the rolled cornified cells. (See Fig. 14.) Since the superficial cells are so thin, they are apt to present all types of folding. The rolled cornified cell is a marked example of the tendency of the cytoplasm of superficial cells to fold over on itself. The only reason this type of cellular form may be confusing is that it must be differentiated from the malignant fiber cell of squamous carcinoma. Two definite characteristics label it as benign: the nucleus is pyknotic, not active as in the cancer cell; the cytoplasm is adequate and careful focusing up and down will reveal the tight folds of the cytoplasm. If there is any doubt about the cell, slight pressure on the cover slip will usually cause the rolled cornified cell to unroll, and then of course identification is simple.

The third group of cells are perhaps more troublesome. They are the precornified cells with extremely active nuclei. (See Fig. 12.) When we speak of active nuclei we are referring to nuclei larger than normal with chromatin in small dense clumps instead of finely granular. The nucleus of the precornified cell may appear larger than usual and the chromatin content may appear increased and irregular. These cells certainly represent some atypicality of the epithelium, but we have not found them to be associated with carcinoma. Whether they represent a precancerous change is impossible to determine at the present time. The nucleus of these cells is certainly atypical in its chromatin arrangement though it is not as distinct as in a cancer-nucleus. The abundance of cytoplasm is of definite aid in the identification of this cell as benign.

General Criteria for the Identification of Superficial Squamous Cells:

1. Cornified cells contain a small pyknotic nucleus showing marked condensation of chromatin material. Cytoplasm is thin, transparent and

abundant and usually stains acidophilic but may occasionally be basophilic.

2. Precornified cells have a vesicular nucleus showing fine granulations of chromatin. Cytoplasm is similar to that of the cornified cell, though a basophilic stain in the cytoplasm is more common than acidophilic.

CHAPTER II

NORMAL CELLS FROM COLUMNAR EPITHELIUM OF ENDOCERVIX AND ENDOMETRIUM

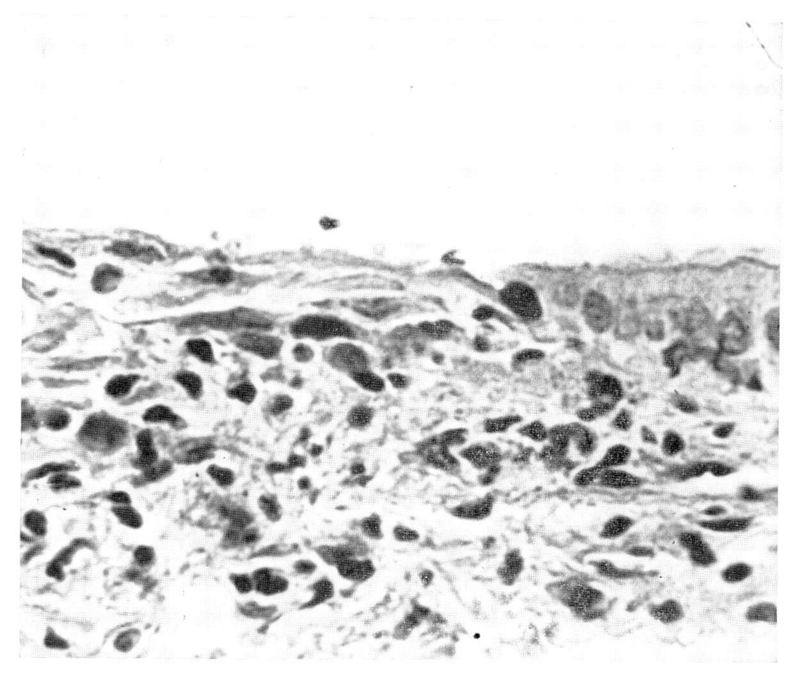

Fig. 16

HISTOLOGIC SECTION: SQUAMO-COLUMNAR JUNCTION

At the squamo-columnar junction the external epithelium of the cervix changes abruptly to columnar epithelium, as illustrated above in the photograph of endocervix. The epithelium of the endocervix is composed of a single layer of columnar cells, some of which are ciliated in the well preserved state. The position of the nuclei is near the basement membrane.

Columnar cells from the endocervix occur occasionally in the vaginal smear. Obviously they are not as frequent in the vaginal smear as in the cervical or endocervical smear. Because of their infrequent occurrence, and very different morphological appearance among squamous cells, they are often erroneously interpreted as suspicious. The endocervical cells should always be considered in the identification of any unusual group of cells.

The cytoplasm of columnar cells degenerates very rapidly. Even if smears are taken directly from the endocervix, the mjaority of the cells will show no cytoplasmic borders. This degeneration often makes the identification of individual cells as endocervical difficult. We have attempted to show in the following plates the various stages of degeneration in endocervical cells from the well preserved ciliated cell to the one with entire loss of cytoplasm.

DESCRIPTION OF ENDOCERVICAL CELLS

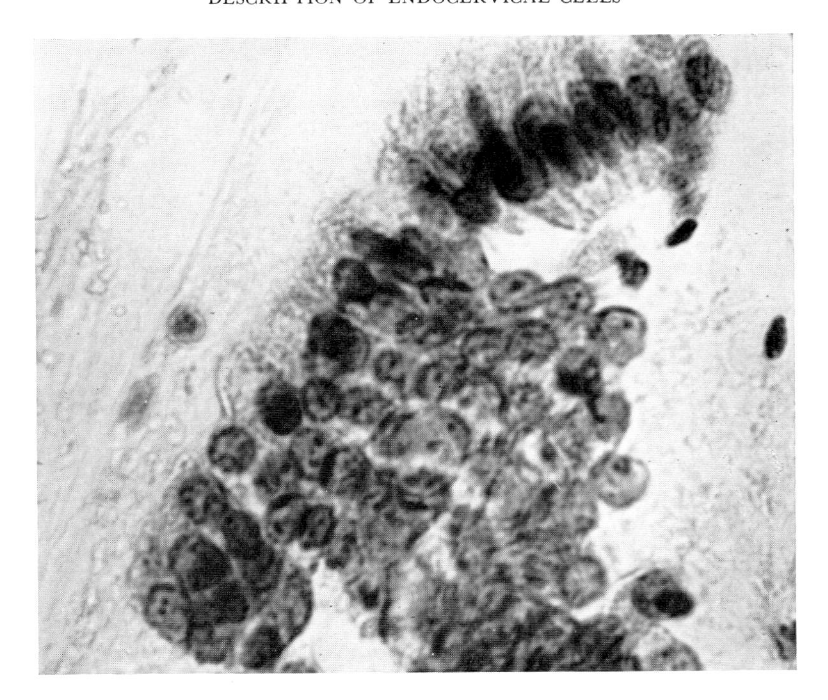

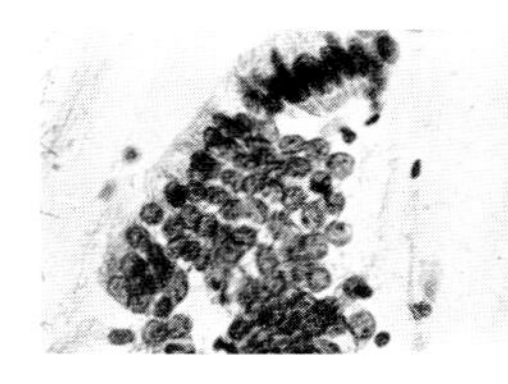

Low Power: Sheet of nuclei, some with tails of cytoplasm.

High Power:

A. Characteristics of Nucleus: 1. Smooth nuclear membrane. 2. The nuclei (see group 3) are smoothly granular with an occasional clump of chromatin. 3. The shape of the nuclei is either round or oval (compare group 1 with group 3) and there is often marked variation in the size. Notice the nuclei in the lower left corner of group 2. Occasionally the nuclei of endocervical cells appear to overlap. (See group 1.)

B. Characteristics of Cytoplasm: 1. Impression of a cell border, but not a true border. 2. The amount of cytoplasm varies greatly, depending on the state of preservation. Group 1 is well preserved and tails of cytoplasm are clearly visible at each end of the nuclei. Often in early degeneration the cytoplasm at one end of the cell is lost, and nucleus appears eccentric in remaining cytoplasm. (See group 1.) Further degeneration leaves only whisps of cytoplasm or completely free nuclei. (See lower left corner of group 2.) If endocervical cells are seen on end (group 3) by focusing up and down, a definite honeycomb appearance is apparent. 3. Staining reaction: Purplish nuclei with orange pink cytoplasm.

C. General Characteristics of Group: Columnar cells arranged either in groups or long sheets. Nuclei are spread evenly and strands of cytoplasm are seen.

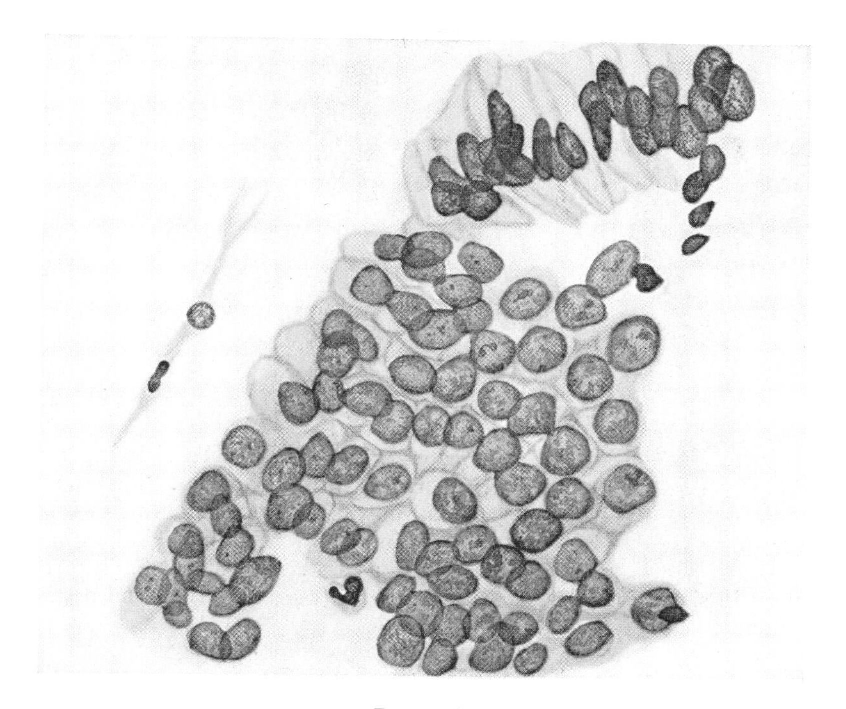

PLATE 3

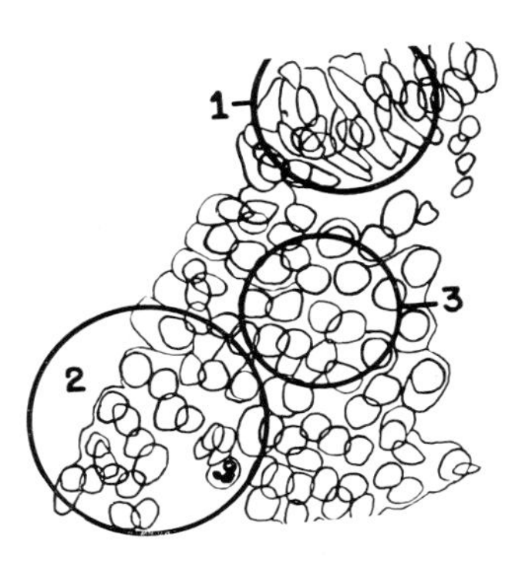

KEY TO ENDOCERVICAL CELLS

1. Group of well preserved endocervical cells in columnar formation. Individual cells are seen. Some overlapping is apparent.
2. Slightly degenerate group of endocervical cells; the upper ones still retain tags of cytoplasm to one side of the nuclei; the ones in the lower corner have lost their cytoplasm and show marked variation in the size of the nuclei.
3. Group of endocervical cells on end, giving honeycomb appearance. Clumps of chromatin are apparent in some of the nuclei.

Fig. 17

ENDOCERVICAL CELLS: GOOD PRESERVATION OF CELLS

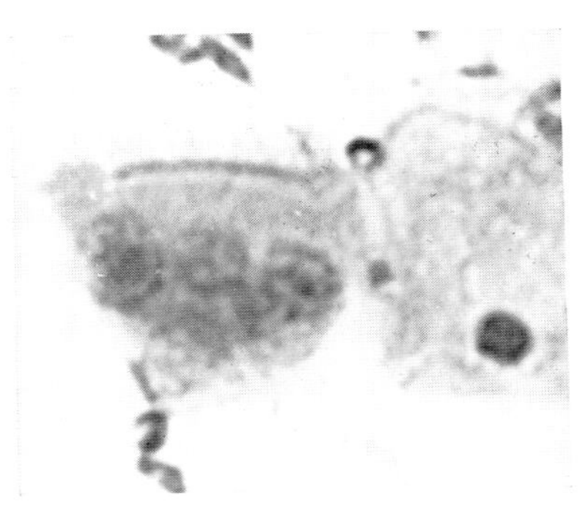

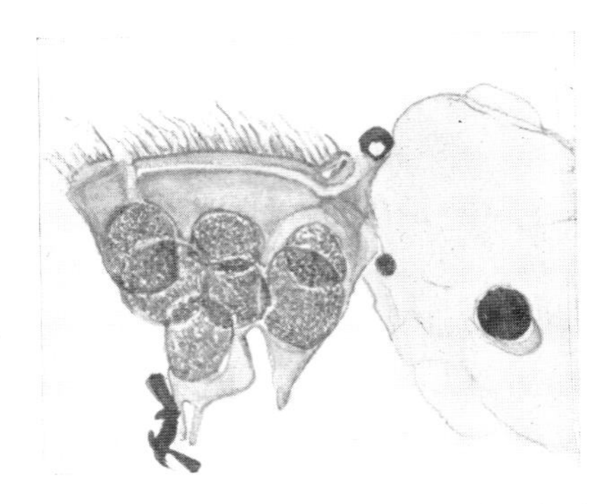

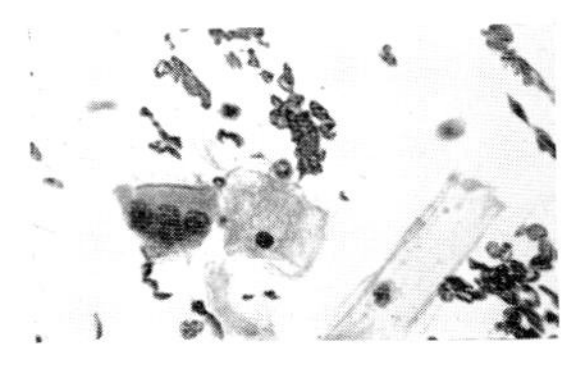

Low Power: Group of columnar cells in classical palisade formation.
High Power: Endocervical cells showing marked preservation of surface cytoplasm. Cilia are present at epithelial border and even the line of insertion can be distinctly identified. Cellular borders, however, are absent. Nuclei are finely granular and vary in size. Notice size variation in two overlapping nuclei at right of group next to cornified cell.

Fig. 18

ENDOCERVICAL CELLS: BEGINNING DEGENERATION OF CYTOPLASM

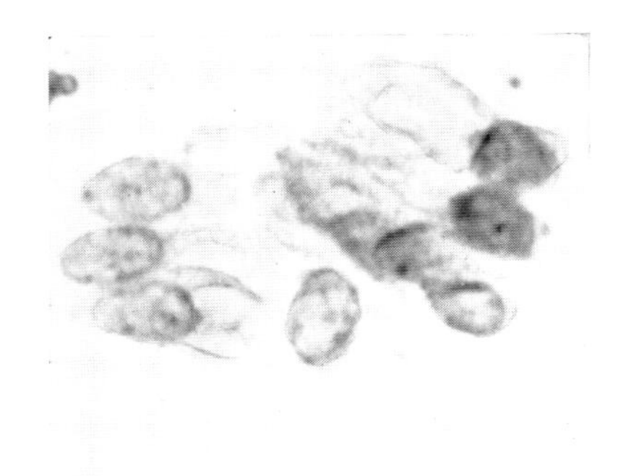

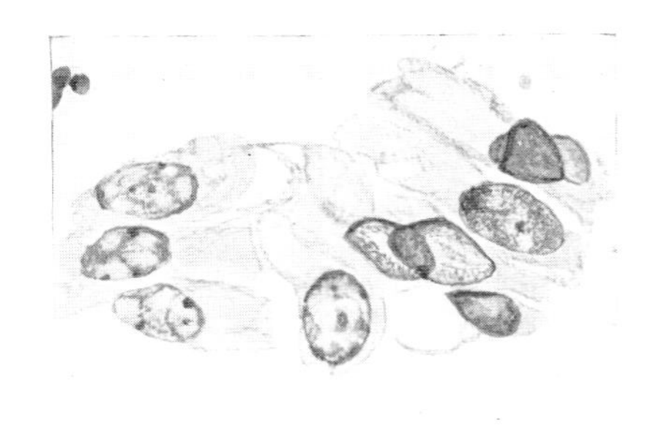

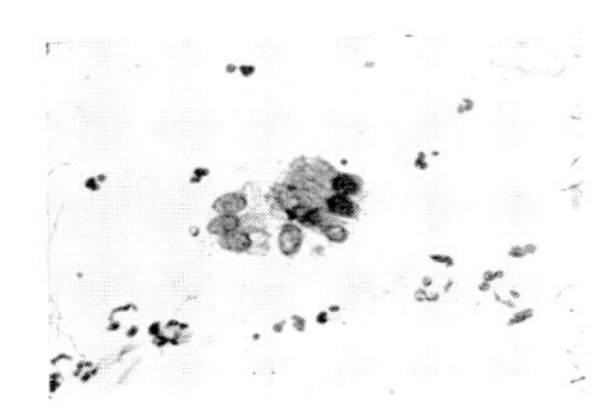

Low Power: Group of columnar cells still retaining palisade formation.
High Power: Ragged edges of cells and loss of cilia indicate beginning degeneration. Nuclei do not show much degeneration, although the lower and upper right nuclei show very little chromatin structure. The cells are still identified as columnar because of the eccentric position of the nucleus. Nuclear variation is in size rather than in shape.

FIG. 19

ENDOCERVICAL CELLS: FURTHER STAGE OF DEGENERATION

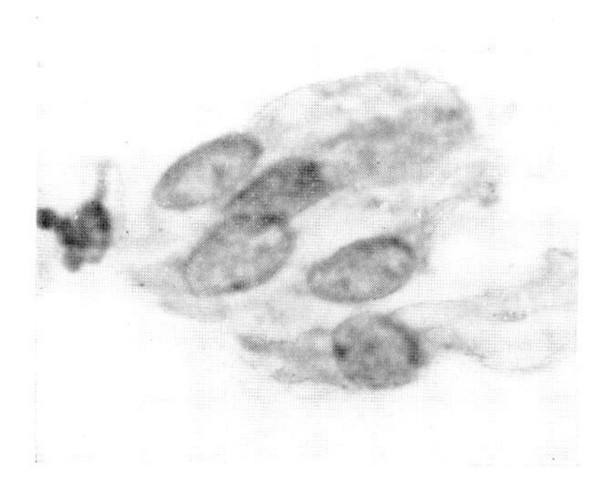

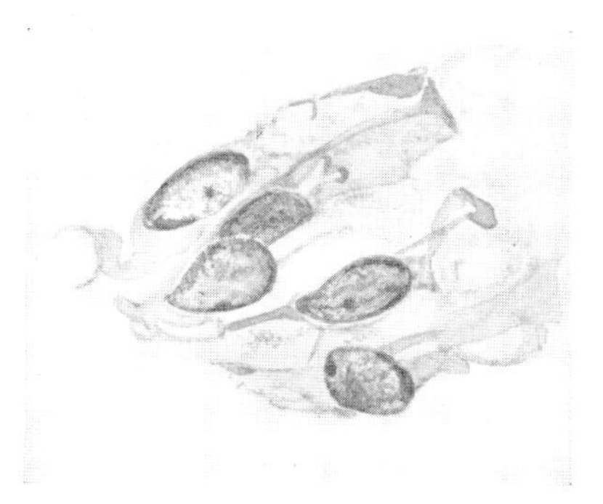

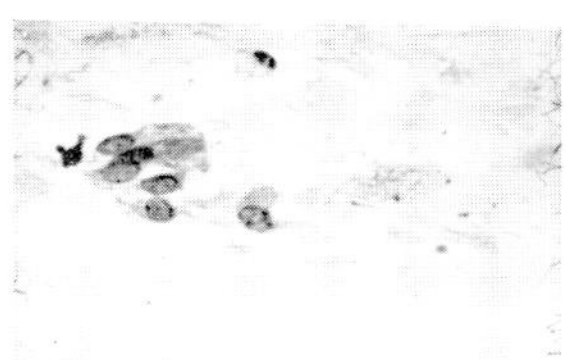

Low Power: Oval nuclei with indistinct cytoplasm.
High Power: The cytoplasm in these endocervical cells has degenerated further, and only the impression of individual cells is left. The cytoplasm is stringy, but still the nuclei are eccentric in the tags of cytoplasm which remain. Nuclei show little evidence of degeneration. Endocervical cells showing this type of degeneration are seen fairly often

FIG. 20

ENDOCERVICAL CELLS: MARKED STAGE OF DEGENERATION

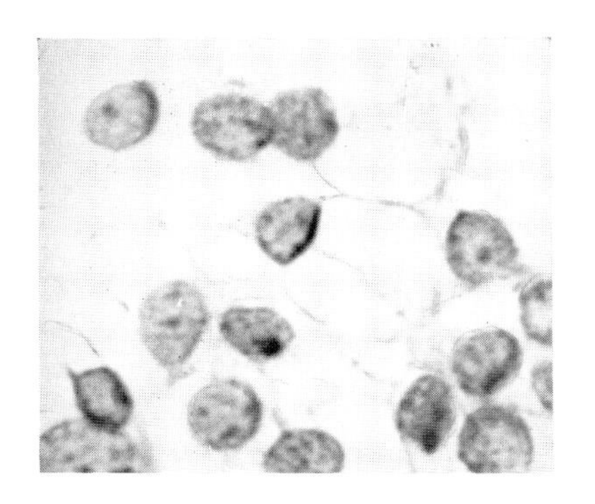

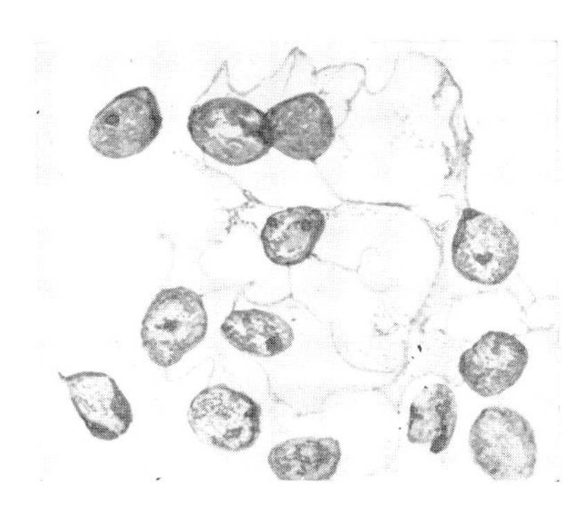

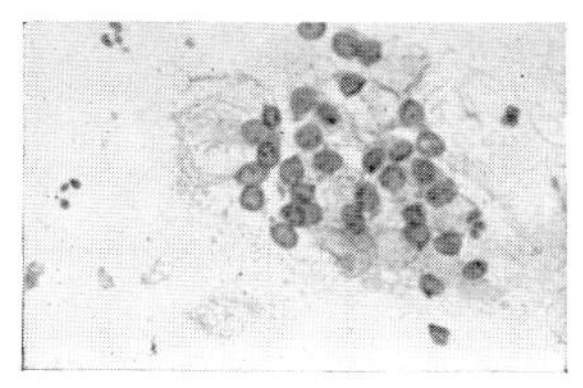

Low Power: Round or oval nuclei with background of cytoplasm.
High Power: These cells are more difficult to identify as endocervicals than those in the figure above. However, careful examination will show that the remaining wisps are not around the nuclei but always at one end, indicating that probably in the well preserved cell the nuclei were eccentric. Nuclei still retain typical pattern.

FIG. 21

ENDOCERVICAL CELLS: MARKED STAGE OF DEGENERATION

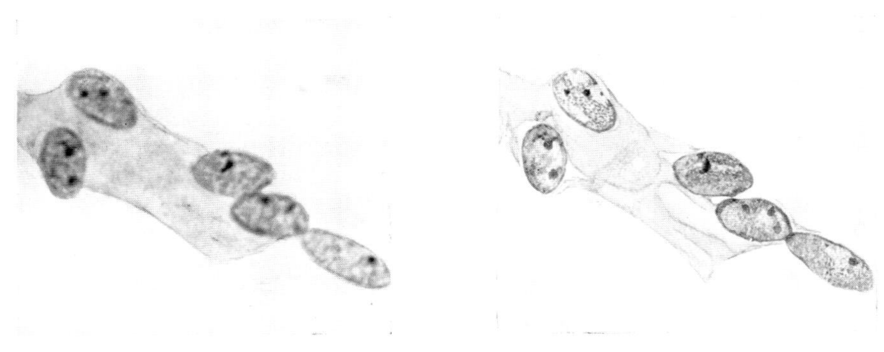

Low Power: Oval nuclei with indistinct cytoplasm. *High Power:* Identification of these endocervical cells depends largely on the nuclei which are still finely granular and vary more in size than shape. Histiocyte nuclei vary more in shape (Fig. 30), basal cell nuclei are round and evenly spaced in a background of cytoplasm (Fig. 2), and endometrial nuclei are smaller and uniform in size (Fig. 26).

FIG. 22

ENDOCERVICAL CELLS: FINAL STAGE OF DEGENERATION

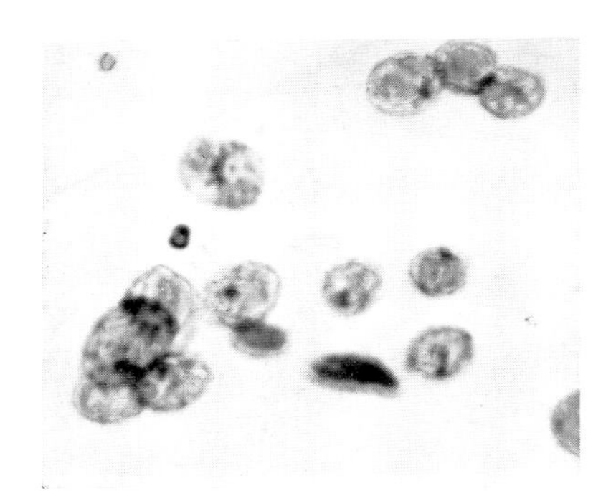

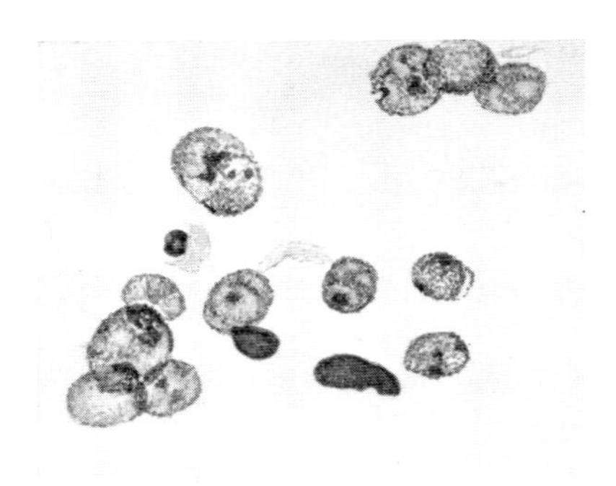

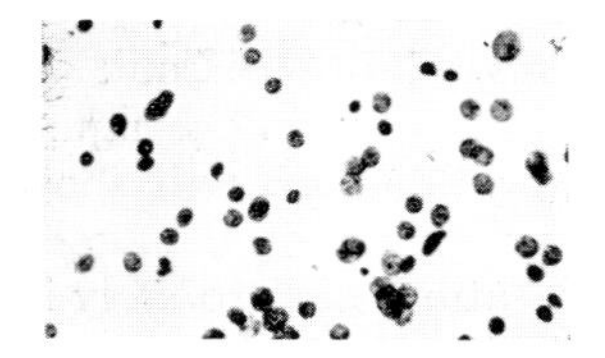

Low Power: Clusters of free nuclei.
High Power: Degeneration has caused total disappearance of cytoplasm. The nuclei show no evidence of degeneration, still being finely granular with one or more prominent clumps of chromatin. Variation is still more in size than in shape. Nuclei, free of cytoplasm, are apt to assume round appearance. In preserved cells the shape is more likely to be oval.

We have found that endocervical cells are misinterpreted as malignant more often than any other single normal cell. This has been true in our own laboratory and in smears sent to us for consultation. The relatively rare desquamation of the cells is partly responsible for the difficulty. The cells from the endocervix appear to be shed less often than any other type of epithelial cell.

We have not been able to find correlation with any definite pathologic condition and the presence of endocervical cells in the smear. We have found atypical endocervical cells in cases of severe chronic endocervicitis, but in the majority of such cases there are no endocervical cells present.

We have stressed the fact that degenerative changes are common in vaginal smears, and that various stages of degeneration must be recognized. In no group of cells are degenerative changes as important as in the endocervical. In general, extreme degenerative changes in squamous cells take place within the nucleus, *i.e.*, karyorrhexis and pyknosis. There is one exception to this—the complete degeneration of the cytoplasm of precornified cells when great numbers of bacteria are present in the smear. In columnar cells, such as endocervical and endometrial, the degeneration is commonly cytoplasmic. The columnar cells do not retain their cellular borders in most instances. If the cells were seen with cellular borders and eccentric nuclei, identification would be relatively simple. However, such preservation of the cells is the exception rather than the rule.

Difficulties in Interpretation: Though the nuclei of both endocervical cells and histiocytes are eccentric in position a distinction may be made between the two types on the basis of nuclear shape and cytoplasmic characteristics. The nuclei of histiocytes vary more in shape than in size. In endocervical cells the reverse is true. The cytoplasm of the endocervical cell is irregular in density or is coarsely vacuolated. Histiocytes show a regular finely vacuolated cytoplasm.

In the rare instances where single endocervical cells desquamate and retain their cellular borders, the cells may assume a round form rather than the typical columnar pattern. They may be distinguished from basal cells by their eccentric nuclei and by the fact that the cytoplasm is not as dense. The cytoplasm may have distinct vacuoles.

As stated above, the identification of occasional endocervical cells as benign is by far the greatest difficulty. In cases of severe chronic endocervicitis, the endocervicals show tremendous variation in size. The size variation between one nucleus and another may be as great as ten times. This type of size variation in free nuclei is apt to be considered as a malignant characteristic. However, in such groups the chromatin variation from nucleus to nucleus will be very slight. The nuclear pattern appears smoothly granular with one or two prominent clumps of chromatin. Such groups have nuclei in which the nuclear border is definite but does not have chromatin condensed at the periphery of the cells, as is common in undifferentiated malignant cells.

Another helpful point is that the nuclei vary little in shape, being either oval or round. Undifferentiated malignant cells present irregularity in shape. These cells can be identified as benign if careful consideration is given to their nuclear structure.

Epithelium of Endocervix and Endometrium

The student should not conclude that all endocervical cells are difficult to identify as benign. If the cytoplasm of the cell has not disappeared completely, their identification is relatively simple, since the eccentric position of the nucleus in the cytoplasm is an extremely helpful point. Too, the size variation described above is seen infrequently and groups of nuclei, larger than endometrial, without cytoplasm, can be classified as endocervical with a fair degree of certainty.

General Criteria for the Identification of Endocervical Cells:

1. Nuclear chromatin is finely granular with several small prominent clumps of chromatin. Nuclear border is distinct. Variation is in size rather than in shape.
2. If cytoplasm is present, it is not dense. Distinct vacuolization may occur, often near the eccentric nucleus. Cellular borders are not distinct.

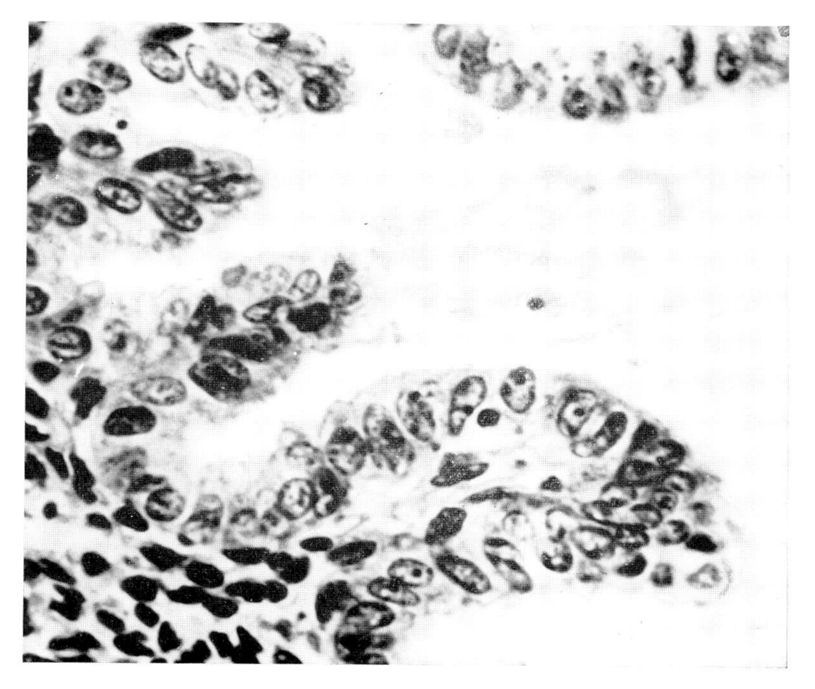

FIG. 23

HISTOLOGIC SECTION: ENDOMETRIUM

Cells from the endometrium are seen frequently in vaginal smears. Unfortunately they seldom are completely preserved. They retain the nuclear morphology, but the cytoplasm degenerates rapidly and it is exceptional to see groups of cells which retain the definite columnar pattern seen in tissue sections, as in the figure above.

The normal occurrence of endometrial cells in the vaginal secretion is during or immediately following the menstrual period. The smear at this time characteristically contains many degenerate superficial cells, many fresh red blood cells and groups of endometrial cells. Since the presence of blood in the vaginal smear is an abnormal finding except during the menstrual period, it is important to have information concerning the cycle of the case before vaginal smears are interpreted. Obviously it is also of importance to know whether the patient is considered to be postmenopausal. If this information is not available, occasionally a smear will be reported as suspicious which actually represents the bleeding period of a normal cycle.

Single endometrial cells occur infrequently and are extremely difficult to identify. Since they are the approximate size of small lymphocytes, the cytologic distinction between the two cells is not an easy one. However, since by far the greatest number of endometrial cells occur in groups, this difficulty is not of any real consequence.

Criteria for identification of endometrial cells and examples of the various types are given on the following pages.

DESCRIPTION OF ENDOMETRIAL CELLS

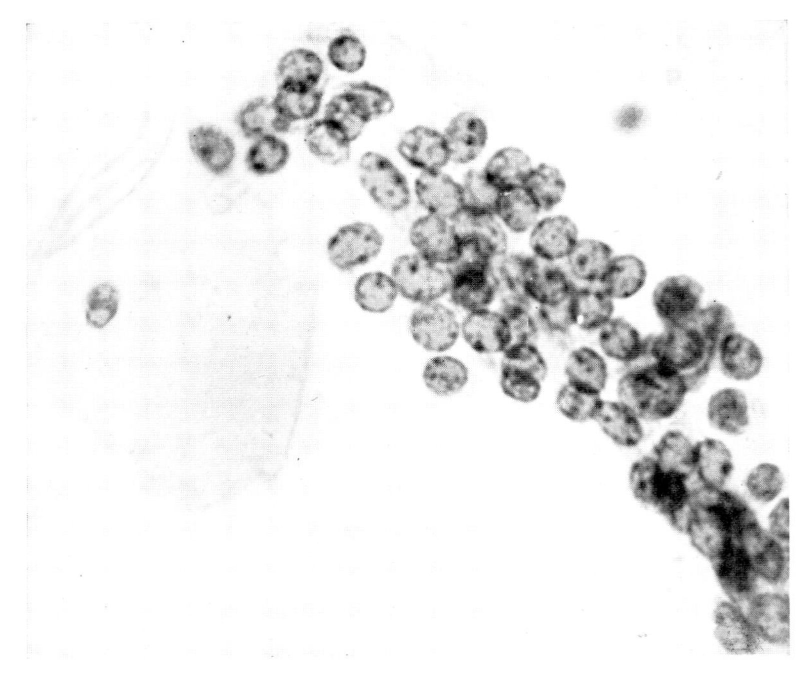

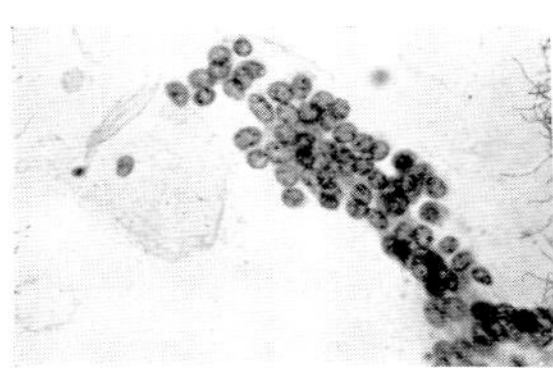

Low Power: Group of nuclei uniform in size and shape.

High Power:

A. Characteristics of Nucleus: 1. Distinct nuclear membrane. 2. All of the nuclei are darker than either basal cell or histiocyte nuclei (see group 3) and the chromatin is arranged in small clumps which give the cells a more active appearance. 3. Endometrial nuclei are relatively small. Compared to size of precornified nucleus in cell 1, they only appear larger when the cells are fresh. Size variation of more than twice is unusual. Contrast the size of the right upper nucleus to the others in group 2. 4. The shape of the nuclei is round or oval, usually round (see group 3), and there is a tendency to overlap. (See group 4.)

B. Characteristics of Cytoplasm: 1. No cytoplasm is visible, but careful examination sometimes reveals a faint degenerate rim with no definite cellular border. 2. The cytoplasm of endometrial cells degenerates rapidly and disappears almost completely by the time the cell reaches the vagina from the endometrium. 3. Staining reaction: Bluish purple nuclei. Cytoplasm, when present, is light green or pink.

C. General Characteristics of Group: The overall picture is uniform, *i.e.*, only slight variation in size and shape and chromatin pattern. The nuclei tend to clump tightly and overlap each other and little or no cytoplasm is seen.

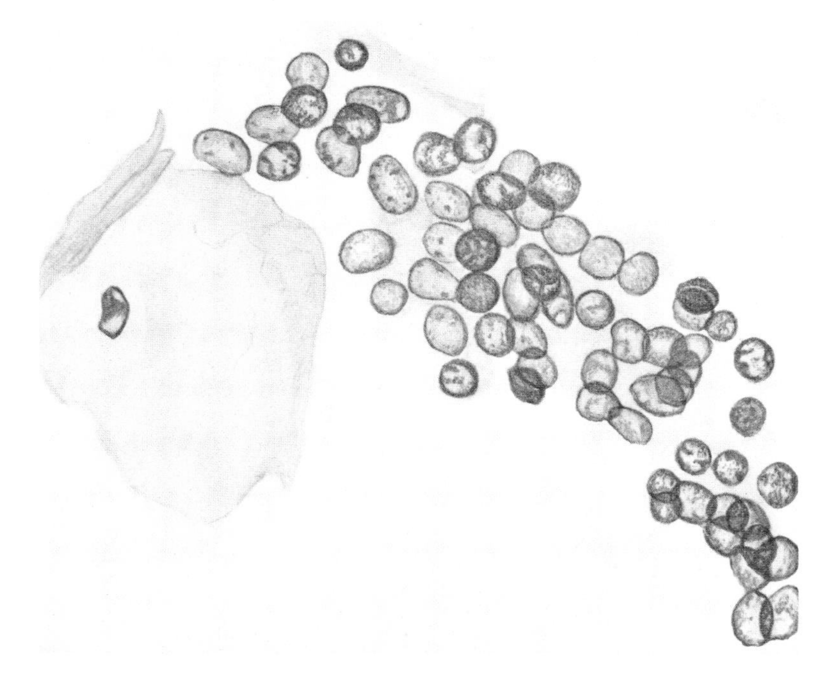

PLATE 4

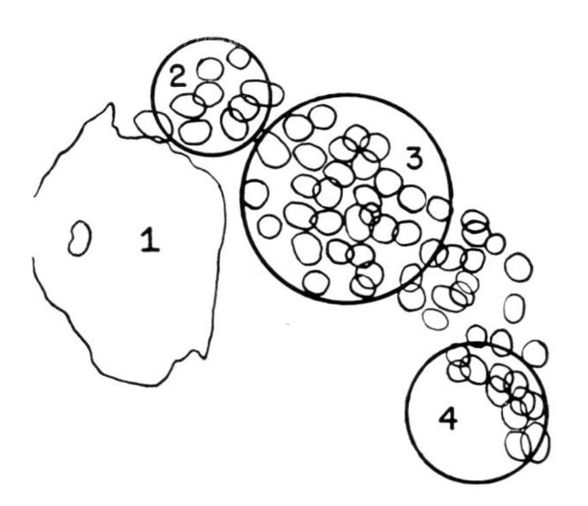

KEY TO ENDOMETRIAL CELL PLATE

1. Precornified cell with a vesicular nucleus and adequate transparent cytoplasm.
2. Round or slightly oval endometrial nuclei, the right upper one being half the size of the rest in the group.
3. Endometrial nuclei, fairly uniform in size and shape with clear nuclear membranes showing a tendency to clumping.
4. Round endometrial nuclei overlapping each other. No cytoplasm is visible.

FIG. 24

ENDOMETRIAL CELLS: WELL PRESERVED

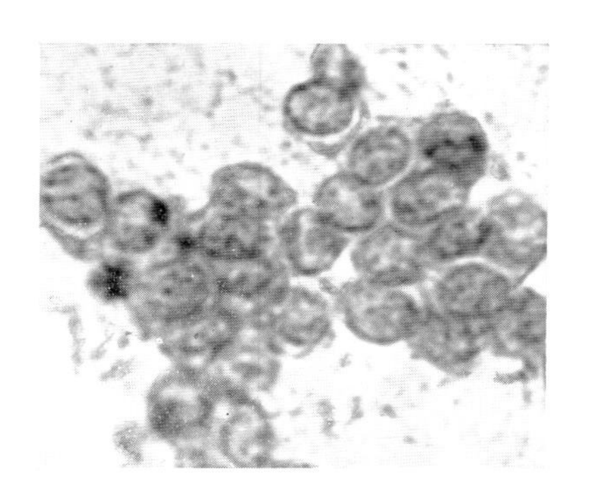

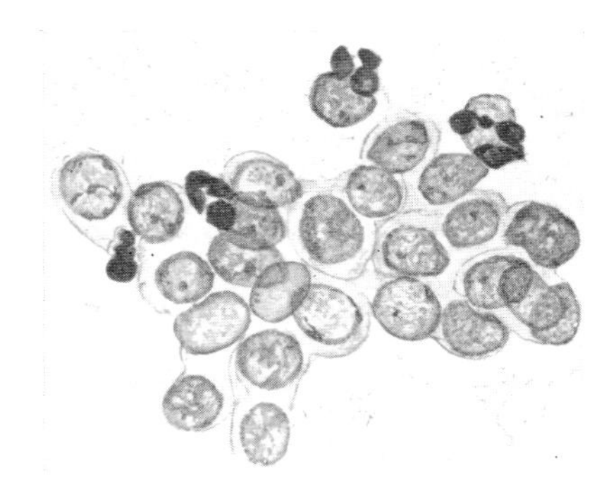

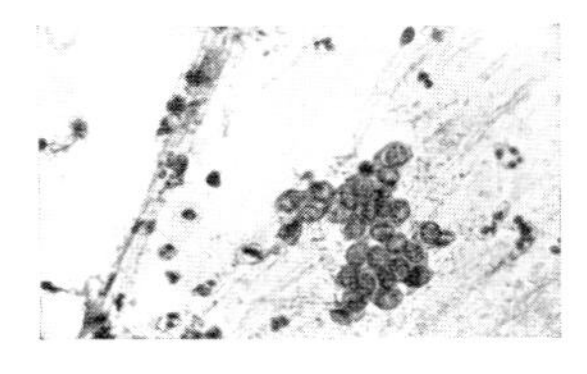

Low Power: Group of round nuclei with little size variation.
High Power: The nuclei have small clumps of chromatin and clear, nuclear borders. The size and shape are constant and some of the nuclei are overlapping. There is evidence of individual rims of cytoplasm around the nuclei, and the cellular outline is reasonably clear, indicating that these cells are fresh and well preserved.

FIG. 25

ENDOMETRIAL CELLS: DEGENERATE

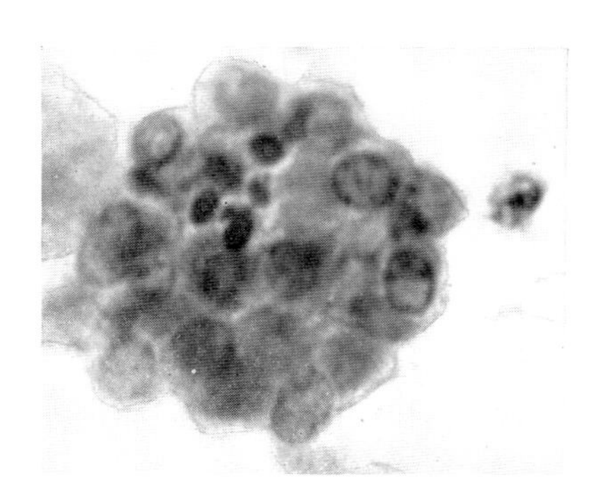

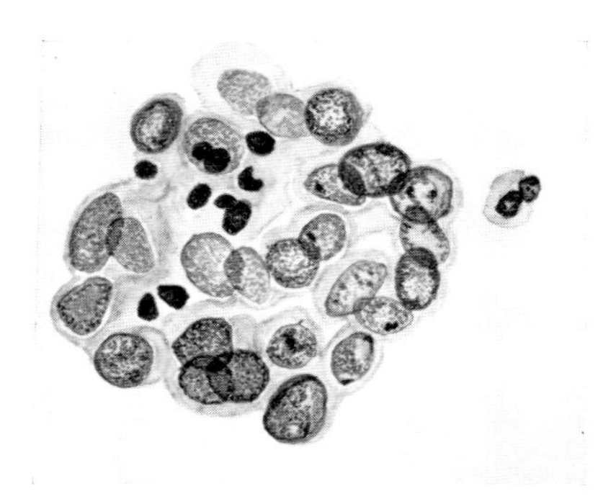

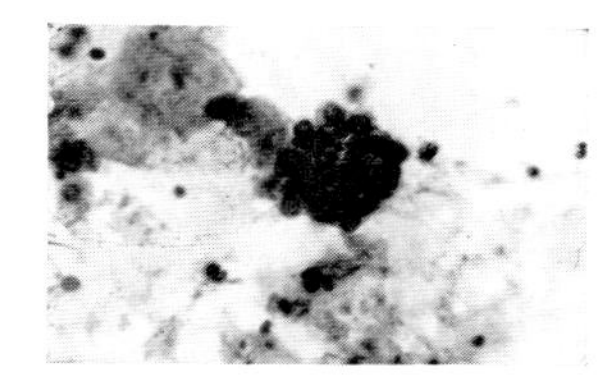

Low Power: Round cluster or granular regular nuclei.
High Power: These endometrial cells show further degenerative changes than those in the figure above. There is a distinct background of cytoplasm but for the most part cellular borders are absent. Nuclei are relatively uniform in size and chromatin is finely granular. There is considerable overlapping of nuclei. Leukocytes and histiocytes are often present in groups of endometrial cells.

FIG. 26

ENDOMETRIAL CELLS: FURTHER STAGE IN DEGENERATION

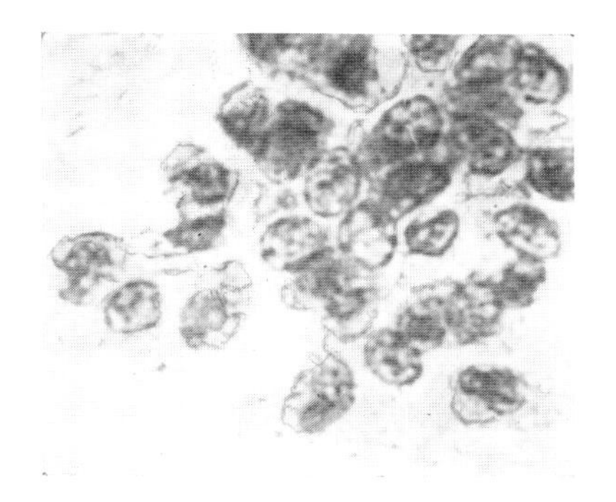

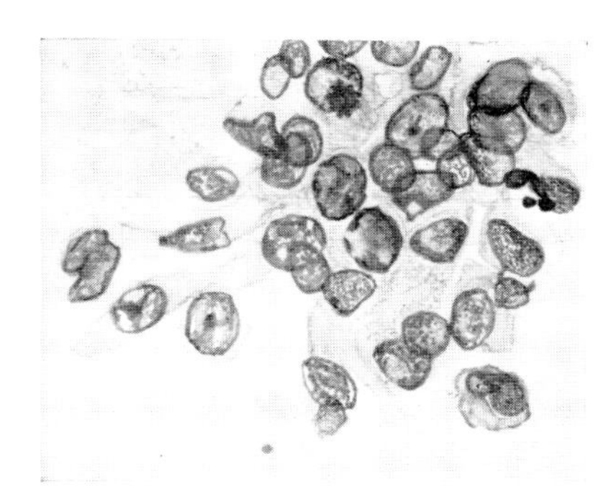

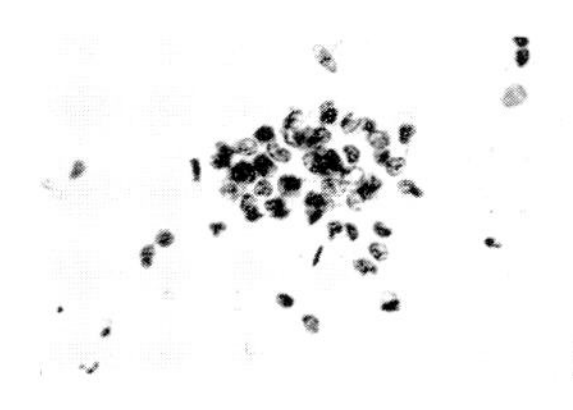

Low Power: Group of nuclei without any cytoplasm.
High Power: The nuclei are degenerate as is shown by the wrinkling and irregular shapes of the clear nuclear borders. (See the cells on the left of the group.) The chromatin structure is consistent and we identify these nuclei as endometrial because of their size and grouping. A small histiocyte is in the lower right-hand corner.

FIG. 27

ENDOMETRIAL CELLS: PYKNOTIC NUCLEI

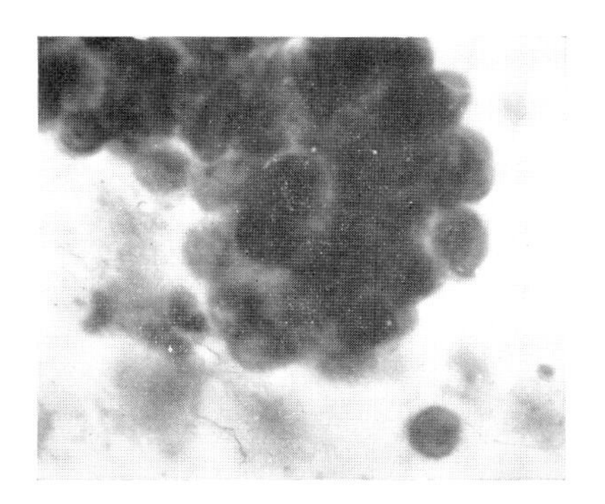

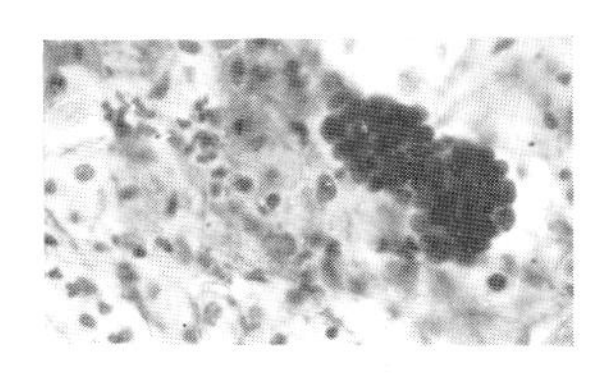

Low Power: Group of tightly clustered, dark nuclei in field of superficial cells.
High Power: These nuclei show very little chromatin structure, the majority of them being smoothly dark. Pyknosis of the nuclei is most common in endometrial cells. They appear hyperchromatic, but the intensity of stain is due to a condensation of chromatin rather than an increase. Nuclei are uniform in size.

Epithelium of Endocervix and Endometrium

The identification of endometrial cells is of real practical significance. Their occurrence in a postmenopausal smear is abnormal and warrants a report of "suspicious" with repeat smears requested. The cells accompanying the endometrial are superficial, indicating some ovarian activity even though the patient is considered clinically to have had cessation of cyclic ovarian function. Since a probable correlation has been established between hyperplasia and carcinoma of the endometrium, any endometrium which is active enough to desquamate cells in a postmenopausal woman should be considered as suspicious until proven otherwise.

As was mentioned in the discussion of endocervical cells, the distinction between endocervical and endometrial depends upon size. Endometrial cells are smaller, and exhibit less variation in size. Obviously definite distinction between the two types is often personal interpretation.

Difficulties in Interpretation: The cells which are desquamated from the endometrial mucosa may occasionally be confused with histiocytes. The nuclei of both types of cells are approximately the same size. If the histiocyte is well preserved with the characteristic foamy cytoplasm, identification is simple. It is when the group is degenerate that the distinction between histiocytes and endometrial cells becomes difficult. The shape of the nuclei is of real aid in this case since nuclei with the characteristic bean shape will usually be present and can be identified as histiocytes. Endometrial cells retain an oval or round shape. Occasionally a group may contain both types of cells clustered together.

The second differentiation which is difficult is that of endometrial cells and undifferentiated malignant cells. The endometrial cells may be identified by their smoothly granular nuclei and their small size and lack of variation. They ordinarily occur in dense clusters. However, in one instance endometrial cells are not easily identified as benign. Cells desquamating from an extremely hyperplastic endometrium are apt to show great variation in size and for that reason appear suspicious. Nevertheless, the nuclei still appear smoothly granular. Often the hyperplasia is due to estrogen administration. It is extremely important that information about any such medication should be available for the cytologist. The smears can then be interpreted much more intelligently.

General Criteria for Identification of Endometrial Cells:

1. Nucleus is finely granular with definite nuclear border. Size of nucleus is approximately one to three times the size of a leukocyte. Nuclei do not vary appreciably in size.
2. Cytoplasm is usually an indistinct background, though occasionally well preserved cells will show a faint cell outline.
3. Cells tend to overlap and to appear in clumps.

CHAPTER III

CELLS NOT ORIGINATING FROM EPITHELIUM OF FEMALE GENITAL TRACT

The vaginal smear contains many cells which do not originate from the epithelium lining the genital tract. The great majority of vaginal smears contain many leukocytes. These are often so numerous that they obscure much of the picture. Red blood cells are seen fairly frequently. Though these should cause no difficulty in interpretation, it is, perhaps, necessary to mention them.

Polymorphonuclear leukocytes are common in all vaginal smears with one exception. A smear indicating a high estrin level, composed of a majority of cornified cells, will in most instances be very clean. There will be very few or no leukocytes present. Evidence of infection cannot be correlated accurately with the number of leukocytes present since their occurrence in great numbers is common. The polymorphonuclear leukocytes are usually very degenerate in the smear preparations. It is usually impossible to classify them, though occasionally well preserved cells are seen and can be identified by the usual staining characteristic of the granules in the cytoplasm. Lymphocytes retain their usual characteristic appearance. (See Plate 5.) They appear as small round cells with deeply staining nuclei and may be identified quite accurately. They may be occasionally mistaken for small endometrial cells, but the density of the nuclear chromatin is greater than that of the endometrial cell and the cell is smaller in size. The presence of leukocytes in smears is very helpful. It is the one cell which remains constant in size. Considerable variation takes place in all normal epithelial cells. The leukocyte is an excellent point of orientation for size differences. If the student will make an effort to compare nuclei to the leukocyte and to determine the relative size a valuable working standard for size relationships can be established.

Blood appears in the vaginal smear in three forms. It may occur as fresh blood, in which case well preserved red blood cells are seen. Fibrin may be present in the smear. It is in long strands of orange staining material. The fibrin should be differentiated from mucus which also occurs in long strands. Mucus stains a deep blue basophilia. The third form is blood pigment, which appears in small granules sometimes collected together in a fernlike arrangement. The blood pigment also stains orange. (See Plate 8.) Occasionally it will be stained a deep green. It is essential to recognize the presence of blood, since its appearance, except at the menstrual period, is an abnormal finding.

Of the cells not originating in the epithelium of the genital tract, the histiocyte is the most important from a diagnostic point of view. It is the only one which gives difficulty in interpretation. There are three types of histiocytes: The large phagocytic cell has obvious large vacuoles filled with ingested material. (See Fig. 29.) The common small histiocytes have oval, round or bean-shaped nuclei arranged eccentrically in finely vacuolated cytoplasm. The phagocytic properties of these small cells may occasionally be observed. The third type we refer to as a foreign body giant cell. These are extremely large cells with many typical histiocytic nuclei arranged peripherally in a loose cytoplasmic network. (See Plate 6.) The figures on the following pages illustrate the various types of histiocytes and their degenerative changes.

DESCRIPTION OF HISTIOCYTES

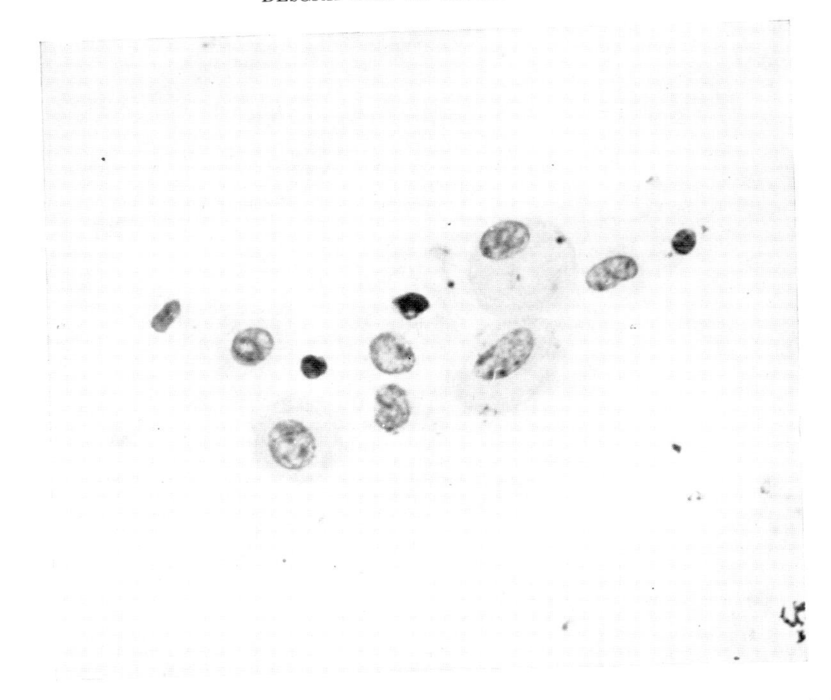

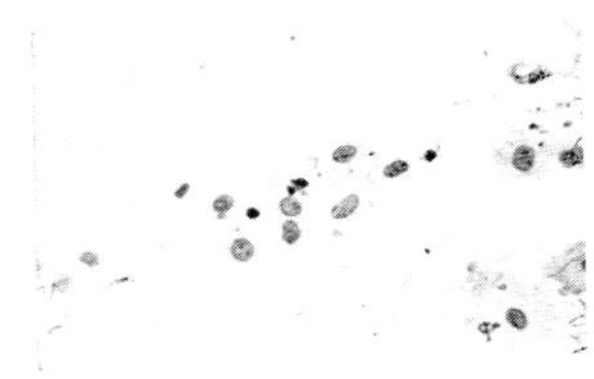

Low Power: Group of cells with light staining cytoplasm and eccentric nuclei.

High Power:

A. Characteristics of Nucleus: 1. Definite nuclear membrane. 2. All of the nuclei appear relatively active because of the small clumps of chromatin which become obvious on the transparent background of finely granular particles. (See cells 1, 3, 6 and 8.) Compare these nuclei with the small dense nucleus of cell 2, a lymphocyte. 3. The nuclei are either oval, round or bean-shaped. Contrast cells 1, 3, 4 and 6 which are round, to cell 8 which is oval; cells 5 and 7 show the characteristic bean-shaped nucleus. 4. The nuclei are usually eccentric in relation to the cytoplasm, as in cells 1, 3 and 4, and may vary in size. Compare cell 1 to 6.

B. Characteristics of Cytoplasm: 1. In well preserved histiocytes there is a cell border (see cells 3, 4 and 6), otherwise, as in cells 7 and 8, the border is fuzzy and indistinct. 2. The cytoplasm may vary in shape and amount (compare cell 5 with 8), but appears foamy or finely vacuolated. 3. Staining reaction: Bluish purple nuclei with faintly staining greenish cytoplasm.

C. General Characteristics of Group: Eccentric oval, round or bean-shaped nuclei with finely vacuolated cytoplasm. The nuclei are relatively small and vary more in shape than in size.

PLATE 5

KEY TO HISTIOCYTE PLATE

1. Small round histiocyte nucleus with faintly staining cytoplasm.
2. Small lymphocyte.
3. Large well preserved histiocyte with a leukocyte superimposed.
4. Round histiocyte with foamy vacuolated cytoplasm.
5. Typical bean-shaped histiocyte nucleus, surrounded by foamy cytoplasm which shows a good cell border.
6. Large round nucleus with obvious clumps of chromatin and adequate finely vacuolated cytoplasm.
7. Small bean-shaped nucleus in faintly staining cytoplasm which has an indistinct cell border.
8. Oval-shaped nucleus, vacuolated cytoplasm, indistinct cell border.

Fig. 28

HISTIOCYTES: VACUOLATED CYTOPLASM

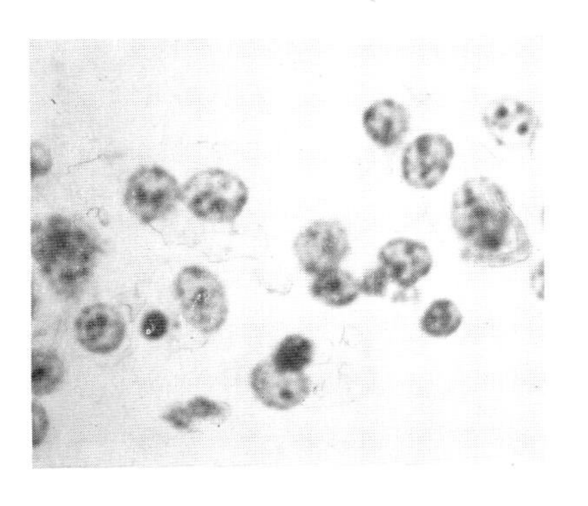

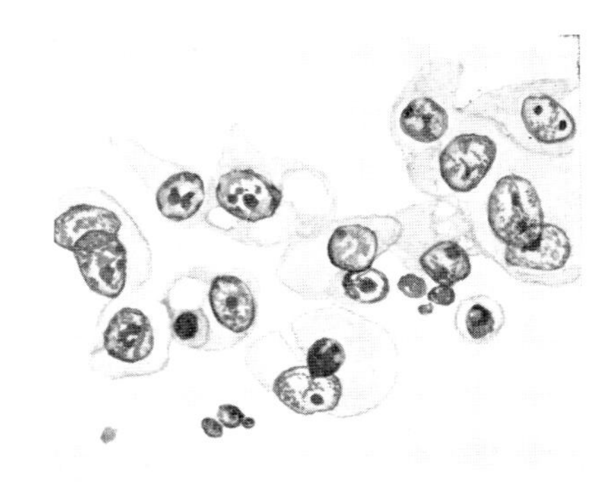

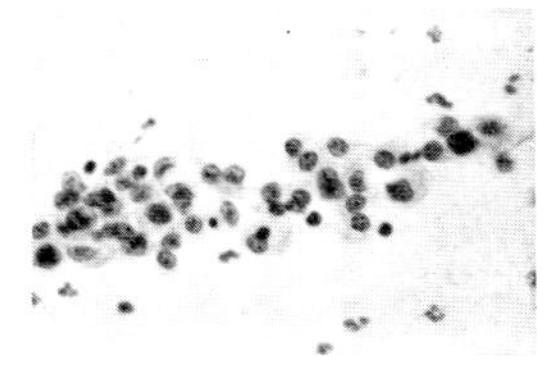

Low Power: Group of nuclei with faintly staining cytoplasm.

High Power: Round or oval nuclei, varying slightly in size with constant chromatin structure, *i.e.*, finely granular particles with obvious clumps of chromatin. The cytoplasm is vacuolated and cellular borders are indistinct, especially in the group of nuclei at the right where it is impossible to see any cellular outline. Several cells appear double-nucleated.

Fig. 29

HISTIOCYTES: LARGE PHAGOCYTIC CELL

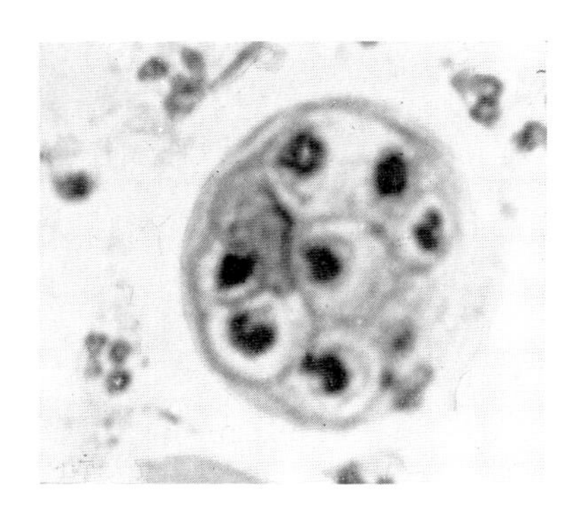

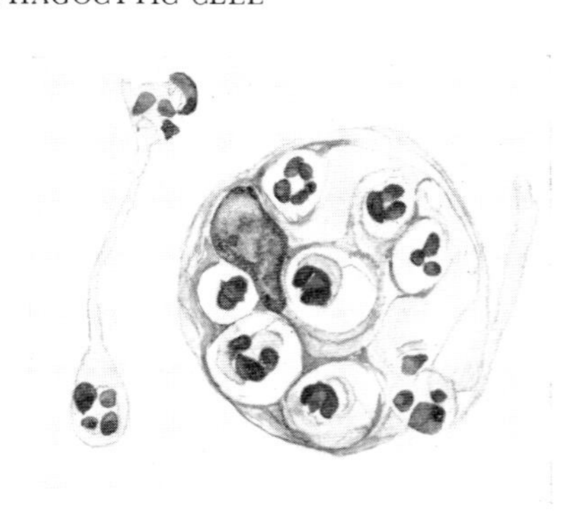

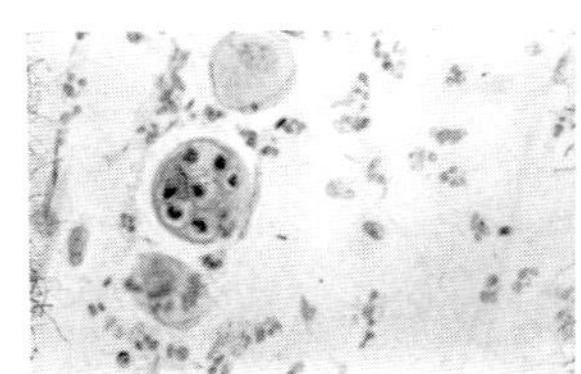

Low Power: Large vacuolated cell with ingested material, outer layer basal cells.

High Power: This large phagocytic histiocyte has seven vacuoles each containing an ingested leukocyte. The nucleus still retains the characteristic chromatin pattern of histiocytes though its shape is somewhat bizarre due to the extreme vacuolization of the cell. Its position is eccentric in the cytoplasm.

FIG. 30

HISTIOCYTES: DEGENERATE

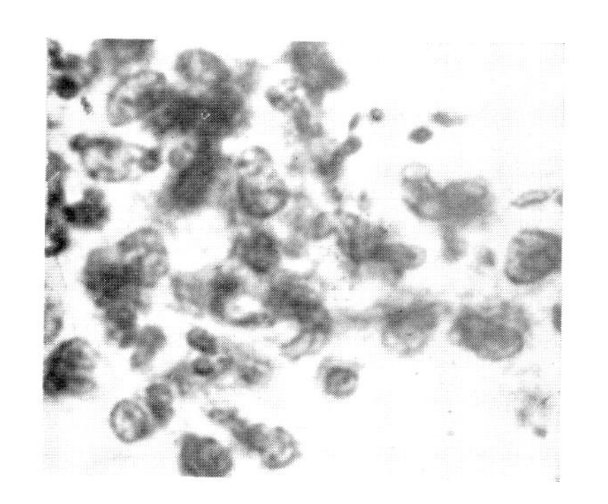

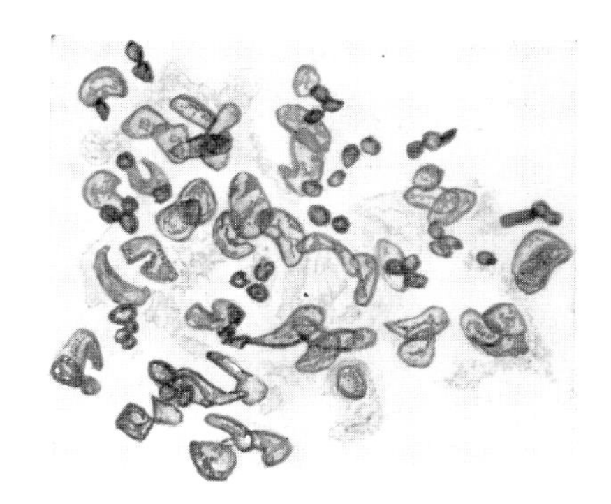

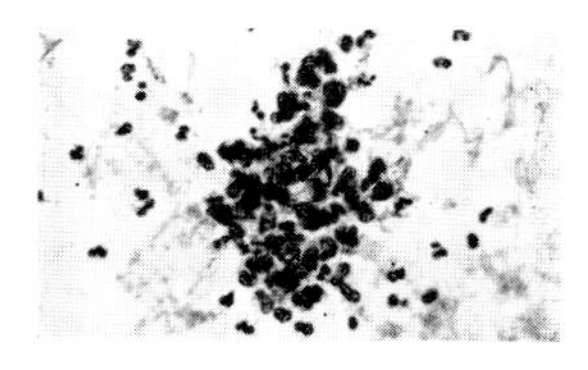

Low Power: Group of dark-staining nuclei with no visible cytoplasm.
High Power: In focusing down on this group it is seen that many of the nuclei are small and degenerate but are typically bean-shaped, which identifies them as histiocytes. The nuclear borders are sharply outlined but the cytoplasm has completely disappeared, because of degeneration. Compare these histiocytes to well preserved ones in the figure below.

FIG. 31

HISTIOCYTES: WELL PRESERVED

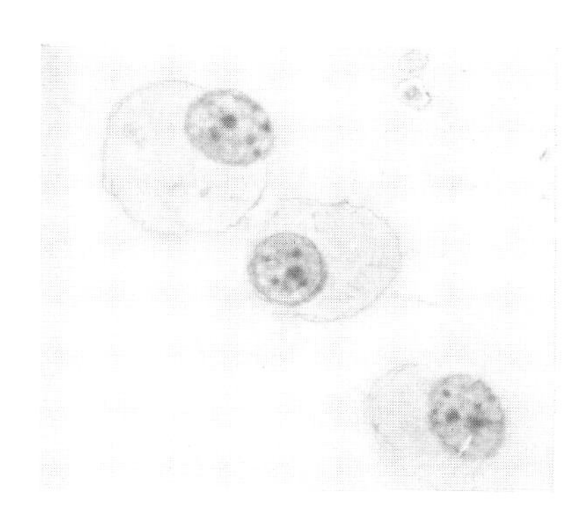

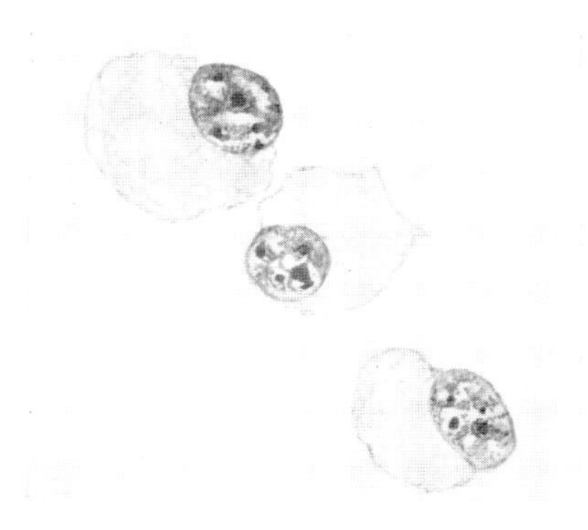

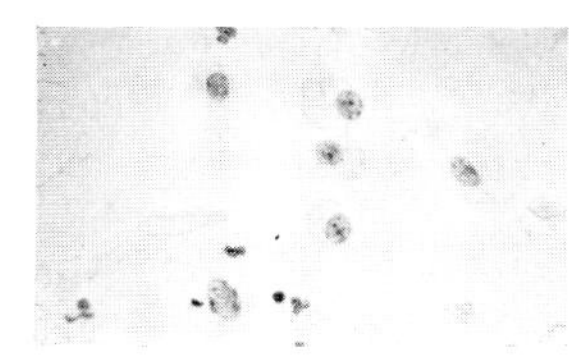

Low Power: Three nuclei, even in size with light-staining cytoplasm.
High Power: These three cells are larger and in a much better state of preservation than ordinarily seen. They illustrate the distinctive criteria for identification of histiocytes: round or oval nuclei with occasional small condensations of chromatin, nuclei eccentric in position and fine vacuolated cytoplasm.

FIG. 32

HISTIOCYTES: MITOTIC FIGURE

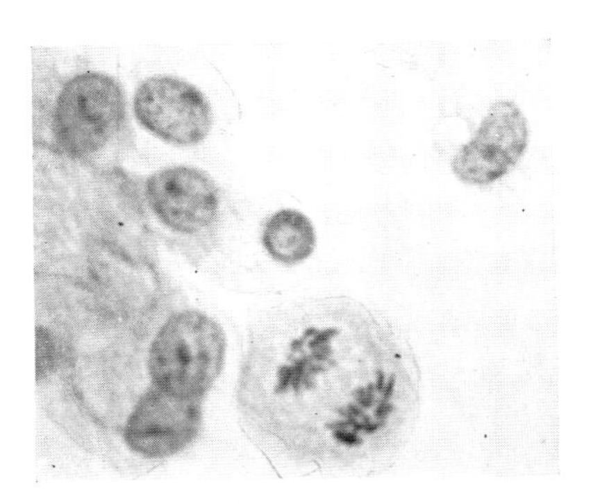

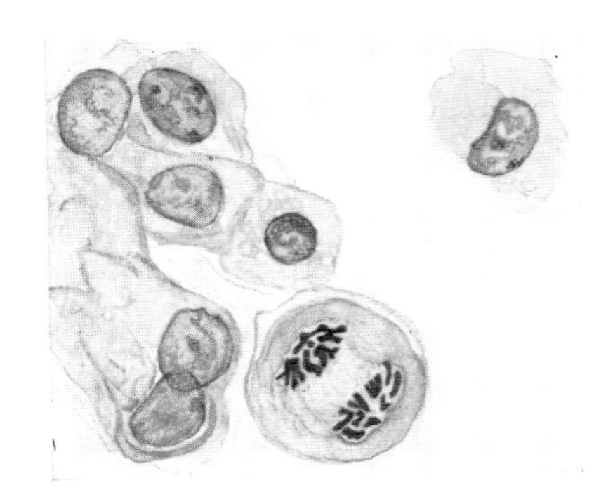

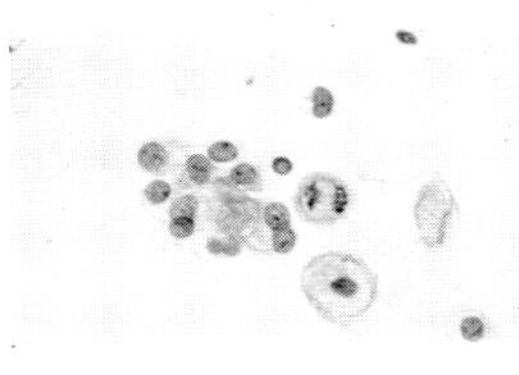

Low Power: Group of histiocytes, one obvious mitosis.

High Power: The nuclei are uniform in size, eccentric in position in the characteristically foamy cytoplasm and an occasional clump of chromatin can be seen. The large histiocyte is in the anaphase stage of mitosis, the chromosomes being clearly visible. Mitotic figures are not indicative of malignancy as they are seen frequently in histiocytes.

FIG. 33

HISTIOCYTES: ELONGATED FORM

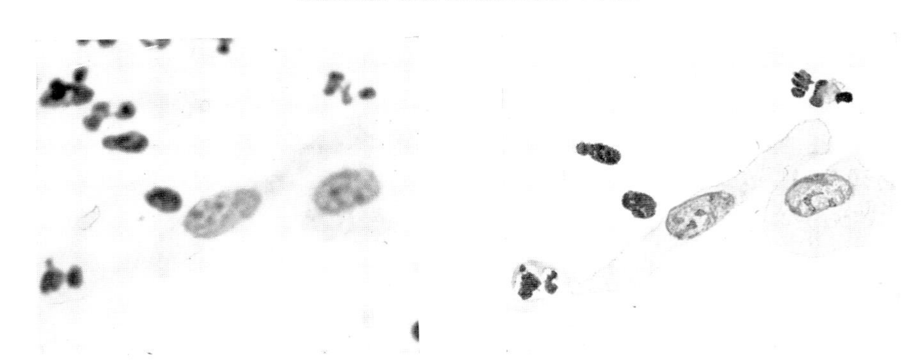

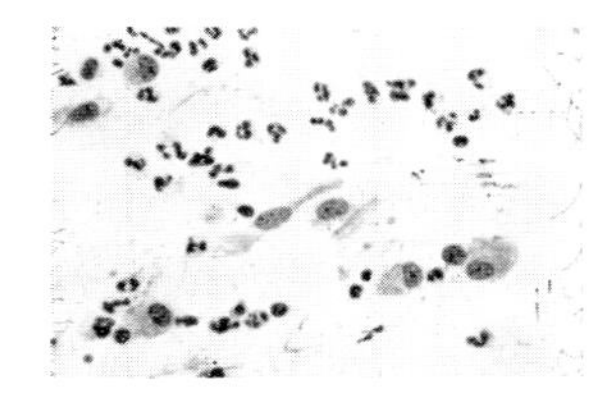

Low Power: Even granular nuclei in field of leukocytes.

High Power: The chromatin particles of these nuclei are evenly distributed with occasional clumps which do not appear hyperchromatic. The cytoplasm of these cells is characteristically light-staining and finely vacuolated, but the elongated shape of the cell on the left is unusual as compared to the typical form of the cell on the right.

The small histiocyte occurs frequently in the vaginal smear. We have not found that it has any specific significance and have not been able to correlate the presence of these cells with any definite pathological process. One would suspect that they occur in cases of chronic infection of the genital tract but this has not always proven to be true.

Difficulties in Interpretation: As in epithelial cells, the intact cell is easily recognized, if careful attention is paid to the criteria for identification of histiocytes. It is only when they are degenerate that difficulty arises. Groups of small histiocytes which have lost their cell borders and are tightly clumped may be mistaken for undifferentiated malignant cells, since the nuclei appear very dark. The density of color is due, not to an increase in chromatin content, but to a condensation of chromatin content. Such groups may be identified as degenerate benign cells for the following reasons: their nuclei are pyknotic, their variation is more in shape than in size and, finally, such groups often have numerous bean-shaped nuclei to assist in the identification. Loss of cellular borders makes the cells more difficult to classify, but if careful consideration is given to nuclear structure, size and shape, the cells should cause minimal difficulty.

Single histiocytes which have lost their cellular borders may also be confusing, especially if the nucleus is oval or round rather than the characteristic bean-shape. Again, attention to the chromatin structure of the nucleus should identify the cells as histiocytes. The nuclei of single histiocytes are not as apt to be degenerate as of those seen in groups. They have the usual smoothly granular chromatin with one or more small condensations of nuclear material, and a distinct nuclear border. They are, in most instances, three to four times the size of a leukocyte. If there is any doubt about the origin of the cell, a search should be made in the smear for typical intact histiocytes and the nuclear structure compared.

A third type of histiocyte which may be troublesome is one in which the cells assume an elongated form. (See Fig. 33.) The cytoplasm is stretched out at either end of the nucleus, so that the nucleus appears to be central in position rather than eccentric. However, these should not be confused with the differentiated malignant cell if careful attention is paid to the nuclear and cytoplasmic characteristics. The cytoplasm is faint, irregular and vacuolated as in the typical form of histiocytes. The nucleus is similar in all respects to those seen in the classical type of histiocyte.

General Criteria for Identification of Small Histiocytes: Small round cell with faint bluish-gray cytoplasm which on close inspection is finely vacuolated. Cellular border is often indefinite and seldom as sharply defined as in squamous epithelial cells. Nuclei are oval, round or bean-shaped, and eccentric in position. Variation is more in shape of the nuclei than in size. The nuclear border is definite, usually more sharply defined than in the squamous epithelial cells. The chromatin of the nucleus is finely granular and often there are small condensations of chromatin. The cells vary in size from about twice to five times the size of a leukocyte.

DESCRIPTION OF A FOREIGN BODY GIANT CELL

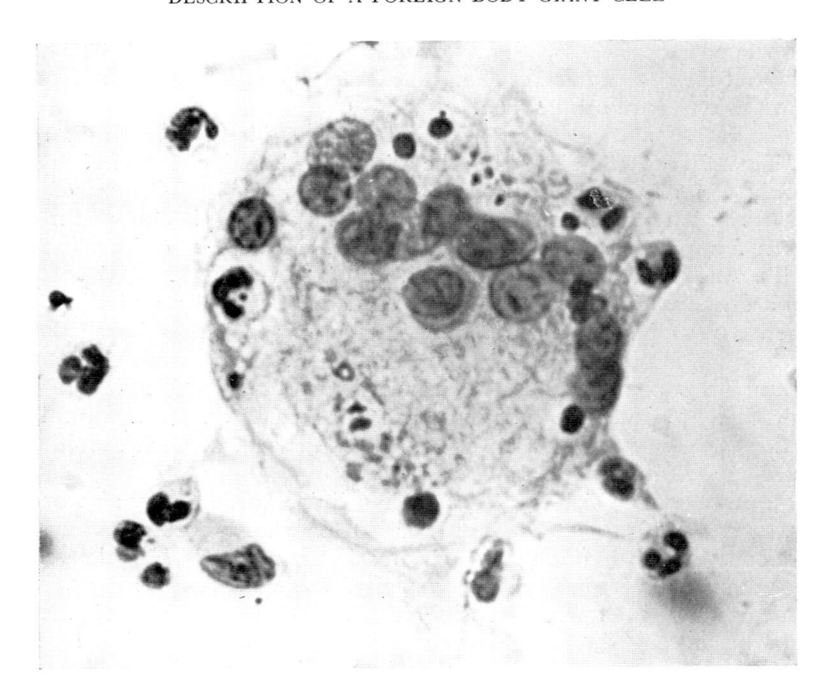

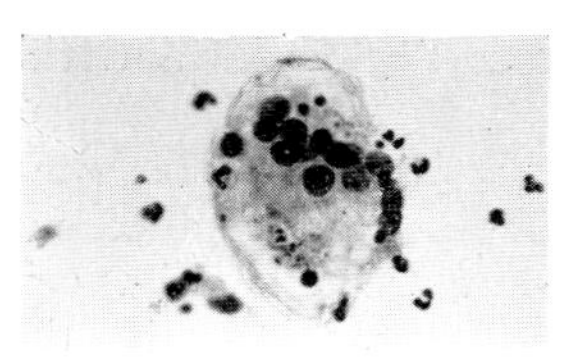

Low Power: Single cell containing many small nuclei.

High Power:

A. Characteristics of Nucleus: 1. The nuclei in group 3 have clear nuclear borders and the same chromatin content, *i.e.*, finely granular particles with obvious clumps of chromatin giving the nuclei an active appearance. The structure of these nuclei is the same as that of cell 1, a small histiocyte with a bean-shaped nucleus and light-staining cytoplasm. 2. The nuclei in foreign body giant cells are either round or oval and there is slight variation in their size. (See group 3.) 3. There are two main characteristics of foreign body giant cell nuclei. The nuclei are always arranged peripherally and tend to overlap each other.

B. Characteristics of Cytoplasm: 1. There is a question as to whether this is true cytoplasm or mucus. However, there are vacuoles present and the leukocytes 2 and 4 appear to be phagocytosed, therefore we believe this to be a true cell with phagocytic qualities. 2. There is no distinct cytoplasmic border. 3. Staining reaction: Bluish purple nuclei with either blue or pink cytoplasm.

C. General Characteristics of the Foreign Body Giant Cell: A large multinucleated cell whose nuclei have the characteristics of histiocytes, *i.e.*, round or oval nuclei with finely granular chromatin and clear nuclear border. The nuclei are arranged peripherally in cytoplasm which has an indistinct border.

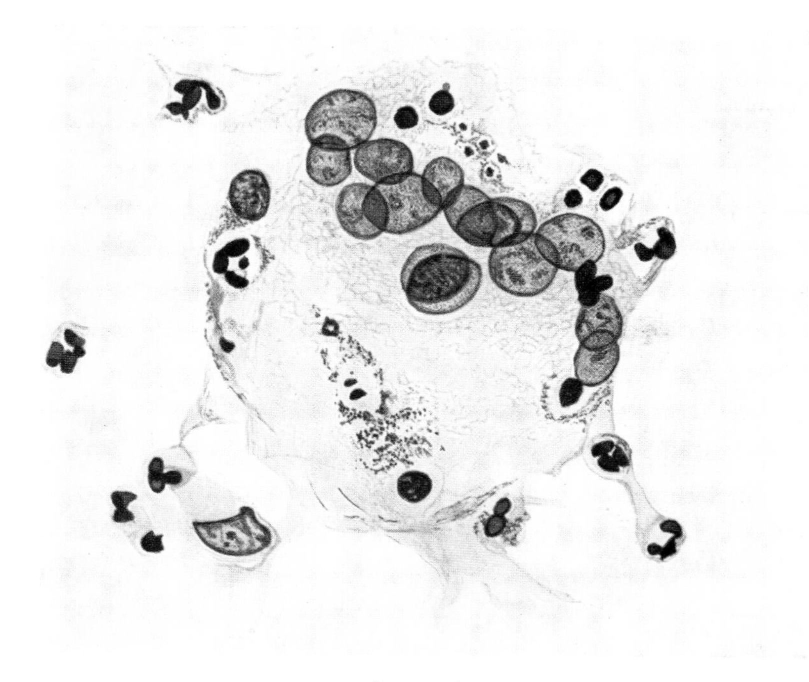

PLATE 6

KEY TO FOREIGN BODY GIANT CELL PLATE

1. Single histiocyte with bean-shaped nucleus and faintly staining cytoplasm.
2. A phagocytosed leukocyte.
3. Group of histiocyte nuclei varying slightly in size with uniform chromatin content. These nuclei are arranged peripherally in the large cell and have tendencies to overlap each other.
4. Phagocytosed lymphocyte.

Foreign body giant cells are encountered only rarely in vaginal smears. They are seen most frequently in postradiation smears, but are also seen when no irradiation has been administered. It is of interest that they were frequent in a smear from a patient with tuberculosis of the endometrium. They have been present in positive smears of patients with carcinoma of the cervix or of the endometrium. No definite correlation has been established between any specific pathologic condition and the presence of these cells in the smear.

Difficulties in Interpretation: Foreign body giant cells cause confusion for only one reason, *i.e.*, they are multinucleated giant cells. The number of nuclei, occasionally as many as forty, impresses the inexperienced as evidence of abnormal division of nuclei. They may be then misinterpreted as malignant giant cells. The lack of variation in chromatin structure, the regularity in shape and size and the peripheral arrangement of the nuclei easily identify them as benign.

Degeneration may cause the nuclei of these giant histiocytes to lose their borders. When the nuclear borders are absent, the nuclei appear fused and assume the form of a giant nucleus. Nevertheless, the cell can still be identified as benign, since the chromatin is finely granular and in this instance stains lightly.

General Criteria for the Identification of Foreign Body Giant Cell:

1. The many nuclei are oval or round in shape, vary little in size and are arranged peripherally in the cytoplasm. Chromatin is finely granular.
2. Cytoplasm is an indistinct background with an indefinite cellular border.

CHAPTER IV

SQUAMOUS CELL CARCINOMA OF THE CERVIX

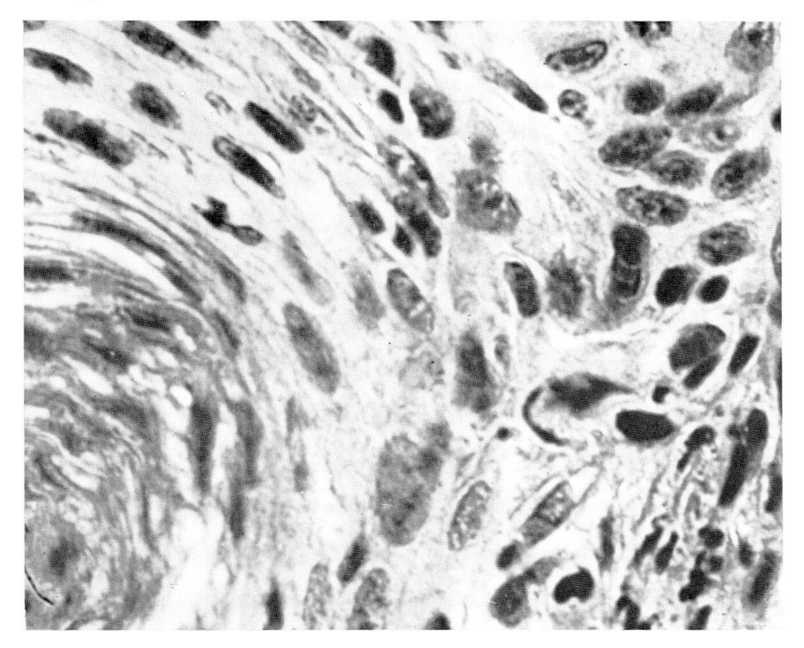

FIG. 34

HISTOLOGIC SECTION: SQUAMOUS CELL CARCINOMA OF CERVIX

Squamous cell carcinoma of the cervix exhibits marked pleomorphism of cellular structure as seen in the section from this type of tumor. When these cells are sloughed from the surface of the lesion, even more marked pleomorphism is seen, since they are not confined by space limitations as in tissue.

For purposes of teaching, we have attempted a rough classification of the types of tumor cells desquamated from squamous carcinoma of the cervix. This classification includes two main groups: undifferentiated and differentiated malignant cells. The basis for distinction between the two groups is presence of cytoplasm and outline of cellular borders. In undifferentiated cells cytoplasm is an indistinct background with absence of cell borders. Differentiated malignant cells have well defined cytoplasm and fairly definite cellular borders. Differentiated malignant cells from squamous carcinoma are of three main types: 1. The fiber cell is a thin elongated cell with a hyperchromatic elongated nucleus. 2. The tadpole cell has, as the name implies, a "head" containing the deep-staining nucleus and a "tail" of cytoplasm. 3. The third type of differentiated malignant cell shows the greatest degree of maturation. It resembles an inner layer basal, with the exception that the nucleus is hyperchromatic and cytoplasmic-nuclear ratio is abnormal. Illustrations and criteria for identification of various types are presented on the following pages.

DESCRIPTION OF UNDIFFERENTIATED MALIGNANT CELLS

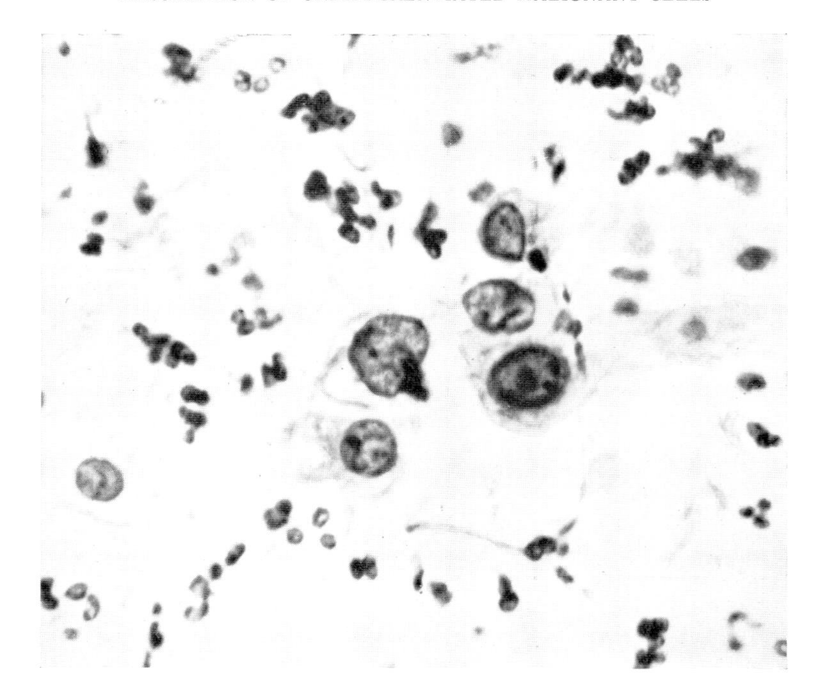

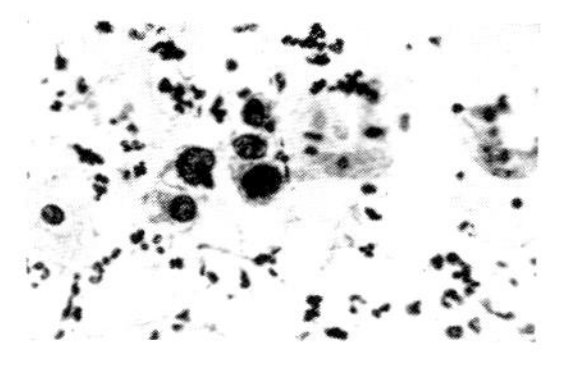

Low Power: Five large active nuclei in a messy field.

High Power:

A. Characteristics of Nucleus: 1. Sharply outlined nuclear borders. 2. All of the malignant nuclei are hyperchromatic, owing to the many irregular clumps of chromatin which have collected mainly at the edge of the nuclei, giving them a darker look. (See cell 5.) Cells 2 and 3 show prominent groups of chromatin not to be confused with nucleoli, which stain red. 3. The nuclei show marked variation in size and shape. Compare cell 1, which is small and round, with cell 2, which is twice as large and triangular in shape. 4. Squamous undifferentiated nuclei are separated from one another, whereas in adenocarcinoma, the nuclei tend to pile or overlap.

B. Characteristics of Cytoplasm: 1. Undifferentiated cells have little or no cytoplasm, but when it is present, the cellular borders are indistinct. 2. Cells 1, 2, 3 and 4 have no visible cytoplasm, but cell 5 shows a faint rim of cytoplasm with a hazy indefinite cell border. 3. Staining reaction: Purplish blue nuclei.

C. General Characteristics of Group: Five large discrete hyperchromatic nuclei varying in size and shape, with sharp nuclear borders and no definite cellular borders.

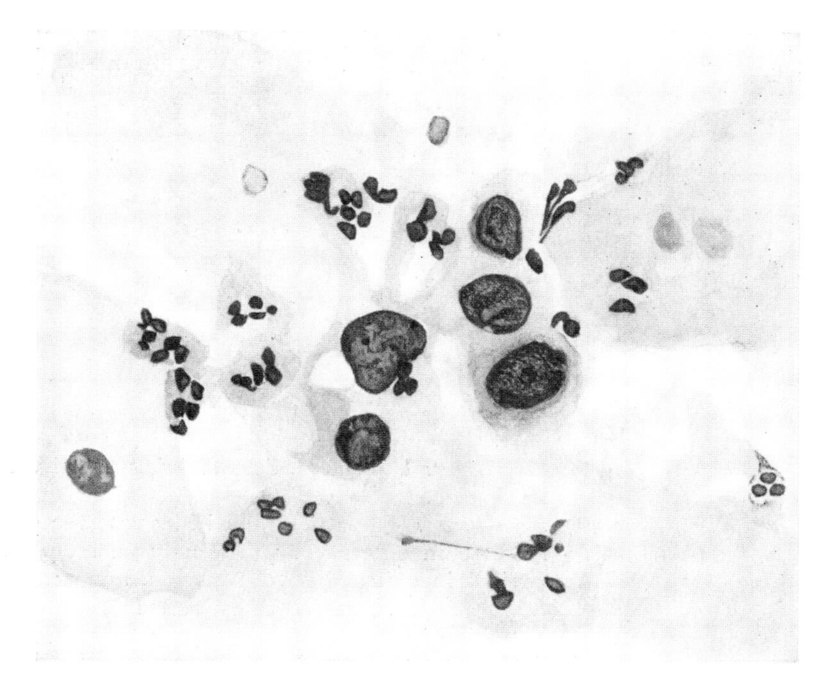

PLATE 7

KEY TO UNDIFFERENTIATED CELL PLATE

1. Round, undifferentiated malignant cell with a clear nuclear border, strands of chromatin and no visible cytoplasm.
2. Large, triangular-shaped undifferentiated cell showing prominent clumps of chromatin and a sharp nuclear border.
3. Undifferentiated cell with two obvious clumps of chromatin. This nucleus is smaller than the rest. There is no cytoplasm.
4. Slightly irregular, but definite nuclear border. Coarse clumps of chromatin and no cytoplasm.
5. Undifferentiated cell with a dense hyperchromatic nucleus and a small amount of cytoplasm with an indistinct cell border. The rest of the field contains cornified and precornified cells and leukocytes.

FIG. 35

UNDIFFERENTIATED MALIGNANT CELLS: EXTREME HYPERCHROMASIA

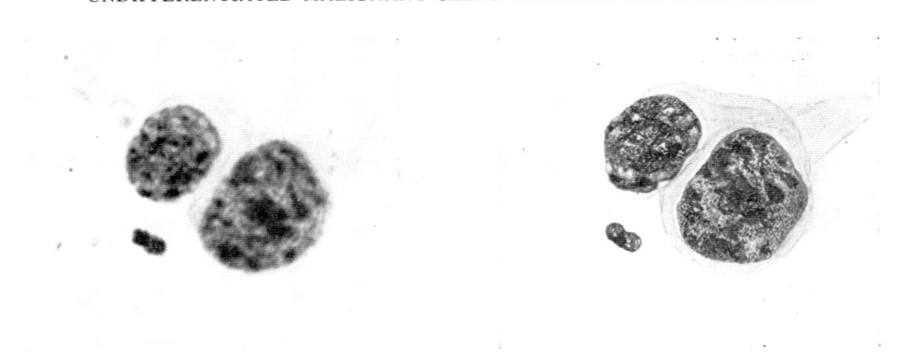

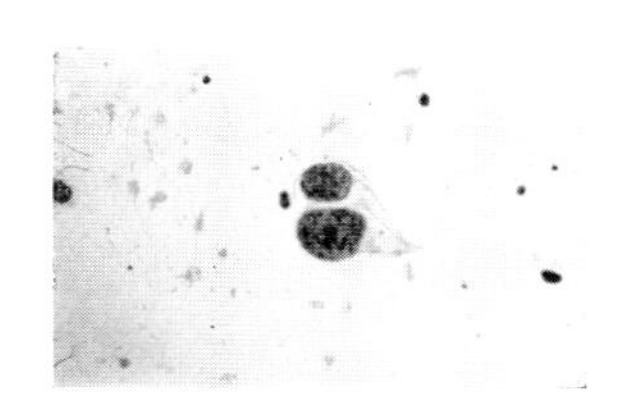

Low Power: Two large, deep-staining nuclei varying in size.
High Power: These two nuclei illustrate the great increase in chromatin which may be seen. The chromatin material is very uneven, appearing in thick strands, dark clumps and fine granules. Variation in the chromatin arrangement is probably the most reliable criterion for the identification of malignant cells. Compare these nuclei with Plate 7.

FIG. 36

UNDIFFERENTIATED MALIGNANT CELLS: VARIATION IN SIZE

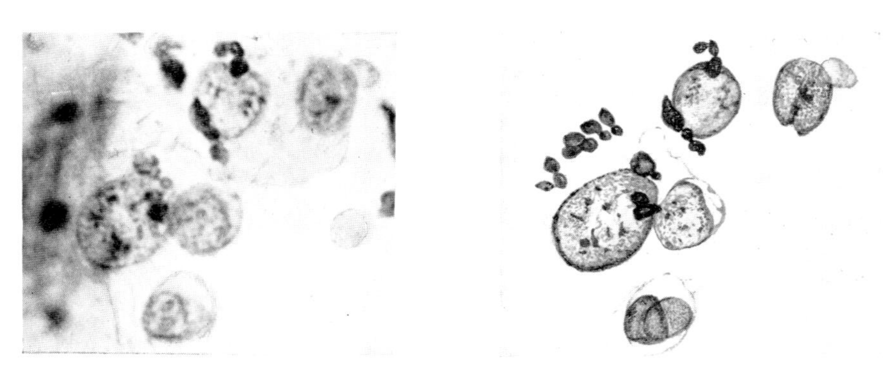

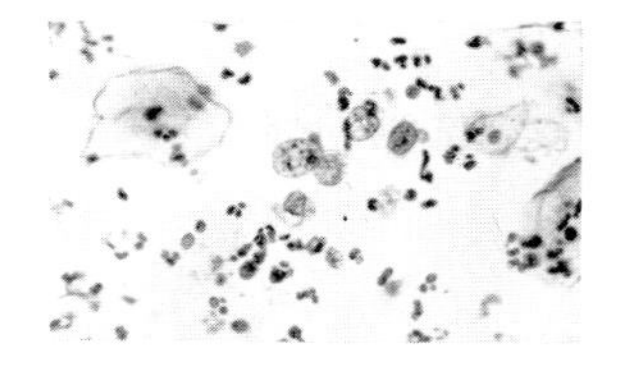

Low Power: Five extruded nuclei varying greatly in size in a field of leukocytes and superficial cells.
High Power: As in Fig. 35 above there is again in these nuclei a marked irregularity of the chromatin. Though the chromatin in these cells does not appear increased, it is distributed unevenly. Nuclear borders are definite but cytoplasmic borders are absent or indistinct. Variation in size is considerable.

Fig. 37

UNDIFFERENTIATED MALIGNANT CELLS: CONDENSATION OF CHROMATIN

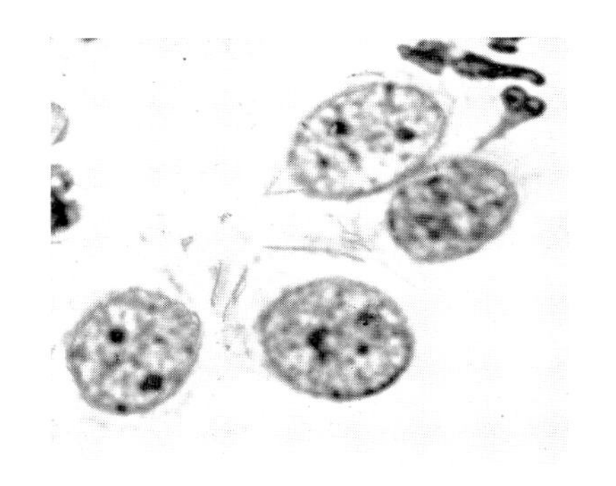

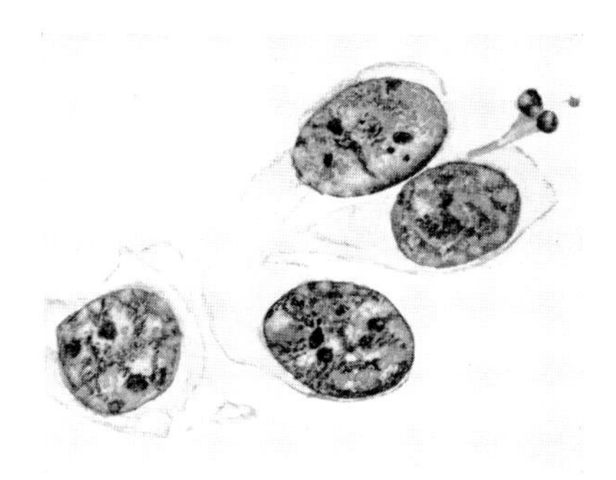

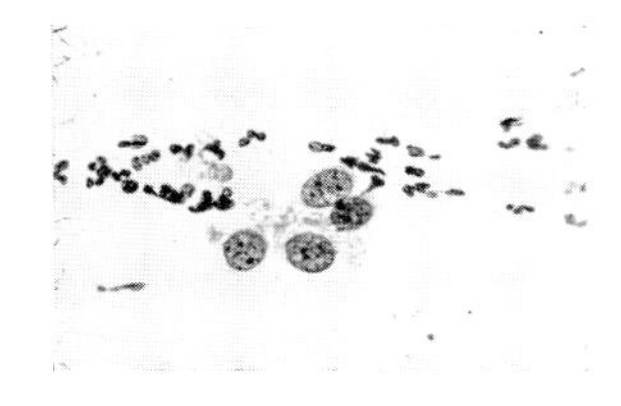

Low Power: Four large free nuclei in a field of leukocytes.

High Power: These four malignant nuclei show marked clumping of the chromatin. Every nucleus has at least three prominent condensations. We interpret these dark-staining borders in the nucleus as condensed chromatin rather than true nucleoli. Such large deeply staining clumps as these are often seen in undifferentiated malignant cells. Cytoplasm is an indistinct background.

Fig. 38

UNDIFFERENTIATED MALIGNANT CELLS: LARGE SINGLE CELL

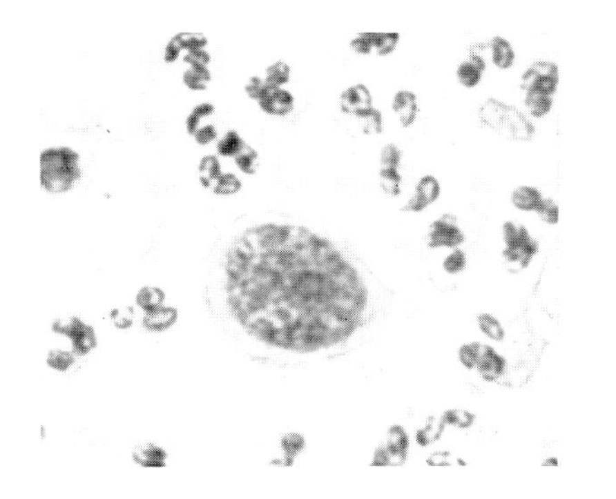

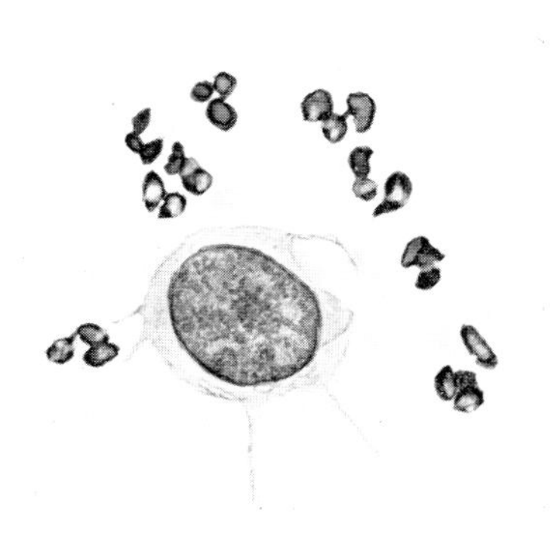

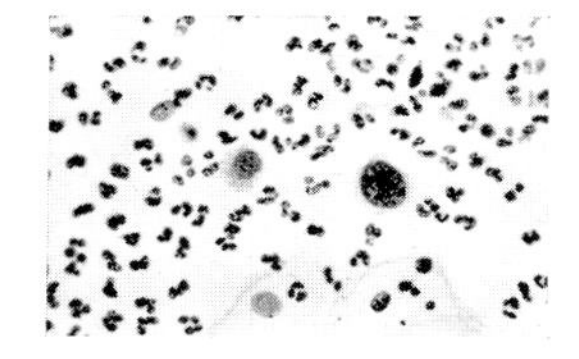

Low Power: Two deep-staining nuclei in field of superficial cells and leukocytes. Notice how the large nucleus is extremely prominent, owing to its increased density as compared with other cells.

High Power: An example of a single cell which can be identified as undifferentiated malignant by the following criteria: increase in chromatin content and irregularity of nuclear pattern, definite nuclear border, scant cytoplasm.

Fig. 39

UNDIFFERENTIATED MALIGNANT CELLS: GIANT CELL WITH NUCLEOLI

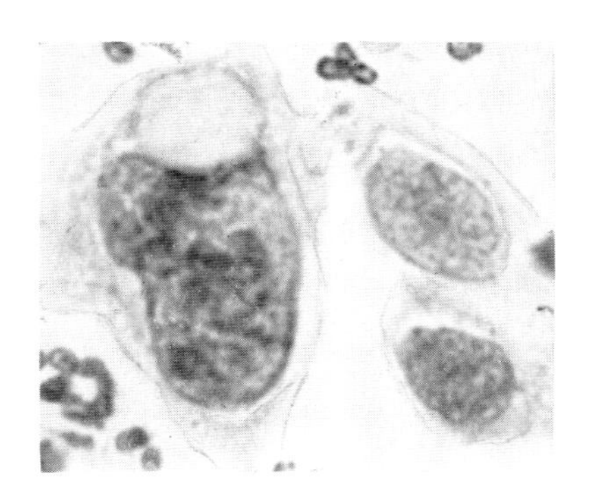

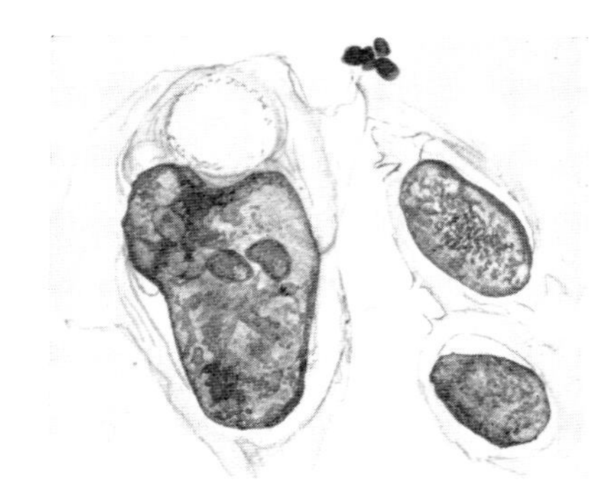

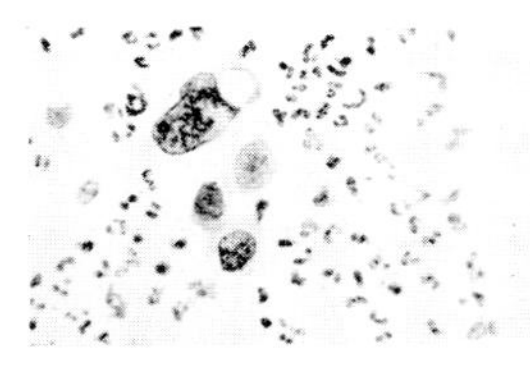

Low Power: Giant nuclei, three fairly even nuclei in field of leukocytes.
High Power: Enlarged nucleus reveals a very sharp nuclear border. The chromatin is extremely irregular and two nuclear bodies are seen which we interpret as nucleoli. There is a definite border to the nucleoli and their appearance is smooth. True nucleoli stain acidophilic instead of the deep basophilia assumed by dense chromatin particles.

Fig. 40

UNDIFFERENTIATED MALIGNANT CELLS: INCREASE IN CHROMATIN CONTENT

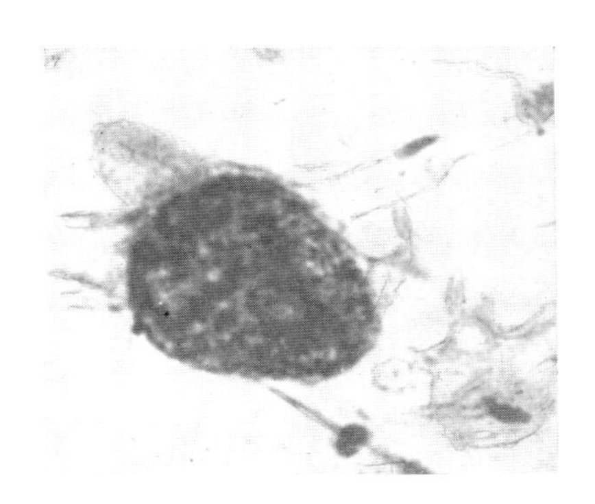

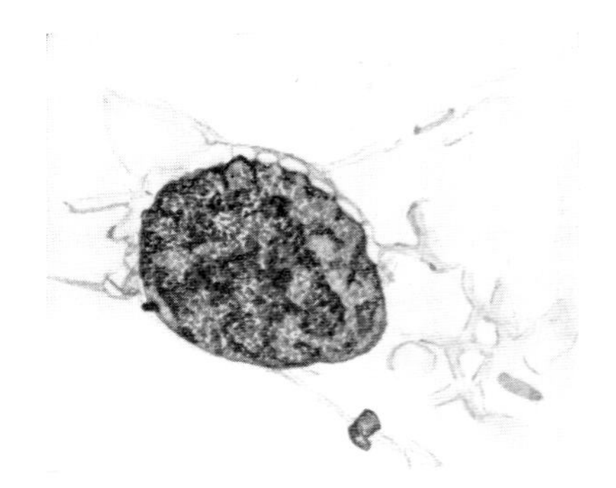

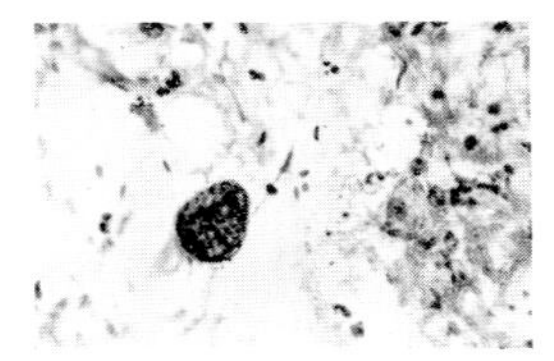

Low Power: Single large densely staining nucleus.
High Power: This enlarged nucleus illustrates the increase in chromatin content which takes place in malignant cells. The irregularity of chromatin pattern may be clearly visualized. Nuclear border is very distinct and irregular. Compare clumps of chromatin in this cell to nucleoli in plate above. Cytoplasm is absent. Blood forms the background.

FIG. 41

UNDIFFERENTIATED MALIGNANT CELLS: PHAGOCYTOSIS

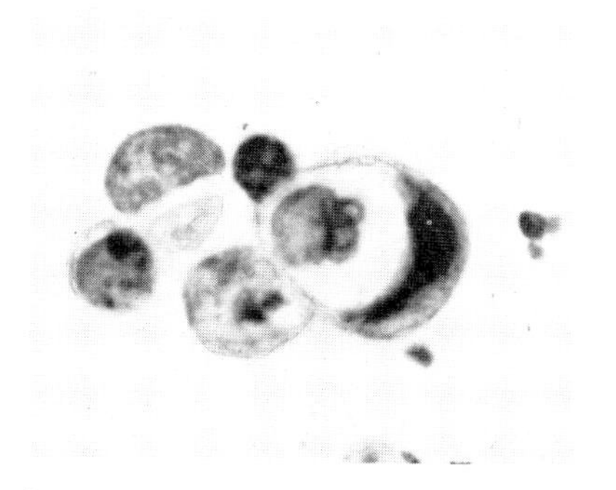

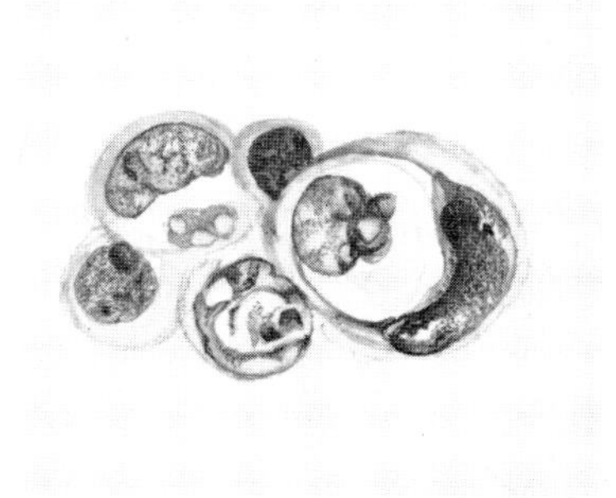

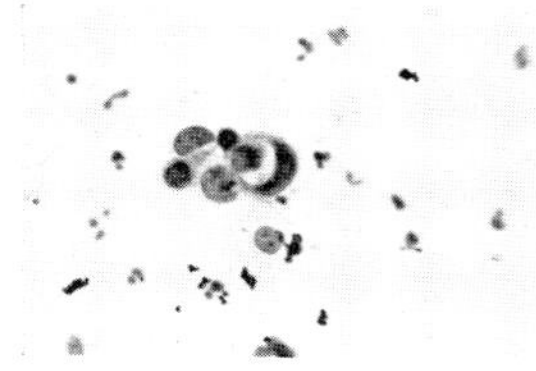

Low Power: Group of free nuclei, one with a large vacuole.
High Power: The vacuolated cell illustrates the occasional phagocytic properties which malignant cells exhibit. The quarter-moon-shaped nucleus has all the characteristics of malignancy and this cell contains a vacuole with an ingested undifferentiated nucleus. The other cells show variation and irregularity in chromatin content plus extreme differences in size. Cellular borders are indistinct.

FIG. 42

UNDIFFERENTIATED MALIGNANT CELLS: MULTINUCLEATED GIANT CELL

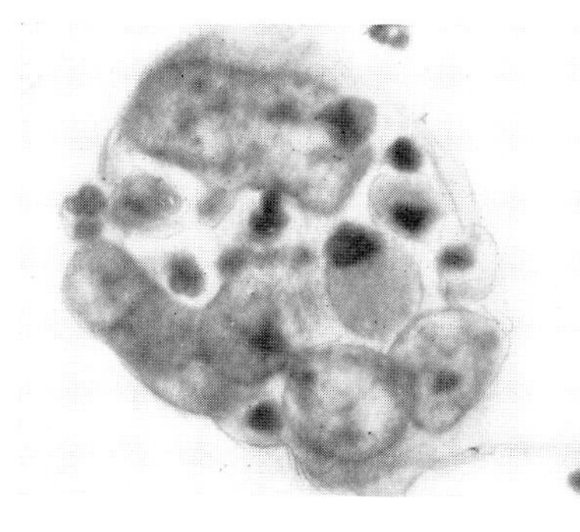

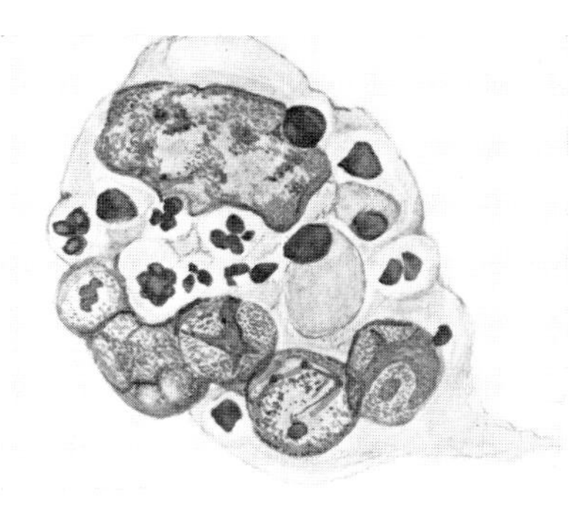

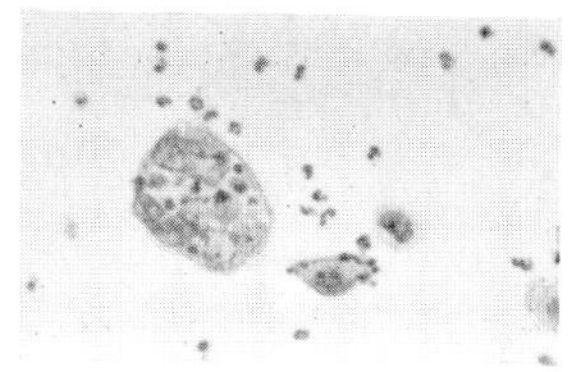

Low Power: One multinucleated cell, a histiocyte and outer layer basal cell.
High Power: The undifferentiated malignant nuclei have clustered to form a giant cell with apparent cytoplasm but indistinct cellular border. Cytoplasm is degenerate, showing vacuolization with leukocytes in the vacuoles. Nuclei vary in size tremendously and retain the characteristics of malignancy. Giant malignant cells are found quite frequently in positive smears.

FIG. 43

UNDIFFERENTIATED MALIGNANT CELLS: VARIATION IN SIZE

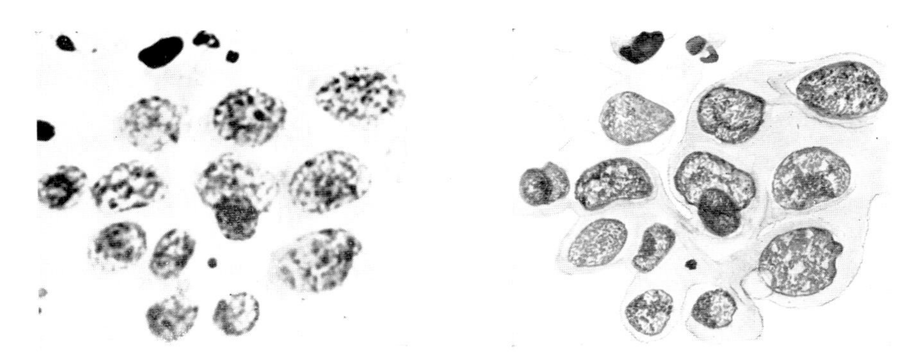

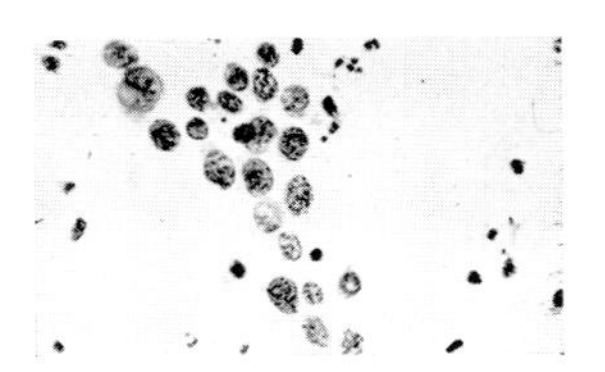

Low Power: Extruded nuclei varying in size with irregular chromatin arrangement.
High Power: There is a background of cytoplasm but no definite cellular borders. Variation in size is greater than variation in shape, though some nuclear borders are irregular. Chromatin varies from coarsely granular to thick strands. Nuclear borders are sharp. Cells are spread quite evenly and there is little overlapping of nuclei.

FIG. 44

UNDIFFERENTIATED MALIGNANT CELLS: DEGENERATION

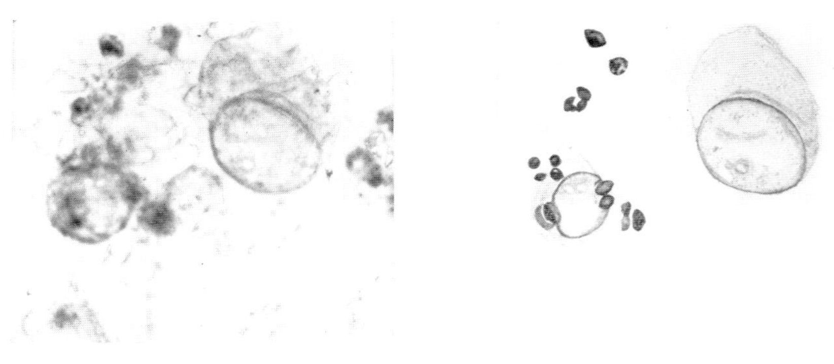

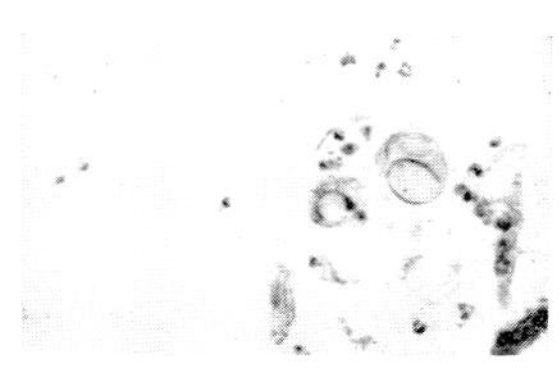

Low Power: Two indistinct nuclei varying in size.
High Power: This is an example of degeneration in undifferentiated malignant cells. The nuclear borders are still very distinct but the chromatin has disappeared entirely. Cytoplasm is indistinct. It should be emphasized that a positive diagnosis should not be made on such degenerate cells as this, but they indicate extensive search of the slide for definite cancer cells.

Undifferentiated malignant cells from squamous carcinoma are the most common tumor cells encountered in vaginal smears. The great majority of cases will show this type of cell. The cells are frequently seen in groups as illustrated in the preceding pages (Figs. 37 to 43). A positive diagnosis is on a more substantial basis if groups are encountered than if only single cells are seen. The reason for this is that identification of single cells must be made upon nuclear structure alone. On the other hand, in groups the additional criteria of variation in size and shape may be used. Groups of undifferentiated malignant cells from squamous carcinoma are more likely to have cells spread discretely (Plate 7) than tightly clustered.

Difficulties in Interpretation: False Positive Reports: Those misinterpreted as tumor cells from squamous carcinoma. Three types of normal cells may give difficulty and be mistakenly identified as undifferentiated malignant cells. They are the histiocyte, endometrial cell and endocervical cell.

Degenerate histiocytes may be confusing as shown in Fig. 30. There are no cellular borders, the nuclei are deep staining and there is variation in shape. However, careful scrutiny will show that the density of the nuclei is not due to an increase in chromatin but to a condensation. The nucleus is pyknotic, not active. The histiocyte nuclei vary more in shape than in size. Malignant nuclei vary both in size and in shape. In such groups of degenerate histiocytes the typical bean-shaped nucleus is usually present. Finally, the nuclei of the histiocyte are much smaller than the nuclei in groups of malignant cells. All of these points are helpful; but by far the most significant of any is the appearance of the chromatin material of the nucleus.

The second group of normal cells which may misinterpreted are the endometrial cells. These should cause little difficulty if attention is paid to the chromatin arrangement of the nucleus. The chromatin is smoothly granular in the fresh endometrial cell (Plate 4), pyknotic in the degenerate cell. Size variation is very slight in cells desquamated from the endometrium.

The third and most important group of normal cells which cause errors are the endocervical cells. We have found these cells to be by far the most confusing. Fortunately they are not frequent in smears from vaginal secretion. These cells cause difficulty because they rapidly lose their cytoplasm and appear as free nuclei, but primarily they are confusing because of their great size variation. The cells from the endocervix vary more in size than any other type of normal cell. Because of the extreme variation the cells are occasionally misinterpreted as malignant. The cytologist must remember that, though malignant cells do show such variation, it is incorrect to interpret nuclei as malignant on size variation alone. Tumor cells must show abnormal nuclear structure in addition. Endocervical nuclei present a smoothly granular chromatin network with occasional small clumps. The nuclei of undifferentiated cells show great irregularity in chromatin network. Compare Fig. 37 with Fig. 22.

Thus, if only cells are interpreted as positive which show definite alterations in the content and distribution of the chromatin, malignant cells may be identified with accuracy.

False Negative Reports: Smears from cases of squamous carcinoma which were read as negative. These errors occur for two reasons. The first is that no cells were shed from the surface of the tumor. This happens very rarely, though we have had cases where actual aspiration from the tumor itself yielded only a rare malignant cell. For some reason a rare tumor seems to have a greater cohesiveness between cells. This is a type of error which represents a limitation of the method.

If cells are present in the smear but either not seen or misinterpreted by the observer, the second type of false negative report occurs. A certain number of this type of error will occur, but theoretically it should be cut to a minimum figure. Every smear should be covered, field by field, under low power and any faintly suspicious cells examined under high power. A brief preliminary examination of the slide under low power often proves profitable and saves time if obvious malignant groups are present. In this connection it might be helpful to point out that in thick smears examination of the edges of the smear is often extremely helpful. However, if no suspicious cells are found on cursory inspection the slide must then be examined methodically.

General Criteria for the Identification of Undifferentiated Malignant Cells from Squamous Cell Carcinoma: The criteria are listed in the order of importance.

1. Irregularity of chromatin network and increase in chromatin content.
2. Sharp nuclear borders.
3. Absence of cellular borders. If single cells, inadequate cytoplasm and no cellular border.
4. Variation in size and shape.

Obviously a diagnosis is more definite if the cells in question fulfill all criteria, and the majority of tumor cells do. The third criterion is essential for the classification of this group of malignant cells as undifferentiated. Variation in size and shape is probably the least reliable criterion. However, it is essential for the student to remember that irregularity of the chromatin network is the criterion that must be fulfilled. A malignant nucleus may not show increase in chromatin content (see Fig. 36), but will show irregularity in distribution.

DESCRIPTION OF DIFFERENTIATED FIBER CELLS

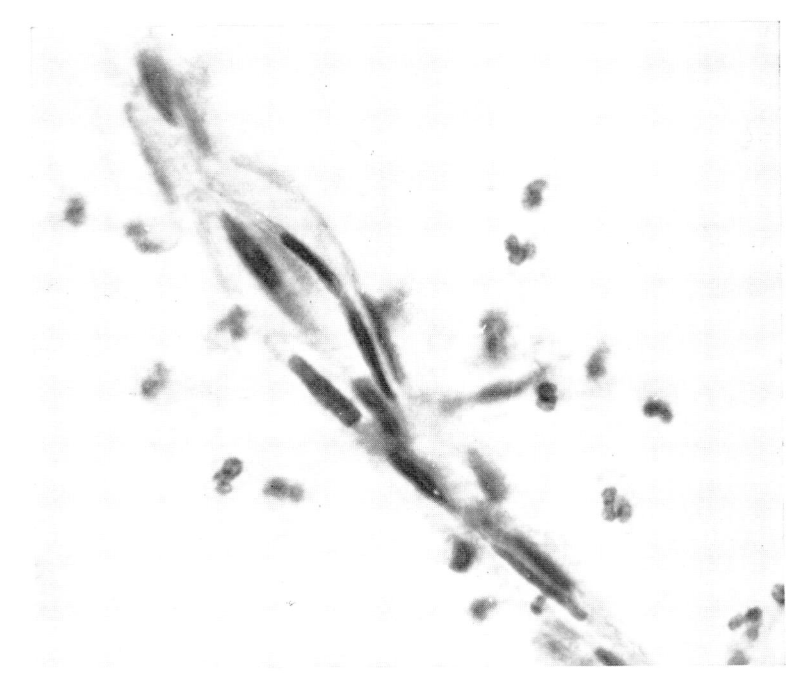

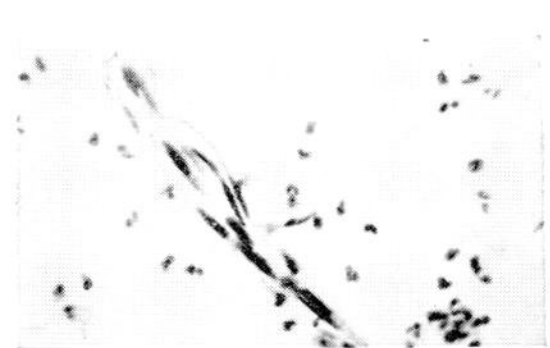

Low Power: Lengthy group of many dense cigar-shaped nuclei.

High Power:

A. Characteristics of Nucleus: 1. Sharp nuclear borders. 2. The malignant nuclei exhibit variation in chromatin content; cell 1 shows several clusters of chromatin, whereas cells 2, 3 and 4 are dense in comparison, because of the increase in chromatin content. 3. The general shape of fiber nuclei is elongated and slightly crooked, but there is great variation in the width and length of the nuclei. Compare in cells 2 and 3 the length of the nuclei and in cells 1 and 4 the width of the nuclei. 4. Cell 5 has two lengthy nuclei situated in a straight line in the cytoplasm.

B. Characteristics of Cytoplasm: 1. Well defined cellular borders. 2. The cytoplasm extends from both ends of the nuclei and is slender and fiber-like in appearance, hence the name fiber cell. Cell 1 has curved, tapered cytoplasm which is very narrow in comparison to cell 6 which has wider and a much longer amount of cytoplasm. 3. Staining reaction: Dark purplish-blue nuclei with acidophilic or in some cases basophilic cytoplasm.

C. General Characteristics of Group: Strands of cytoplasm varying slightly in width and length with elongated, cigar-shaped nuclei also showing variation in width, length and chromatin content.

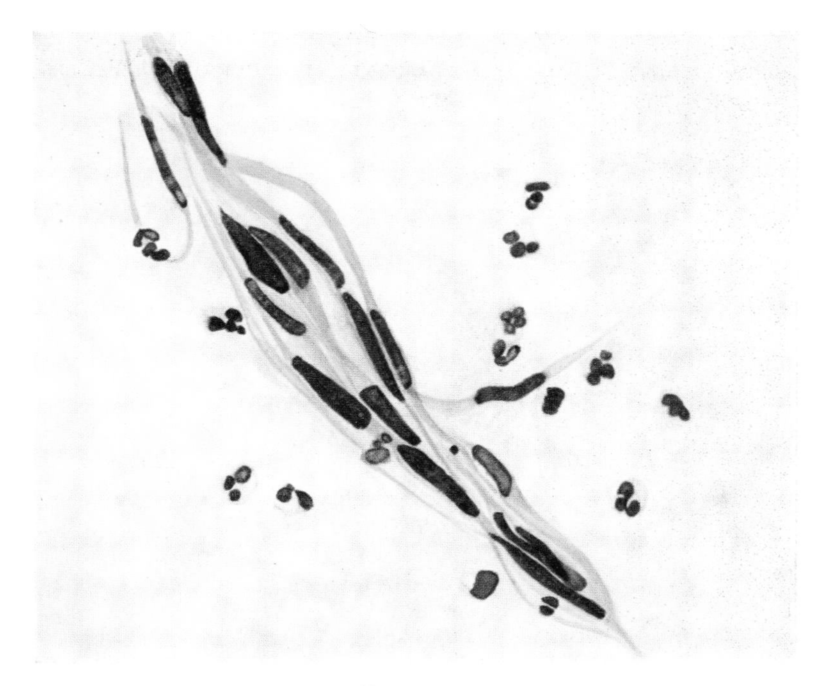

PLATE 8

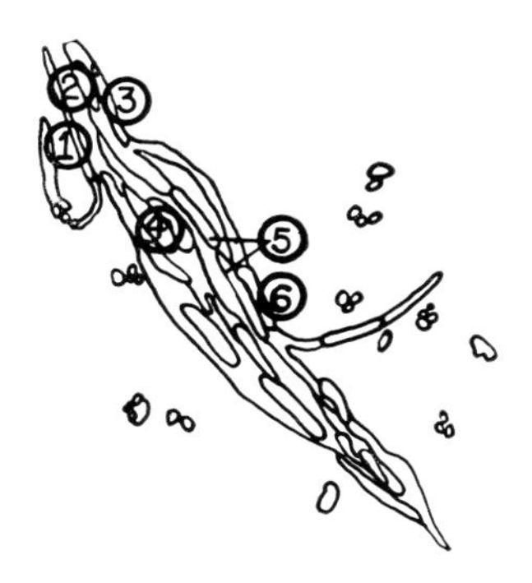

KEY TO DIFFERENTIATED FIBER CELL PLATE

1. Differentiated fiber cell showing several clumps of chromatin in the nuclei and thin, curved cytoplasm.
2. Fiber cell with a short, dense nucleus and small amount of cytoplasm.
3. Fiber cell with an elongated, dense nucleus and indistinct cytoplasmic borders.
4. Fiber cell with a wide hyperchromatic nucleus.
5. Differentiated fiber cell with two long nuclei centered in a straight line in the cytoplasm.
6. Fiber cell with extremely long tail of cytoplasm.

Fig. 45

DIFFERENTIATED MALIGNANT CELLS: TYPICAL GROUP OF FIBER CELLS

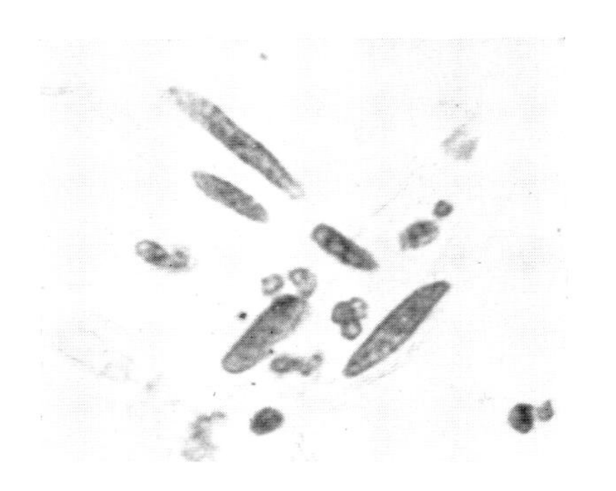

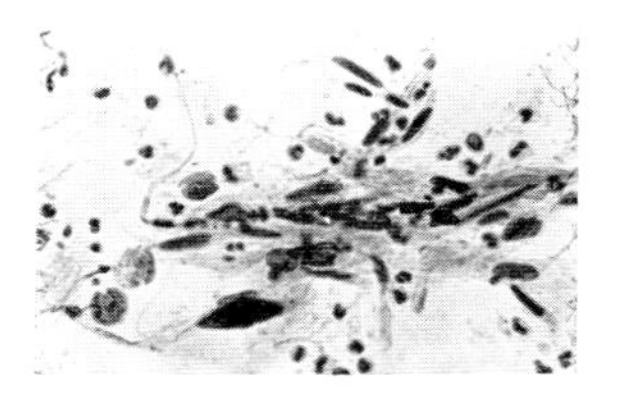

Low Power: Field of many irregularly arranged cigar-shaped nuclei.
High Power: Five elongated hyperchromatic nuclei showing various degrees of length and width. Each nucleus has a sharp border and numerous clumps of chromatin which have stained darkly. The cytoplasm is also elongated and is visible at each end of the nuclei. The disorderly pattern of these cells is typical of malignant fiber cells.

Fig. 46

DIFFERENTIATED MALIGNANT CELLS: FIBERS IN CHAIN PATTERN

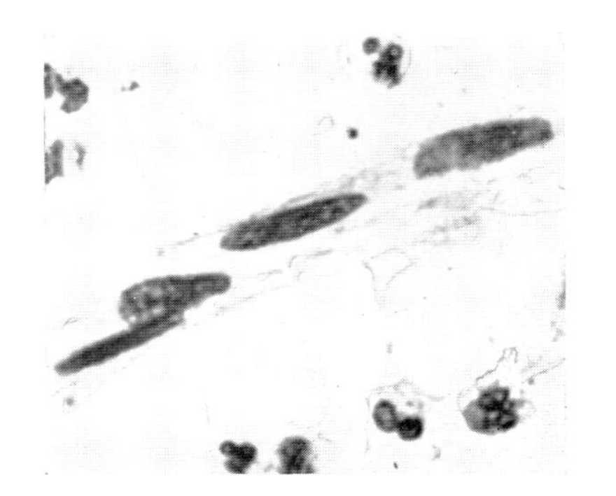

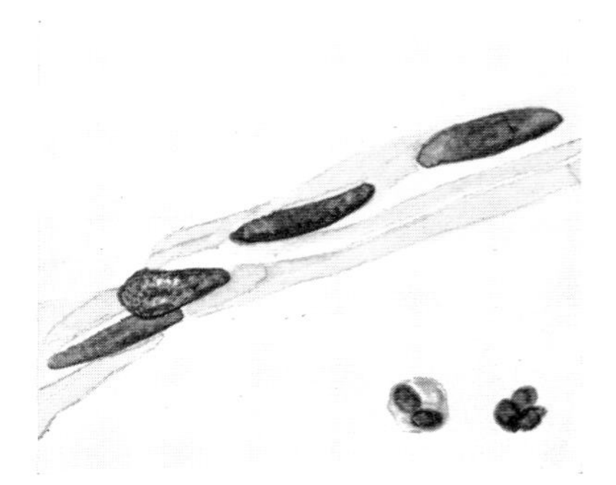

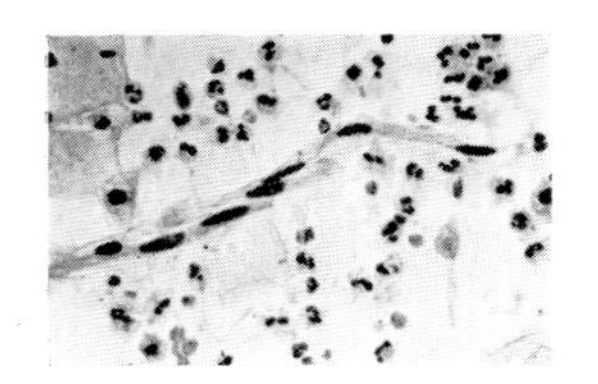

Low Power: Long string of dark oval nuclei extending the length of the field.
High Power: These malignant nuclei vary slightly in shape and size and appear almost pyknotic owing to the intensity of the chromatin particles. The cell borders are fairly definite and the cytoplasm is strung out in the same fashion as the nuclei. Compare these cells to the rolled cornified cell in Fig. 14.

FIG. 47

DIFFERENTIATED MALIGNANT CELLS: DEGENERATE FIBER CELLS

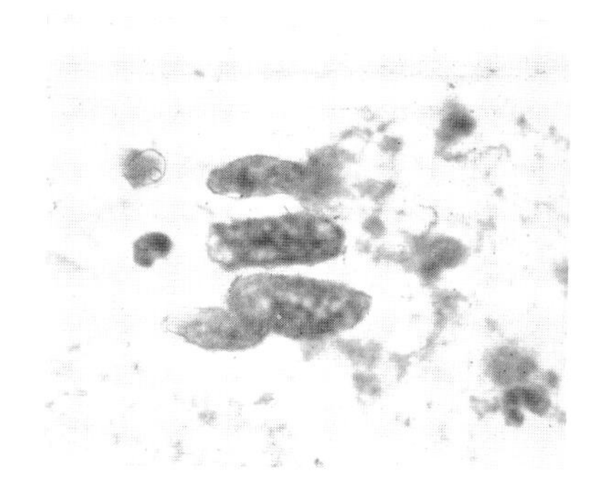

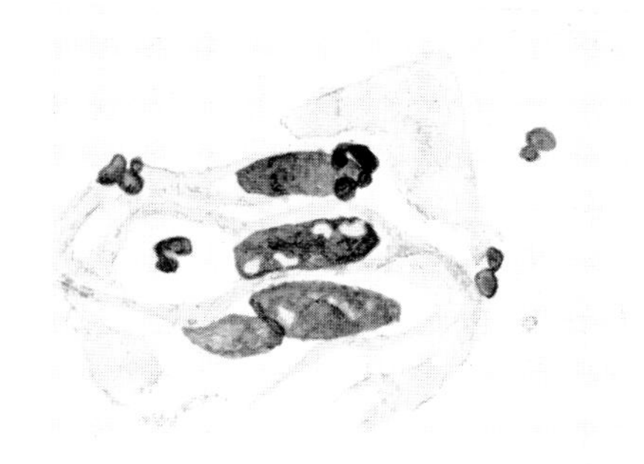

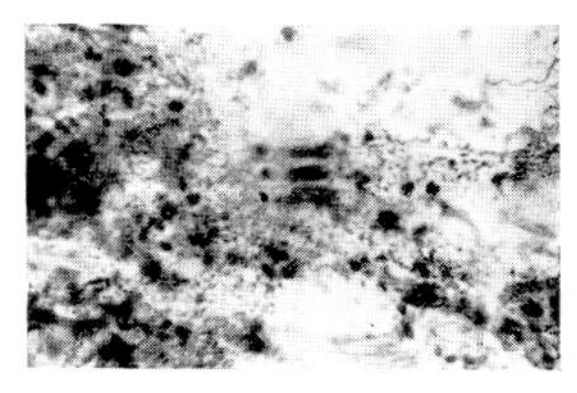

Low Power: Group of elongated nuclei with no visible cytoplasm.
High Power: We identify these nuclei as malignant fiber cells instead of undifferentiated cancer cells only because of their shape and position. They are not as hyperchromatic as usual because of the degenerate changes; causing vacuoles in the nuclei, indistinct nuclear borders and no evidence of any cellular border. Compare these cells to figure below.

FIG. 48

DIFFERENTIATED MALIGNANT CELLS: INDEFINITE MATURATION

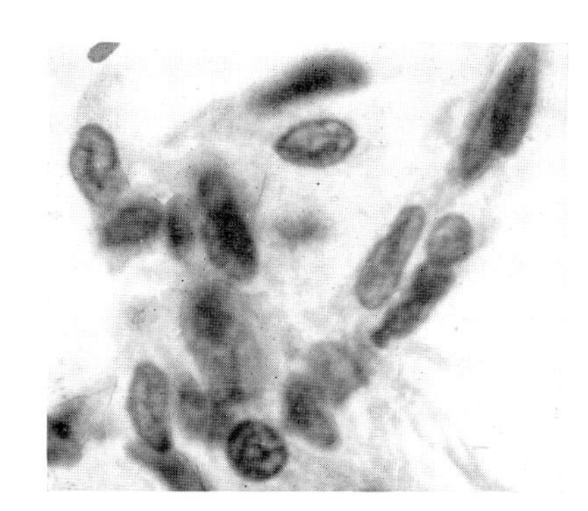

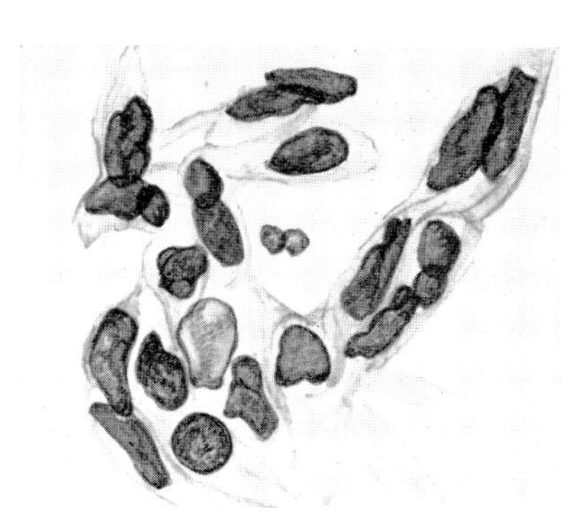

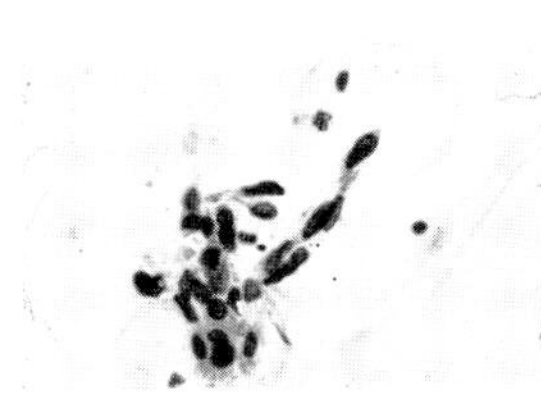

Low Power: Group of hyperchromatic nuclei, some appearing elongated.
High Power: These cells fulfill the criteria for malignancy, *i.e.*, irregularity in size and shape, increase in chromatin content and sharp nuclear borders. However, the elongated forms of some of the cells merely suggest some differentiation, since even an impression of cellular borders is absent. Cells such as these indicate further search for definite fiber cells.

DESCRIPTION OF DIFFERENTIATED TADPOLE CELLS

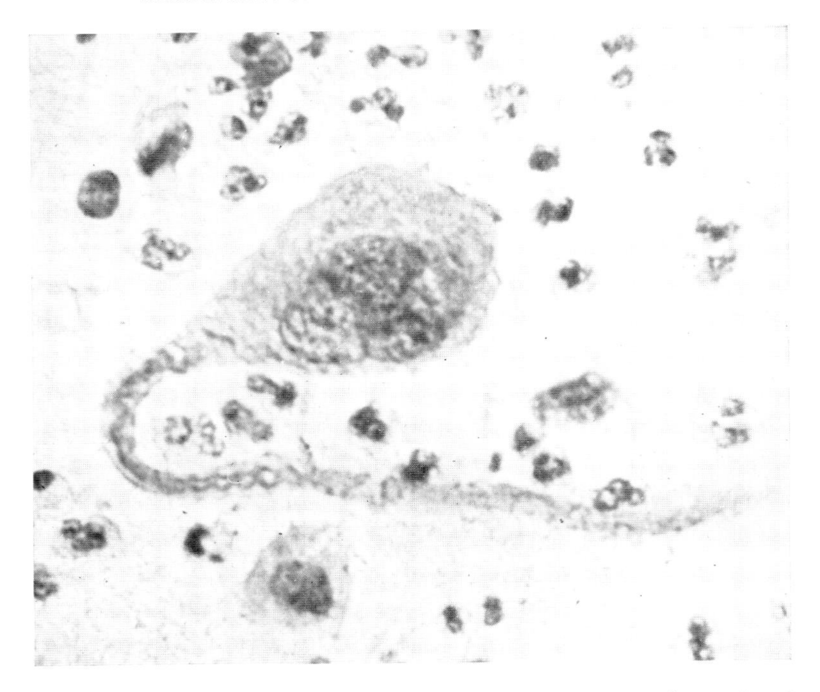

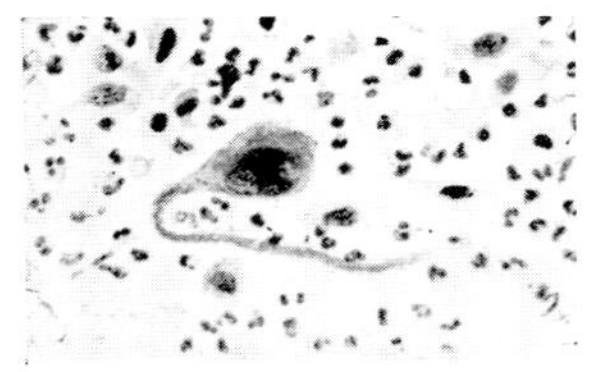

Low Power: Discrete cell with a large head and nucleus and a long tail of cytoplasm.

High Power:

A. Characteristics of Nucleus: 1. Sharp nuclear border. 2. The nucleus in the head of cell 2 is extremely large and there are indentations along the nuclear border. Compare the chromatin content of this malignant nucleus to those of cells 1 and 3. Notice all have definite clumps or strands of chromatin particles which are unevenly grouped, leaving transparent spaces. 3. The nucleus in cell 2 is eccentric in position, irregular in shape and abnormally big, being about seventeen times larger than a leukocyte.

B. Characteristics of Cytoplasm: 1. Cells 1 and 2 have clearly defined cellular borders. 2. Cell 2 has a long sharply outlined tail of cytoplasm stretching almost the whole length of the high-power field. This cell is called a tadpole because of the shape of the cytoplasm and it is also a differentiated cell because there is adequate cytoplasm with a good cell border. 3. Cell 3 is undifferentiated because there is no evidence of cytoplasm. 4. Staining reaction: Bluish-purple nuclei with acidophilic or basophilic cytoplasm.

C. General Characteristics of Group: Single cell with well defined cytoplasm in the shape of a tadpole with a long tail and a head containing one large malignant nucleus.

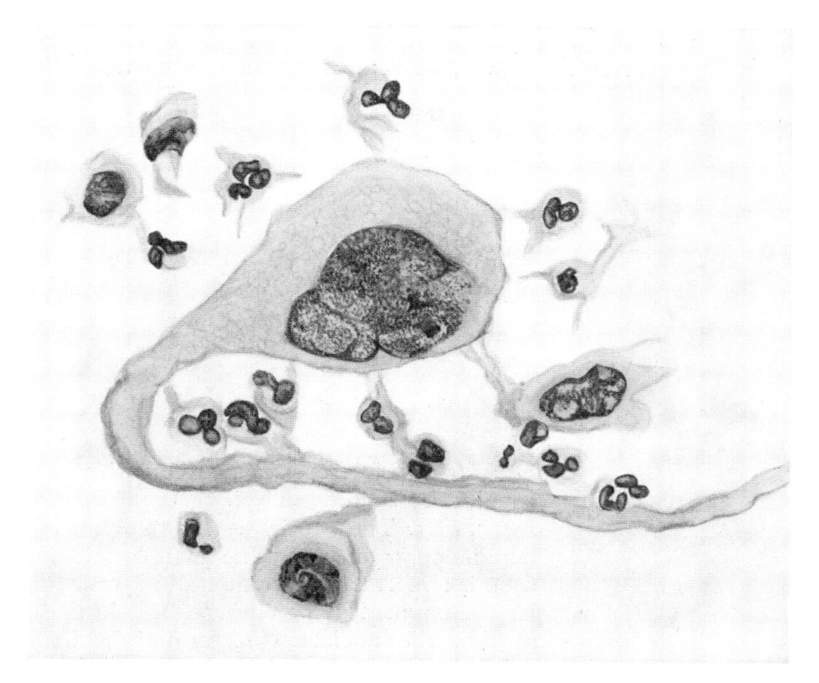

PLATE 9

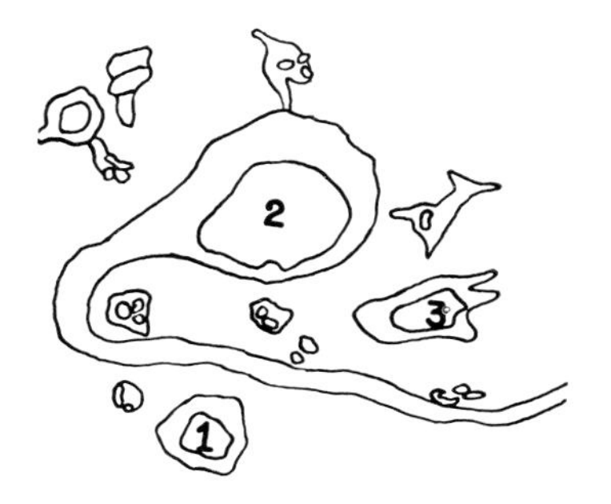

KEY TO DIFFERENTIATED TADPOLE CELL PLATE

1. A third type of differentiated cell with a large hyperchromatic nucleus and adequate, well defined cytoplasm.
2. Large, differentiated tadpole cell with clearly defined cytoplasm with a head containing one malignant nucleus and a long tail almost filling the high-power field.
3. An undifferentiated malignant nucleus with a sharp, wrinkled nuclear border, many clumps of chromatin and no clear cellular border.
4. Gray-blue background is old blood pigment.

FIG. 49

DIFFERENTIATED MALIGNANT CELLS: DOUBLE-NUCLEATED TADPOLE

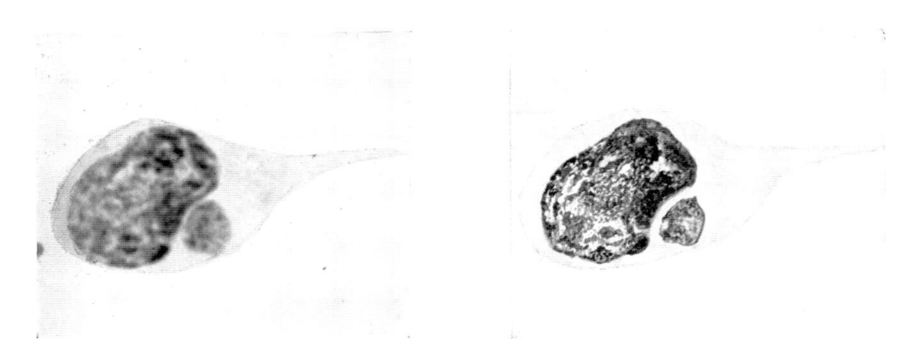

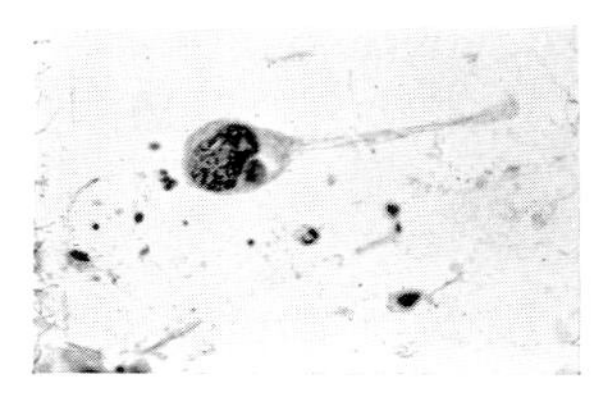

Low Power: Bizarre elongated cell with two, unequal, deep-staining nuclei.
High Power: The two hyperchromatic nuclei vary tremendously in size. The cytoplasmic nuclear ratio is abnormal. The chromatin is condensed in large amounts leaving empty spaces in the nuclei so that the nuclear configuration is irregular. Nuclear borders are extremely sharp and the cellular border is fairly well defined.

FIG. 50

DIFFERENTIATED MALIGNANT CELLS: SMALL MULTINUCLEATED TADPOLE

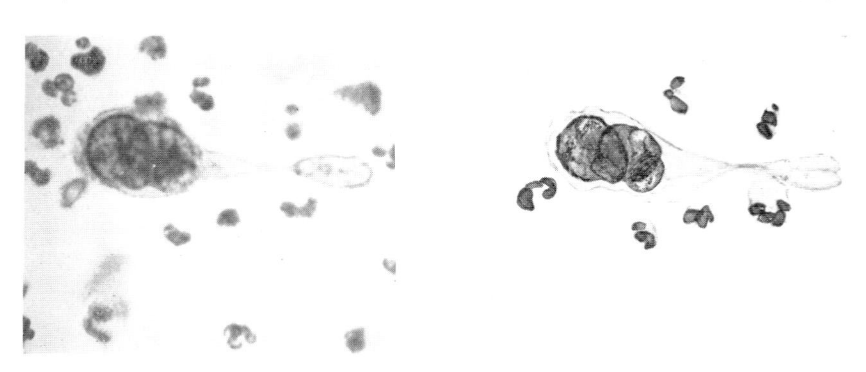

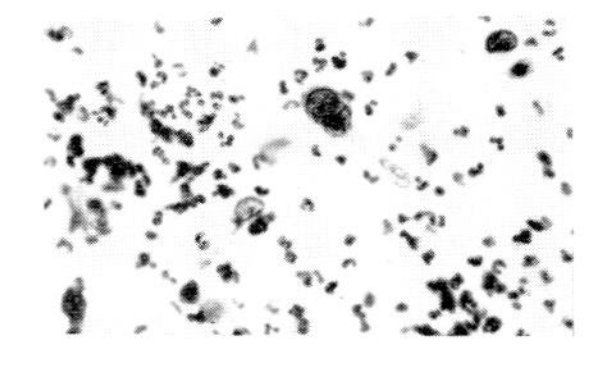

Low Power: Group of three hyperchromatic nuclei.
High Power: Closer examination reveals a well defined cellular border with a short, wasp-waisted tail of cytoplasm, typical of the tadpole shape. At the head of the cell there are three malignant nuclei, about eight times larger than a leukocyte with definite clumps of chromatin especially collected at the nuclear borders giving them a very definite outline.

FIG. 51

DIFFERENTIATED MALIGNANT CELLS: TADPOLE WITH BROAD NUCLEUS

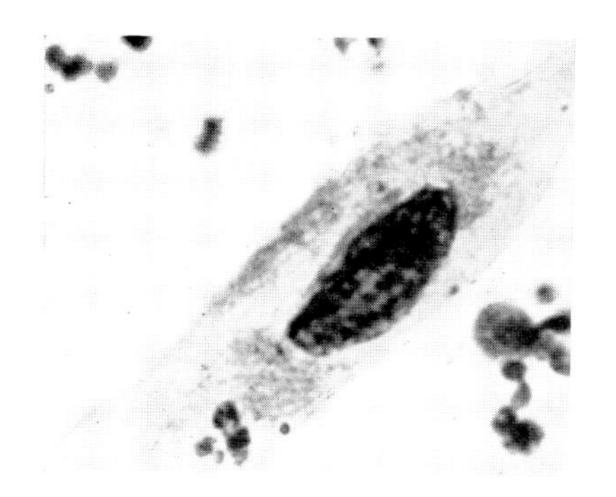

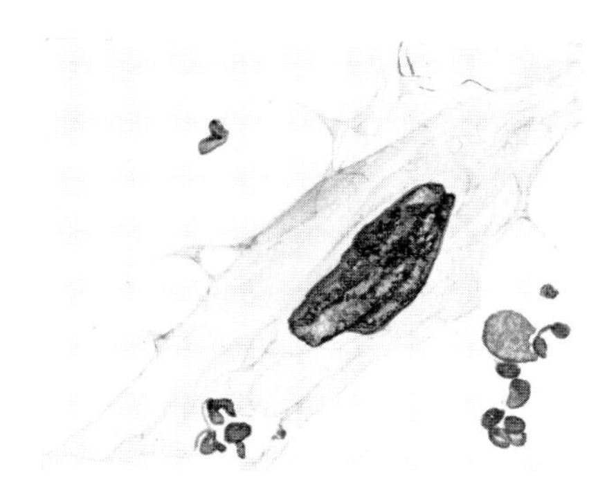

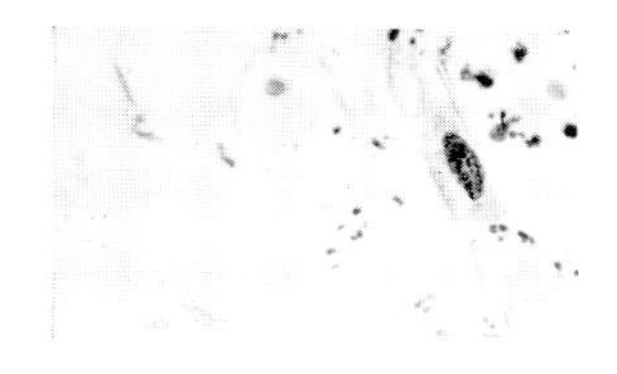

Low Power: Elongated cell with deep-staining nucleus.

High Power: This cell illustrates the difficulty in any exact classification. The elongated nucleus would suggest a fiber cell. The long tail of cytoplasm which is greater at one end of the cell suggests the form of a tadpole. Obviously borderline cells such as this occur and rigid classification is unimportant as long as it is identified as a differentiated malignant cell.

FIG. 52

DIFFERENTIATED MALIGNANT CELLS: IRREGULAR TADPOLE

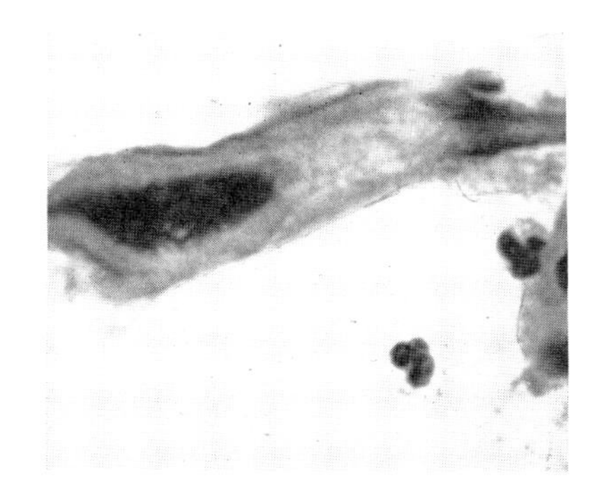

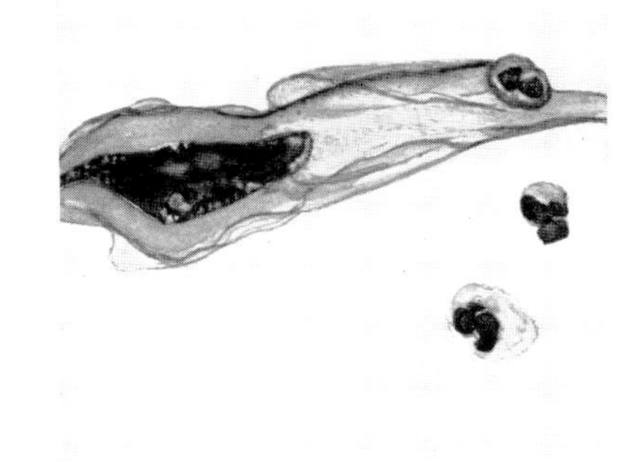

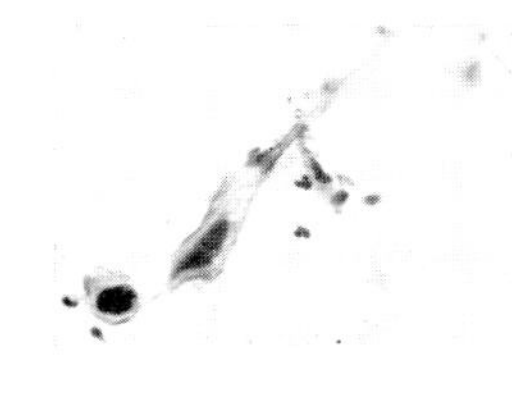

Low Power: Two cells with large black nuclei. One cell shows tail of cytoplasm.

High Power: The irregular-shaped nucleus is large and the hyperchromatism is so intense that it is difficult to distinguish individual particles, though some irregularity of chromatin structure may be seen. The position of the nucleus is extremely eccentric. The cytoplasm of the cell is abnormally distributed.

FIG. 53

DIFFERENTIATED MALIGNANT CELLS: MULTINUCLEATED TADPOLE

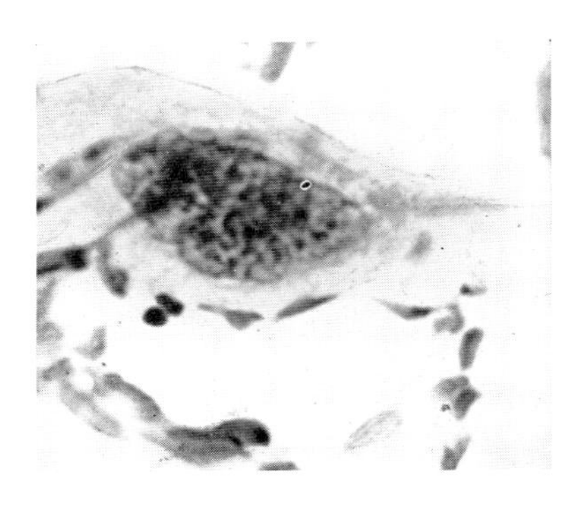

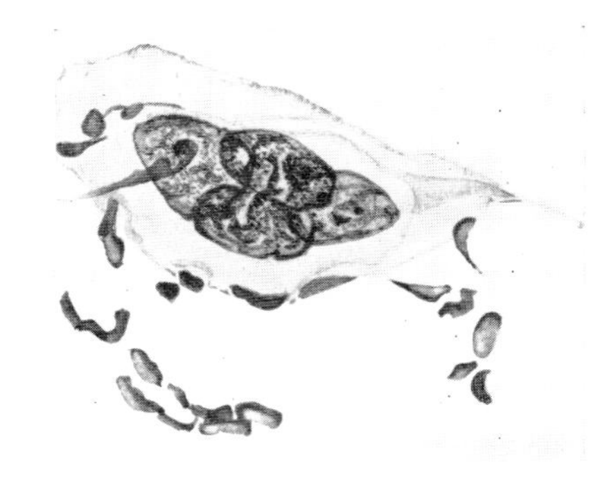

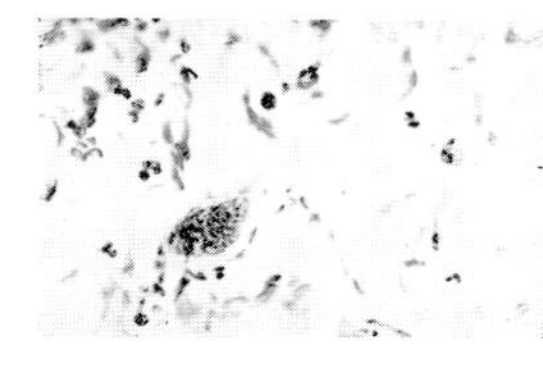

Low Power: Large granular nucleus in field of leukocytes and blood.

High Power: This differentiated malignant cell has four nuclei which show irregularity in chromatin structure, sharp nuclear borders and a fairly definite cellular border. Cytoplasm is present but abnormally distributed. A thin tail of cytoplasm extends from the round portion of the cell. Tadpole cells are often multinucleated.

FIG. 54

DIFFERENTIATED MALIGNANT CELLS: TADPOLE WITH CENTRAL NUCLEUS

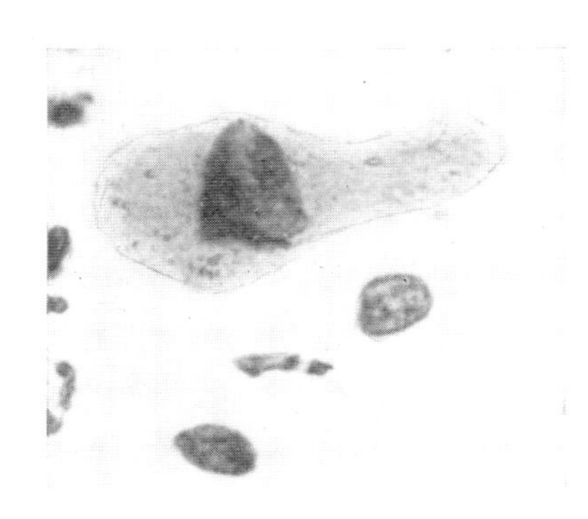

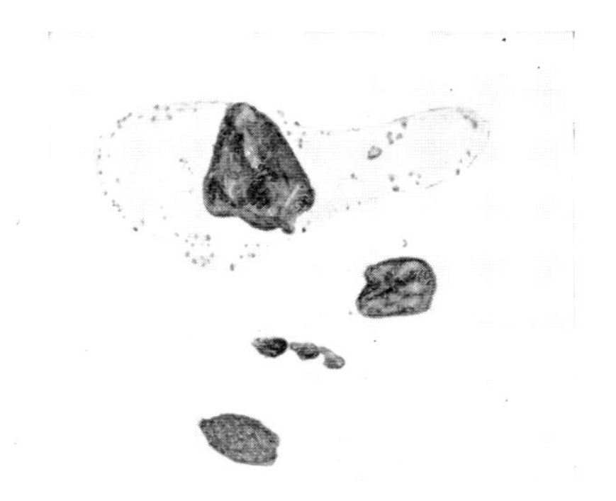

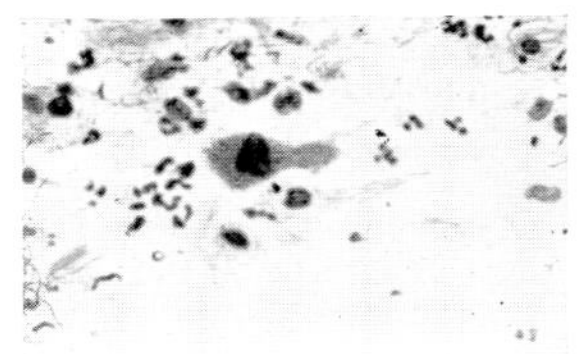

Low Power: Field of large extruded nuclei, one with pear-shaped cytoplasm.

High Power: Three malignant nuclei showing irregular chromatin structure. Two of the nuclei have no cytoplasm and are undifferentiated carcinoma cells. The triangular nucleus has aberrant cytoplasm which classifies it as a differentiated malignant cell. The nucleus is centrally located, but the shape of the cytoplasm identifies this cell as a tadpole.

The fiber cell occurs frequently in smears from cases of squamous carcinoma, and is a more common type of differentiated malignant cell than the tadpole. These cells are occasionally seen in a pearl formation. (See Plate 12.) In that instance the cells are arranged in concentric circles retaining their nuclear characteristics which identify them as malignant.

Difficulties in Interpretation: The two types of normal cells which may be confused with the malignant fiber cell are the tightly rolled cornified cell and the connective tissue cell. The reasons for identifying the rolled cornified cell as a superficial cell are given in the discussion of squamous epithelial cells. Cells originating from connective tissue are present only rarely in the vaginal smear. Because of their extremely infrequent occurrence, they may be erroneously regarded as suspicious, if it is not remembered that cells of connective tissue origin may desquamate into the vaginal secretion. The cells occur both singly and in clusters. They may be identified as benign on the appearance of the nuclear material. It is very finely granular. The cytoplasm of this thin elongated cell usually shows the wavy pattern characteristic of connective tissue. To illustrate how rarely this type of cell is found, it should be recorded that it has been seen only four times in our laboratory.

General Criteria for Identification of Fiber Cells: 1. Thin, elongated cell with fairly distinct cellular borders. If cellular borders are not clear, the observer usually gains a definite impression of the width and length of the cell with the exception of groups which show complete degeneration of cytoplasm. In this latter instance, identification is made on the elongated hyperchromatic nuclei. 2. Deep-staining elongated nuclei. The hyperchromasia of these nuclei is more intense than in any other type of malignant cell. However, careful observation will still identify definite alterations in nuclear pattern. 3. Variation both in size and shape of nuclei.

The Tadpole Cell: The differentiated malignant cell classified as a tadpole cell is the least frequent of any type of cancer cell from squamous carcinoma. Its morphology is so distinctive that there should be no difficulty in identifying it in smears from untreated squamous carcinoma. Occasionally radiated epithelial cells may be confused with tadpole cells. (See chapter on radiation changes in cells of the vaginal secretion.) However, this is a special instance. In the routine diagnosis of vaginal smears the identification of tadpole cells may be made with a great degree of accuracy.

General Criteria for Identification of Tadpole Cells: 1. Cytoplasm is present but abnormally distributed. The cell has a "head" containing nucleus or nuclei (these cells are often multinucleated) and a long tail of cytoplasm. Cellular borders are distinct. 2. Nuclei fulfill criteria for malignant cells, *i.e.*, increase in chromatin content, irregularity of chromatin pattern and sharp nuclear border.

DESCRIPTION OF THIRD TYPE DIFFERENTIATED CELLS

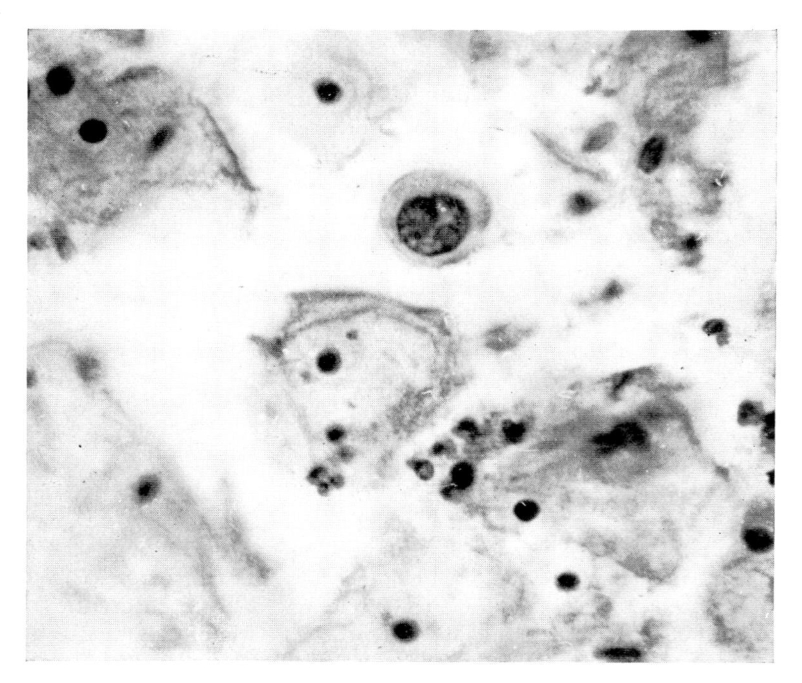

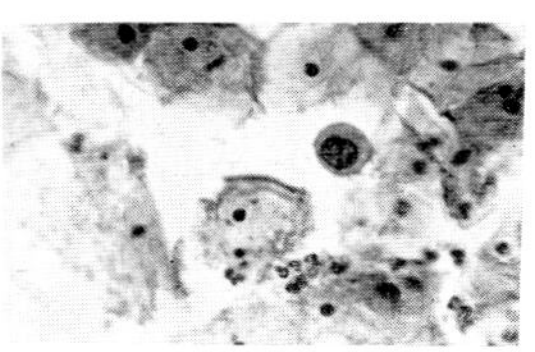

Low Power: Round cell with hyperchromatic nucleus in a field of cornified and precornified cells.
High Power:

A. Characteristics of Nucleus: 1. Sharp but uneven nuclear border. 2. The chromatin content of cell 1 exhibits all the characteristics of a malignant nucleus. Notice the prominent dark-staining clusters of chromatin which are distributed irregularly throughout the nucleus, giving it an active uneven appearance. Compare to cell 2, a precornified nucleus. 3. The general shape of the nucleus is round but the nuclear border may be slightly indented or wrinkled in appearance. 4. The nucleus is usually the size of a basal cell nucleus or larger. Notice that the nucleus is much too big for the surrounding cytoplasm producing an abnormal nuclear-cytoplasmic ratio.

B. Characteristics of Cytoplasm: 1. Well outlined cellular border. 2. As in cell 1, the amount of cytoplasm is moderate and usually surrounds the nucleus. 3. Staining reaction: Basophilic or acidophilic.

C. General Characteristics of Group: A round cell the size of an inner layer basal cell which shows definite cellular border. Nucleus is large for amount of cytoplasm present and exhibits the usual characteristics of a malignant nucleus, *i.e.*, increase and irregularity of chromatin material, sharp nuclear border. Nucleus often appears wrinkled or indented.

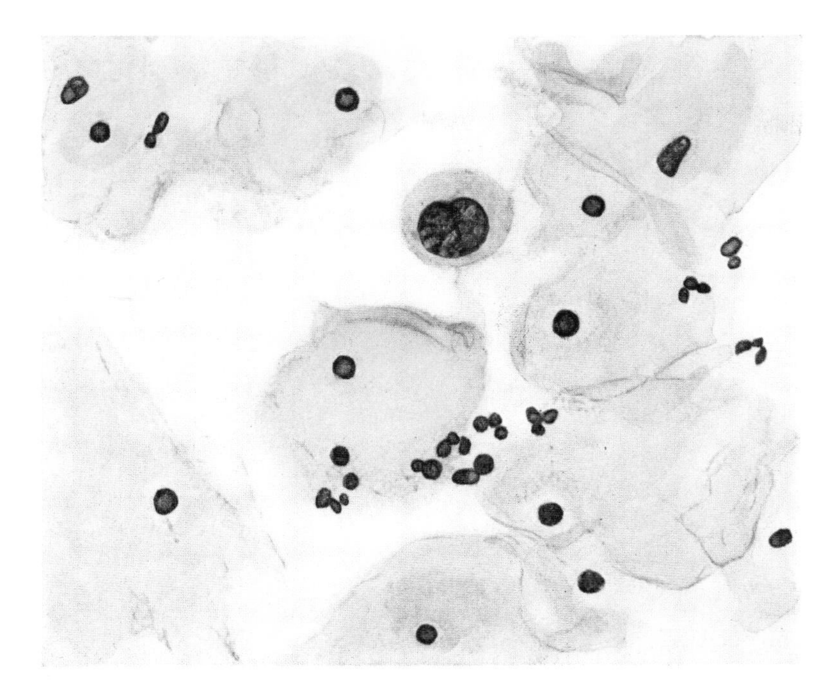

PLATE 10

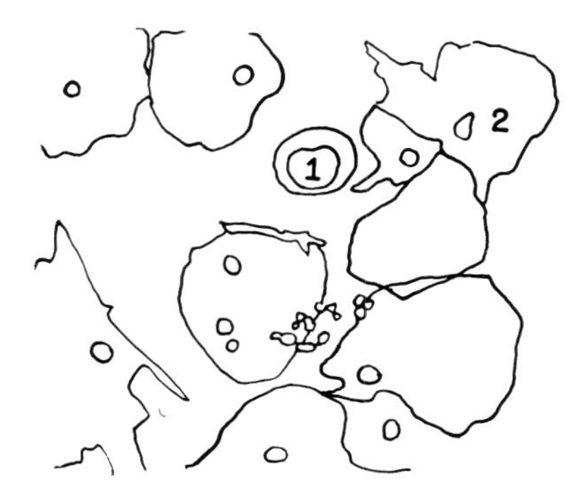

KEY TO THIRD TYPE DIFFERENTIATED CELL PLATE

1. Third type differentiated malignant cell with an active hyperchromatic nucleus and slightly indented periphery with a clear nuclear and cytoplasmic border.
2. Precornified cell with a small even nucleus and a large amount of folded transparent cytoplasm.
3. All other epithelial cells in the field are degenerate cornified cells.

FIG. 55

THIRD TYPE DIFFERENTIATED: LARGE NUCLEUS

Low Power: Single cell with a huge hyperchromatic nucleus.
High Power: A typical malignant third type differentiated cell showing a large nucleus with a clear nuclear border and many clumps of chromatin. two clusters being obvious. Compare the size of this nucleus to the leukocyte in the same field. Note the nucleus is too big for the amount of well defined cytoplasm around it.

FIG. 56

THIRD TYPE DIFFERENTIATED: WRINKLED NUCLEUS

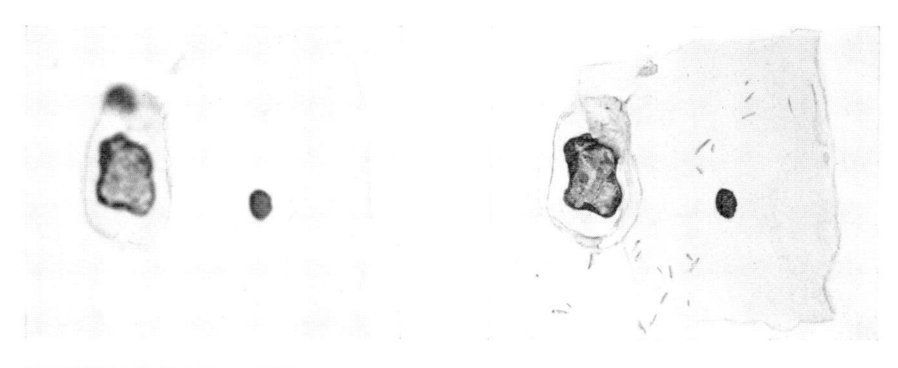

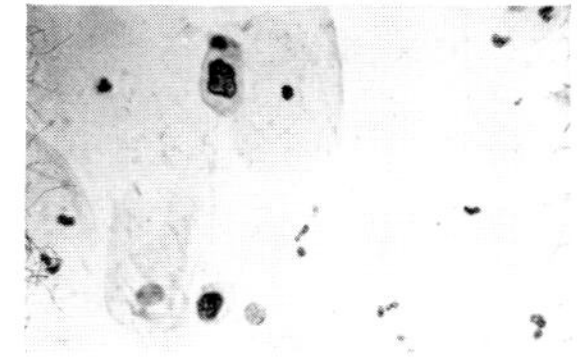

Low Power: Squamous epithelial cells and two cells with active nuclei.
High Power: This third type differentiated cell has a large nucleus and shows the characteristic wrinkling of the nuclear border, which is well defined because of the condensation of the chromatin particles at the edge. There is adequate well outlined cytoplasm. Bacteria are superimposed on a cornified cell.

Fig. 57

THIRD TYPE DIFFERENTIATED: IRREGULAR CELL BORDER

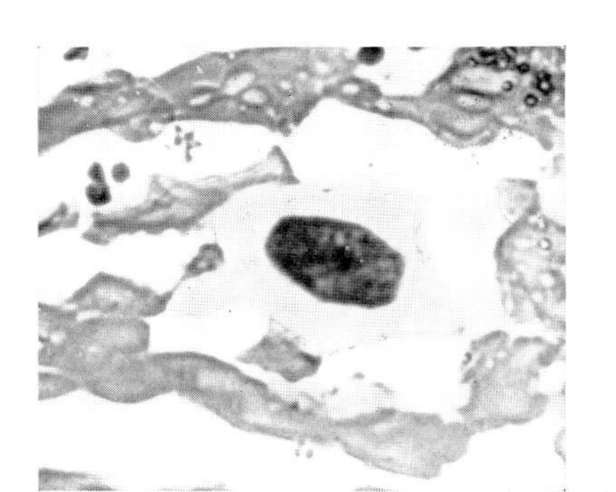

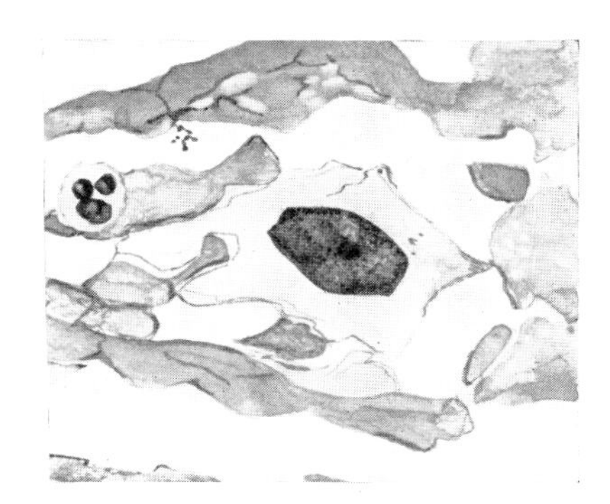

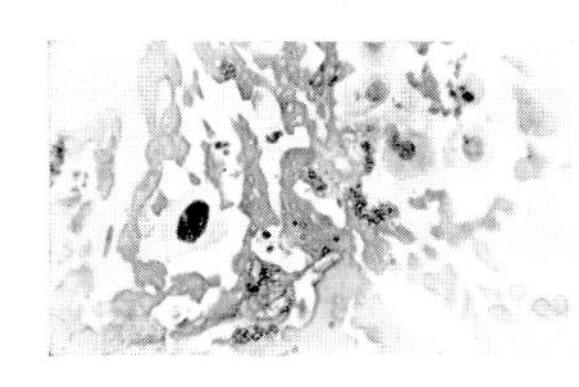

Low Power: Basal epithelial cells and one discrete cell with large dense nucleus.
High Power: The cell border is irregular in shape but clearly outlined, as are all differentiated malignant cells. The nucleus is centrally located and appears dense owing to the many clumps of chromatin. Part of the cytoplasm is obscured by blood, which is frequently seen in malignant smears.

Fig. 58

THIRD TYPE DIFFERENTIATED: MULTINUCLEATED

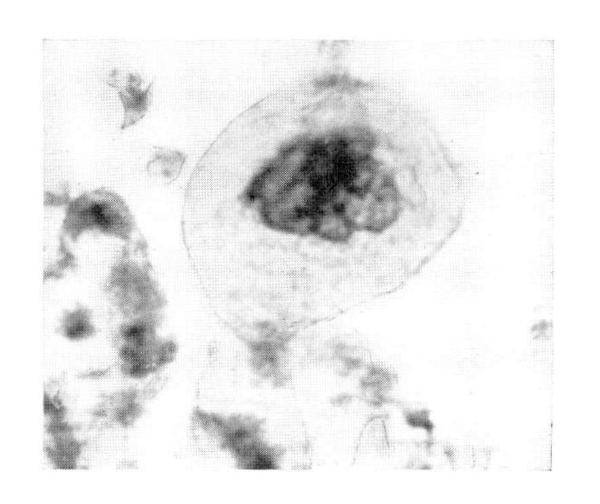

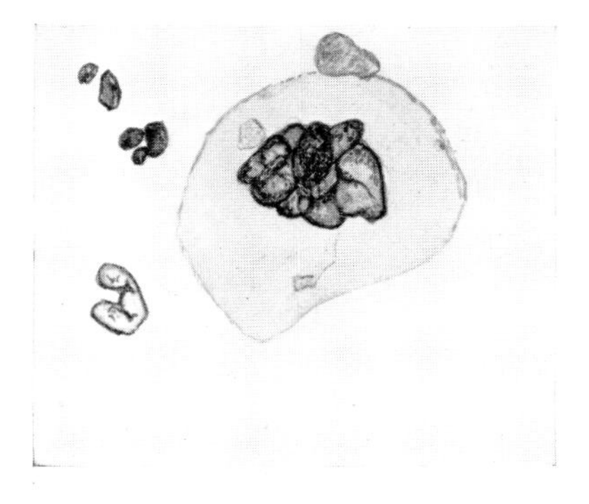

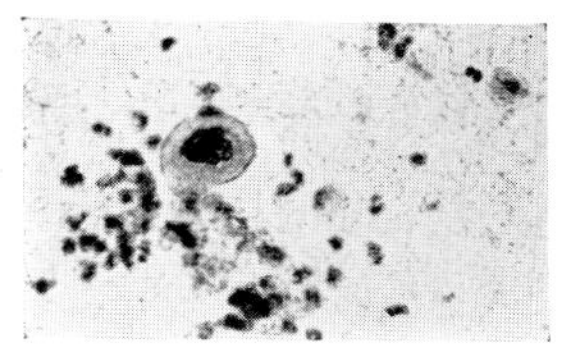

Low Power: Large single cell in a field of histiocytes.
High Power: Closer examination reveals a clear cell border and eight malignant nuclei, varying in size and shape, with sharp nuclear borders. Compare the chromatin content to the histiocyte on the left. In malignant giant cells the nuclei are grouped centrally in the cell, whereas in foreign body giant cells the nuclei are arranged peripherally.

FIG. 59

THIRD TYPE DIFFERENTIATED: GROUPS

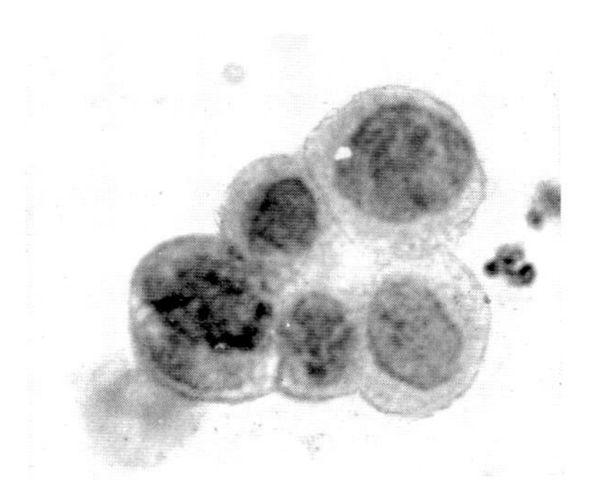

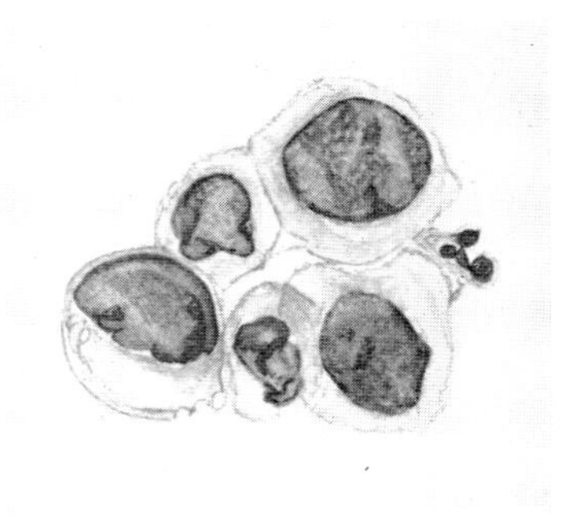

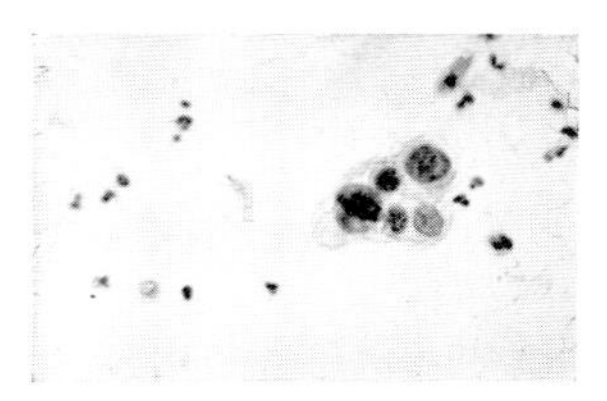

Low Power: Cluster of five cells with deep-staining nuclei.
High Power: This group of third type cells show well defined wrinkled nuclear borders, variation in size and shape and definite cytoplasm with clear cell borders. All cells have an abnormal cytoplasmic-nuclear ratio. It is much more common to find third type cells singly than in groups. Compare this group to basal cell in Plate 1.

FIG. 60

THIRD TYPE DIFFERENTIATED: DOUBLE-NUCLEATED

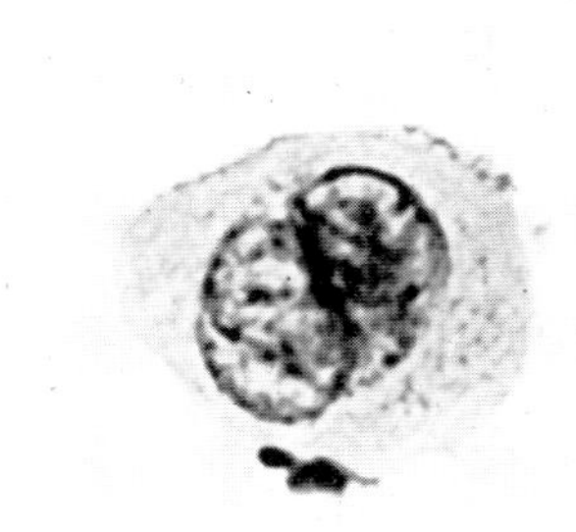

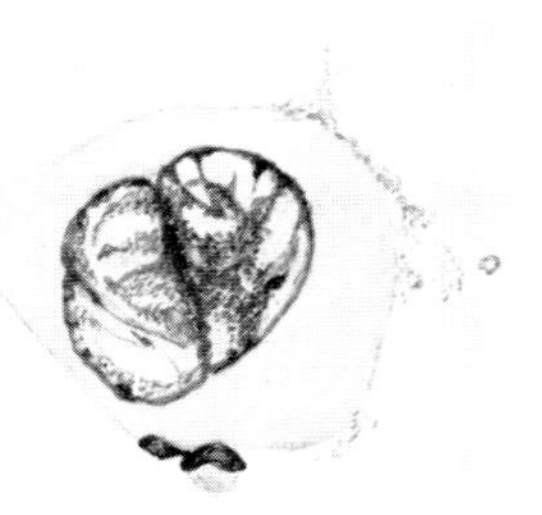

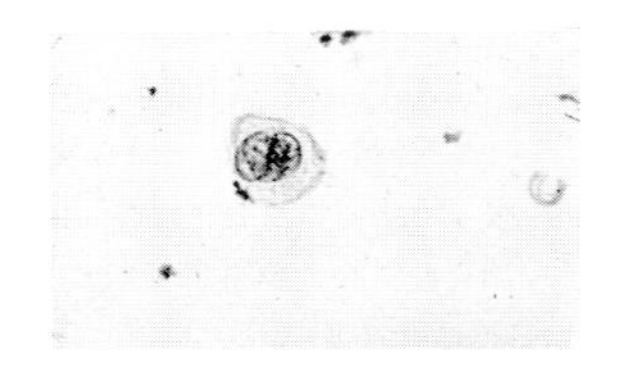

Low Power: Discrete cell with two enlarged nuclei.
High Power: The chromatin particles are unevenly distributed in dense clumps giving the nuclei a transparent quality where there is no chromatin. The cytoplasm is smoothly granular and the cell wall is visible. These malignant nuclei have the same structure as undifferentiated cells. Classification of these cells as differentiated is based on cytoplasm and a good cellular outline.

The term "third type differentiated" is certainly not descriptive. Perhaps the only attribute of the term is that it gives the number of types of differentiated malignant cells from squamous carcinoma. Though the name of the cell is a poor one, the cell itself is extremely important, since it is the type of malignant cell which occurs in early carcinoma of the cervix. These third type differentiated cells may occur in any positive smear from a case of squamous carcinoma in addition to other cells such as undifferentiated and fiber cells. However, when *only* this type of cell occurs the majority of cases show carcinoma-in-situ histologically rather than invasive carcinoma. It should be emphasized that we do *not* attempt a diagnosis of carcinoma-in-situ cytologically. The term carcinoma-in-situ necessitates an examination of the architecture of the lesion. Nevertheless, it appears that there is some definite correlation between early non-invasive carcinoma of the cervix and the presence of third type differentiated cells exclusively. This should not be interpreted as meaning that all cases of carcinoma-in-situ show only this type of positive cells. Many of them contain every type of cancer cell. On the other hand, in those smears in which only third type differentiated cells are found, the tumor is usually non-invasive. Because these cells do occur in early carcinoma, their identification is essential. They are perhaps misinterpreted as benign more often than any malignant cell, since they are the most differentiated and resemble normal cells more closely than any other cancer cell.

Difficulties in Interpretation: The normal cell which is confused with the third type differentiated cell is the inner layer basal cell. The distinction between the benign basal cell and the differentiated malignant cell depends upon nuclear structure. As has been emphasized in the criteria for all types of cells from squamous carcinoma, the identification of a cell as malignant relies on aberration in the nucleus. In no other cell are nuclear changes so important for identification as in the third type differentiated, since they are the *only* criteria upon which to base a decision. The nucleus of basal cells has finely granular chromatin. Occasionally there are one or more small clumps of chromatin. The nuclear border is not sharp. (See Plate 1.) The chromatin in the nucleus of the third type differentiated cell is irregular. There are large clumps of chromatin, heavy strands and fine granulations. The surface of the nucleus often appears wrinkled (See Plate 10.) The nuclear border is very definite. As in identification of all malignant cells, if the nuclear changes are scrutinized with care, the third type differentiated cells should be recognized with accuracy.

The distinction between a third type differentiated cell and an undifferentiated malignant cell rests upon the presence of a cellular border. It is difficult at times to be certain whether a border exists or not. It is safer to classify the cells as undifferentiated if there is any question as to whether a true border exists. Undifferentiated carcinoma cells commonly occur in groups, while it is infrequent to see third type differentiated cells in groups. The most exact way of determining whether cell borders exist is to see if the entire circumference of the cell is definitely outlined. If one tries to distinguish definitely where one cell ends and another begins, impression of presence of cellular outlines will become a certainty.

General Criteria for Identification of Third Type Differentiated Malignant Cells:

1. Nucleus has an irregular chromatin network. There is an increase in chro-

matin content. Nucleus often appears wrinkled. Border of nucleus is very sharp.

2. Cellular border is distinct. The presence of a definite cellular outline identifies the cell as a differentiated malignant one. Cytoplasmic-nuclear ratio is often abnormal but this is not a reliable criterion, since cells may occur in which the ratio is within normal limits.

SUMMARY

In reporting positive smears we attempt to determine the type of tumor present. If we are able to do so with some certainty the report reads positive, "consistent with squamous cell carcinoma." If only undifferentiated malignant cells are present, it is impossible to tell the type of tumor. However, very few smears from squamous carcinoma contain only undifferentiated cells. These cells are certainly in the majority in most instances, but careful searching of the positive smear will usually present others which may be identified as fiber, tadpole or third type differentiated cells. *Only* if differentiated cells of one or more of these three types are present may a smear be called "consistent with squamous cell carcinoma."

We have found that positive reports which also give the suspected type of tumor are extremely helpful to the clinician, especially in the unsuspected cases. Such reports focus attention toward the portion of the uterus where the tumor is most likely to be. For example, during the period when we reported smears as positive only and did not try to determine the type of tumor, an unsuspected case which was thought to be squamous carcinoma by smear had a dilatation and curettage but no biopsy of the cervix. The reverse has also been true; a biopsy of the cervix obtained but no curettage when the smear contained positive cells interpreted as being desquamated from an adenocarcinoma. For this reason we feel that positive reports of smears should give the type of tumor suspected if at all possible. With experience the well trained cytologist should be able to identify the tumor in the majority of positive smears.

CHAPTER V

ADENOCARCINOMA OF THE ENDOMETRIUM

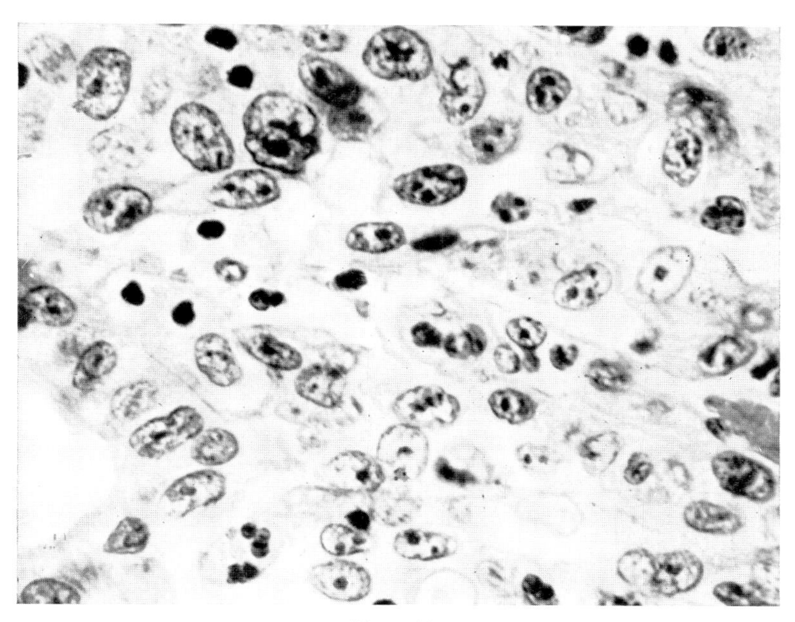

FIG. 61

HISTOLOGIC SECTION: ADENOCARCINOMA OF THE ENDOMETRIUM

Cells of adenocarcinoma of the endometrium do not show the great variation so evident in squamous carcinoma. For this reason cells from adenocarcinoma are not as easy to recognize cytologically as those from squamous cell carcinoma.

Malignant cells from adenocarcinoma of the cervix exhibit the same characteristics as those from carcinoma of the endometrium. We have classified the cells from adenocarcinoma into two groups: differentiated and undifferentiated, basing the classification on the presence of cellular borders. The differentiated cells of adenocarcinoma have an apparent cell border, not sharp but giving an impression of cellular outline. The nuclei are large, about eight to ten times the size of a leukocyte, and round or oval. They are eccentric in position and present the usual characteristics of malignant nuclei. The nuclei are large in relation to the amount of cytoplasm present and the nuclear-cytoplasmic ratio is abnormal. The cytoplasm does not stain evenly and often shows extreme vacuolization. These cells may be distinguished from the third type differentiated cell of squamous carcinoma by three criteria. The nucleus is eccentric rather than central in location. The cytoplasm has an irregular appearance and is often vacuolated. The cellular border is not as sharp.

The undifferentiated cells of adenocarcinoma occur in tight groups with a tendency to piling. Their nuclear structure is characteristic of malignancy. The nuclei vary more in size than in shape, but variation is never extreme. A background of cytoplasm is often vacuolated, but cellular borders are absent.

DESCRIPTION OF DIFFERENTIATED MALIGNANT CELLS FROM ADENOCARCINOMA

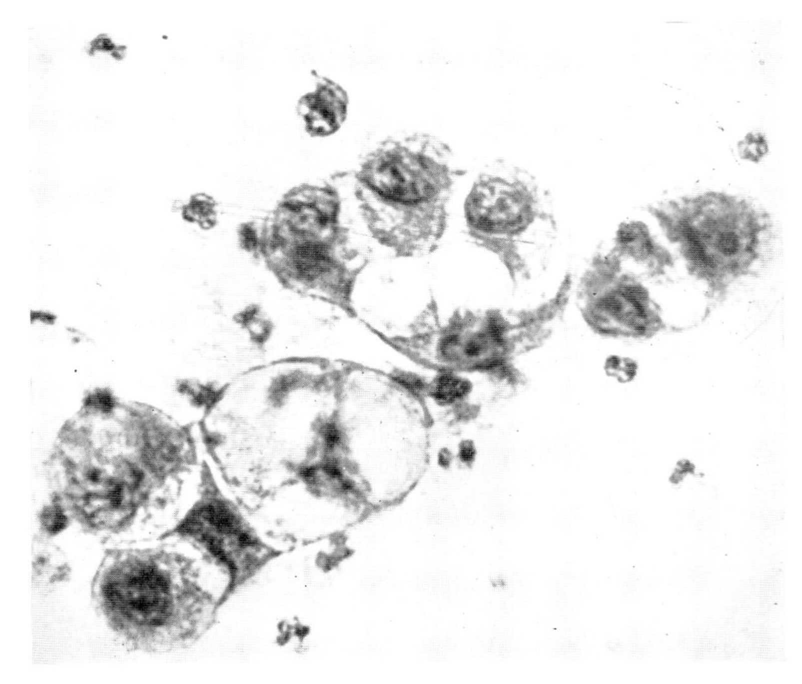

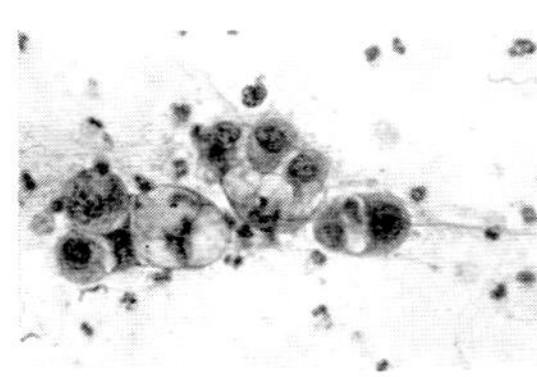

Low Power: Group of dark-staining nuclei with vacuolated cytoplasm.

High Power:

A. Characteristics of Nucleus: 1. Sharp nuclear borders. 2. All of the nuclei exhibit prominent clumps of chromatin (see cell 1 and group 4), but the intensity of hyperchromaticism varies according to the stage of degeneration of the nucleus. For example, cell 2 shows clumps of chromatin, but they have stained lightly and the nuclear border appears hazy. 3. The nuclear variation in adenocarcinoma is more in size than in shape. The nuclei as a whole are fairly constant in shape (see group 4), whereas the variation in size is more obvious. Compare the small lower nucleus in group 5 with others in the field.

B. Characteristics of Cytoplasm: 1. Distinct cellular borders. 2. Each nucleus has an adequate amount of clearly visible cytoplasm which may be intact (see cell 1) or show many pouchy vacuoles (see cell 3), which tend to enlarge and distort the shape of the cellular border. 3. Often several cells are conglomerated and the edges of the cytoplasm and vacuoles form a definite line around the cells. (See groups 4 and 5.) 4. Staining Reaction: Purplish blue or pink cytoplasm.

C. General Characteristics of Group: Collection of cells with eccentric active nuclei, varying more in size than in shape, with the large cytoplasmic vacuoles which are characteristic of adenocarcinoma.

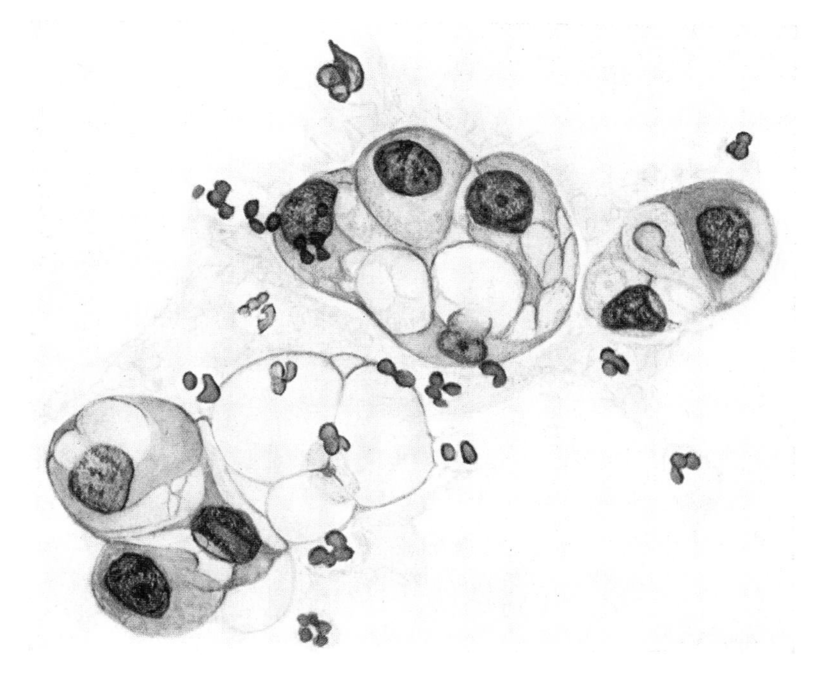

PLATE 11

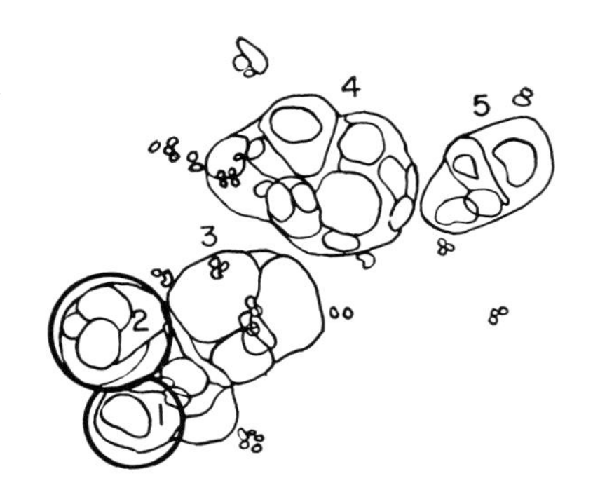

KEY TO ADENOCARCINOMA CELL PLATE

1. Differentiated malignant cell with a dark-staining nucleus and adequate well defined cytoplasm.
2. Light-staining differentiated cell with vacuolated cytoplasm and a degenerate malignant nucleus.
3. Faintly-staining nucleus in cytoplasm which is entirely vacuolated.
4. Group of four differentiated malignant cells with fairly uniform nuclei and vacuolated cytoplasm. Notice the cell borders are jointed together forming a continuous line.
5. Two differentiated cells showing variation in nuclear shape and vacuolated cytoplasm.

FIG. 62

ADENOCARCINOMA OF ENDOMETRIUM: MITOSIS

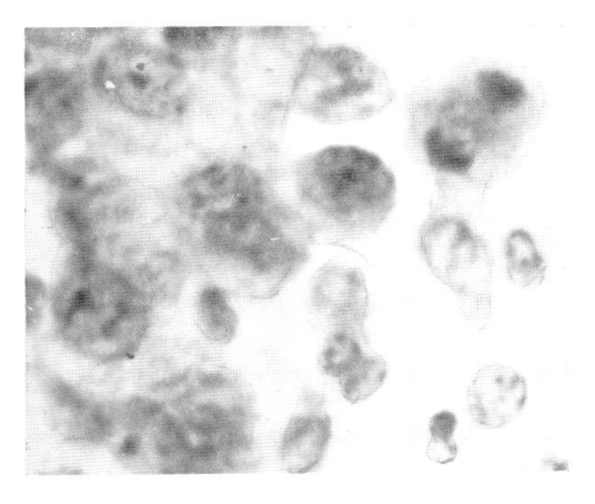

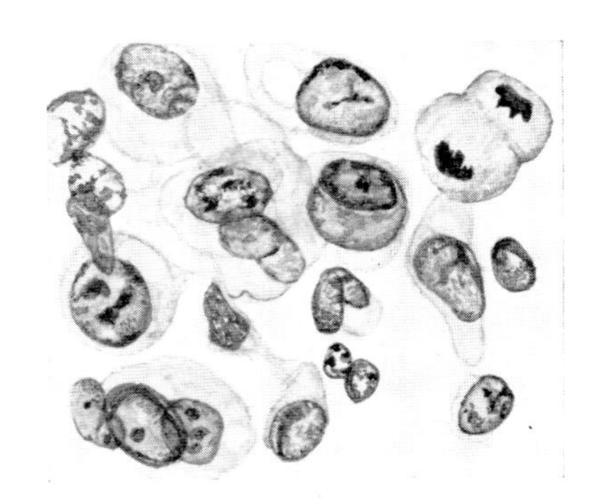

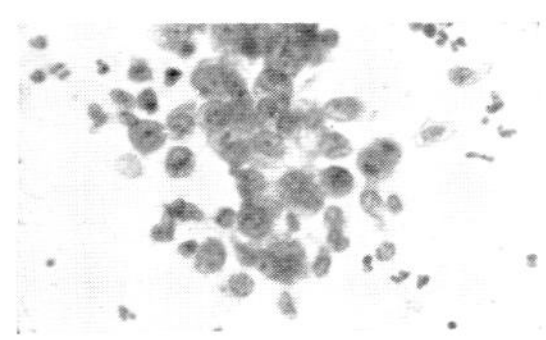

Low Power: Field of uneven nuclei showing distinct granulations.
High Power: Nuclei show extreme variation in size. The chromatin is arranged in dense irregular clumps, thick strands and fine granules. A mitosis is present. Compare this group of cells with that of histiocytes illustrating a mitosis. (See Fig. 32.) The two groups show how identification of a cell in mitosis depends on the cells which accompany it.

FIG. 63

ADENOCARCINOMA OF ENDOMETRIUM: DIFFERENTIATED MALIGNANT CELLS

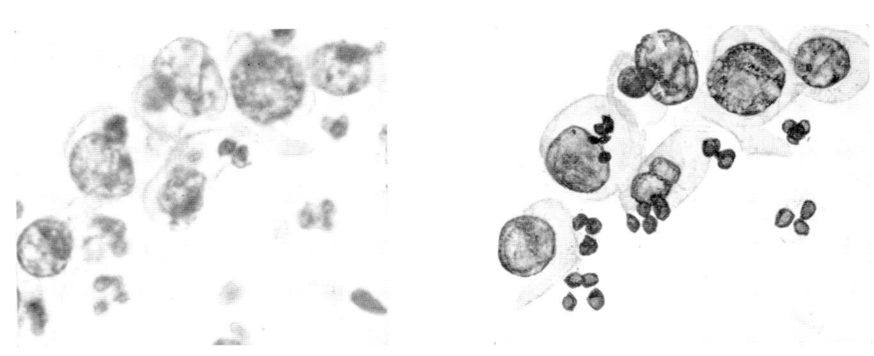

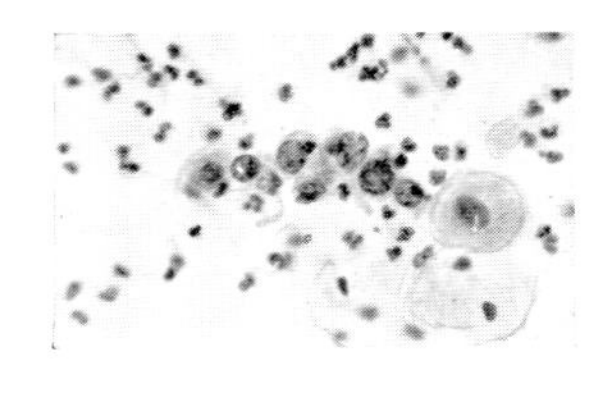

Low Power: Group of nuclei much deeper stained than nucleus of outer layer basal.
High Power: The distinction between these cells and the third type differentiated (see Fig. 59) is based on three criteria. First, the nuclei are eccentric, not central in position. Second, cellular borders are not distinct. Third, the cytoplasm does not stain evenly. It is extremely light in spots, dense in others.

Fig. 64

ADENOCARCINOMA OF ENDOMETRIUM: DEGENERATE MALIGNANT CELLS

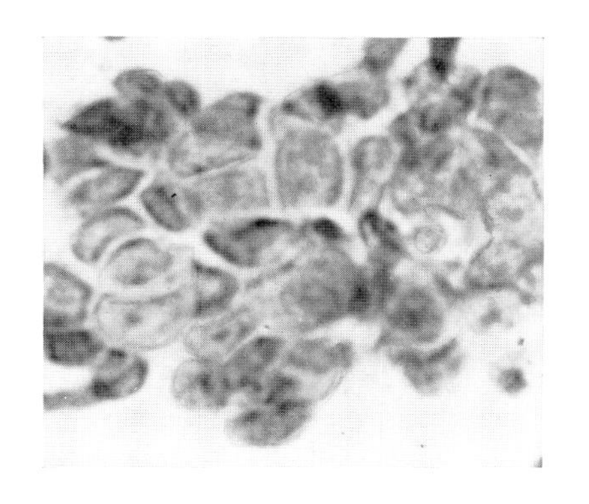

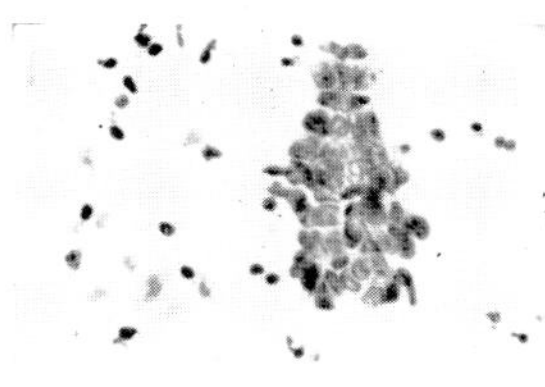

Low Power: Dense cluster of deep-staining nuclei. *High Power:* These nuclei are somewhat degenerate. However, variation in chromatin pattern is still very marked. In this group the undifferentiated malignant cells show marked variation in both size and shape. Cytoplasm is an indistinct background. Tight clumping of large numbers of malignant cells is more common in adenocarcinoma than in squamous carcinoma where cells tend to be arranged discretely.

Fig. 65

ADENOCARCINOMA OF ENDOMETRIUM: SINGLE VACUOLATED MALIGNANT CELL

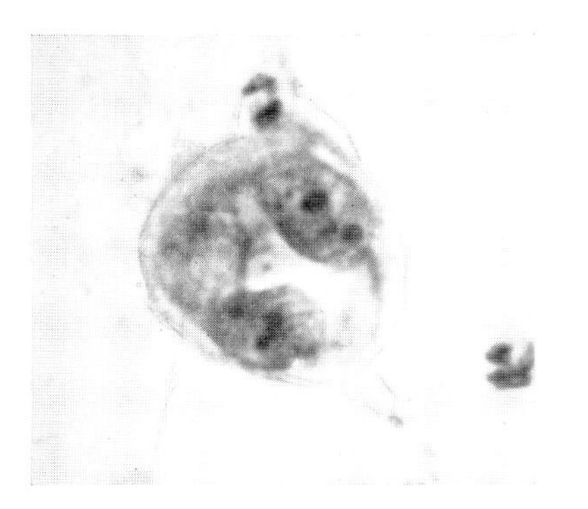

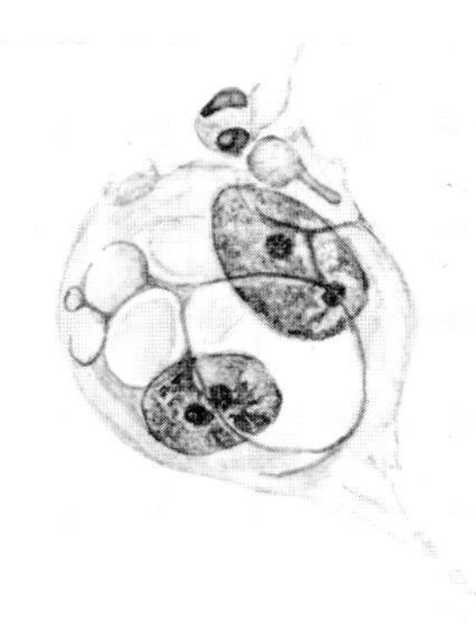

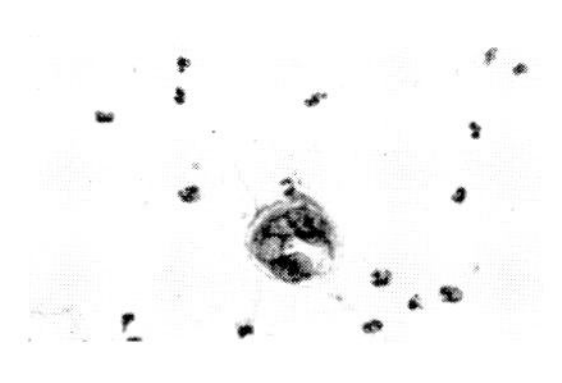

Low Power: Vacuolated, double-nucleated cell. *High Power:* This is an example of a single cell which can be identified as desquamating from adenocarcinoma. The nuclei differ in size and the chromatin content exhibits the marked irregularity of structure which is so typical of a malignant cell. The cellular border is fairly distinct and shows extreme vacuolization. One vacuole overlies part of each nucelus.

FIG. 66

ADENOCARCINOMA OF ENDOMETRIUM: VACUOLATED GROUP

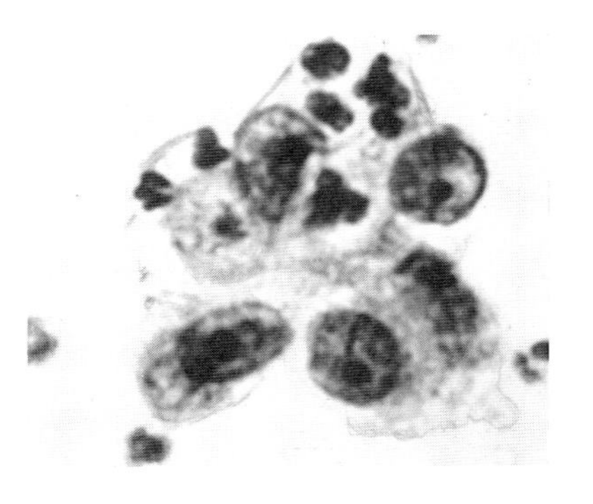

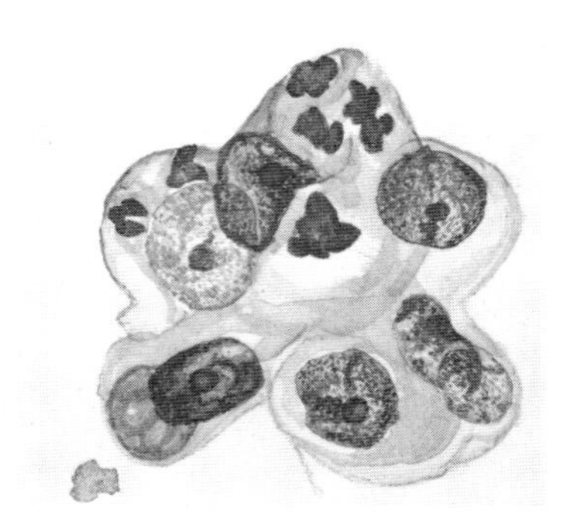

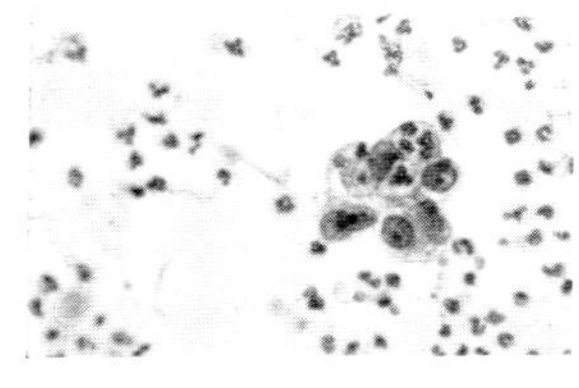

Low Power: Large clump of irregular nuclei with vacuolated cytoplasm.

High Power: The majority of the cells in this field show prominent nucleoli. The nuclear borders are sharp and chromatin is irregular in appearance. The cytoplasm shows some vacuolization with inclusion of polymorphonuclear in the vacuoles. No definite cellular borders are present except at the edges of the group, but nuclei appear to be eccentric.

FIG. 67

ADENOCARCINOMA OF ENDOMETRIUM: PHAGOCYTOSED MALIGNANT CELL

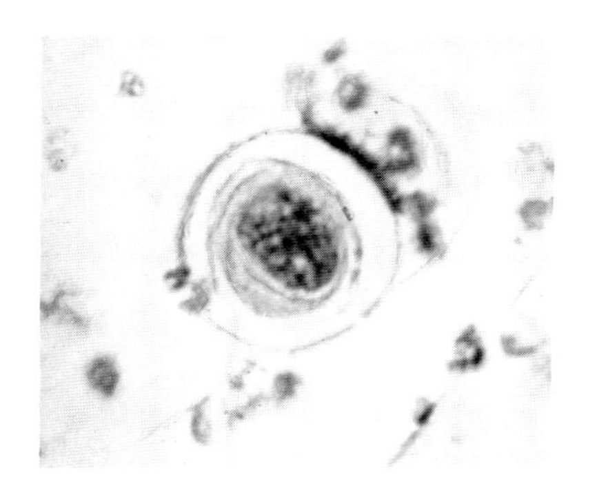

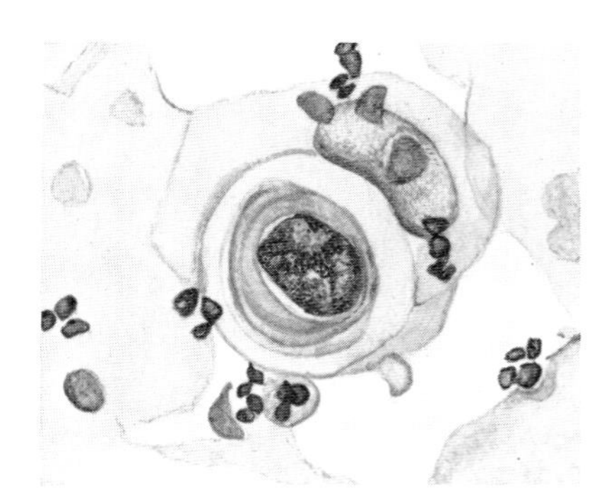

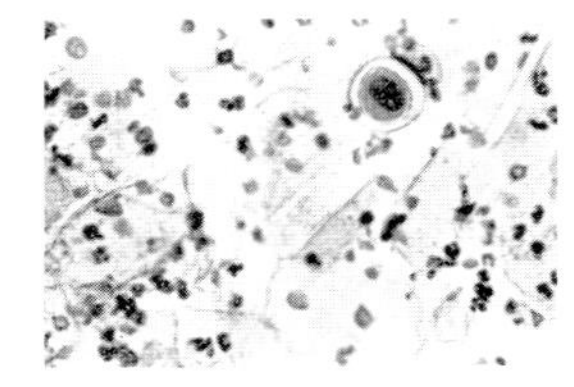

Low Power: Large engulfed cell in field of squamous cells.

High Power: This is a rare type of cell found in adenocarcinoma, and illustrates the occasional phagocytic property of malignant cells. The phagocytosed cell has a slightly eccentric hyperchromatic nucleus. The nucleus of the vacuolated cell contains a large dark body which might be interpreted as a nucleolus, since it is homogenous and has a smooth outline.

FIG. 68

ADENOCARCINOMA OF ENDOMETRIUM: UNDIFFERENTIATED MALIGNANT CELLS

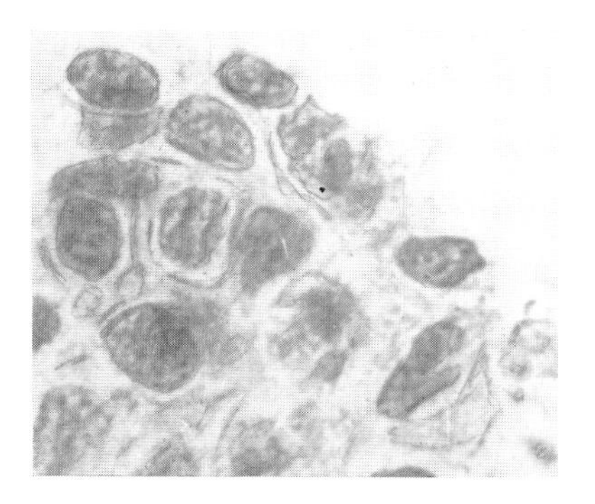

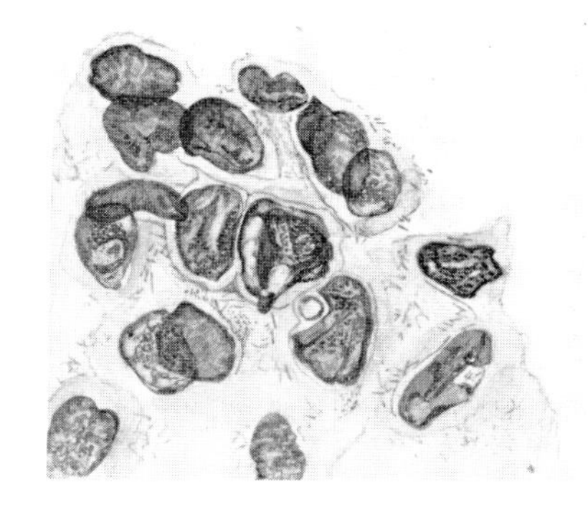

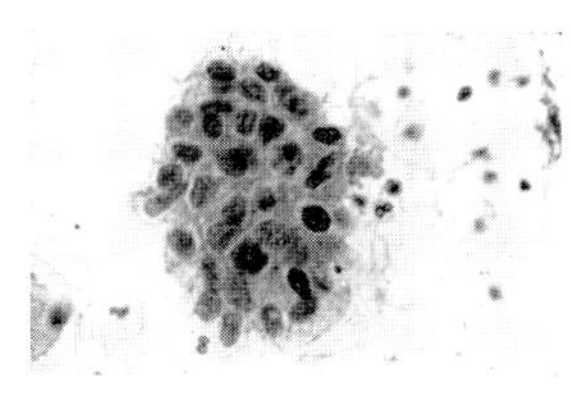

Low Power: Dense cluster of hyperchromatic nuclei.

High Power: These cells have no characteristics to identify them as adenocarcinoma. There are no cellular borders, no vacuolization. They are typical undifferentiated malignant cells, quite tightly grouped, showing marked differences in both size and shape and extreme variation in the chromatin content. Chromatin is concentrated at the border of the nuclei and nuclear borders are irregular.

FIG. 69

ADENOCARCINOMA OF ENDOMETRIUM: GIANT MALIGNANT CELL

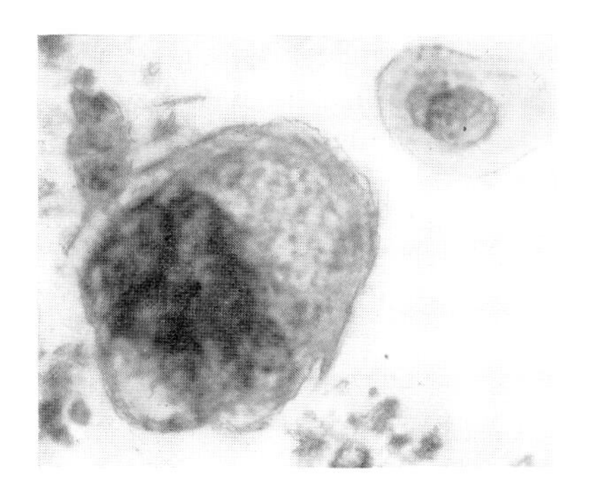

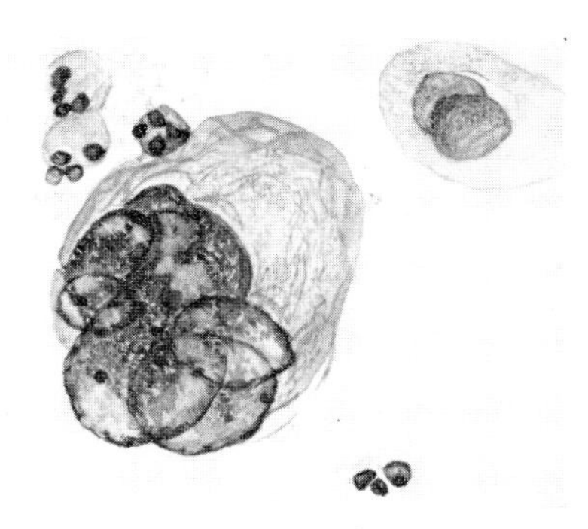

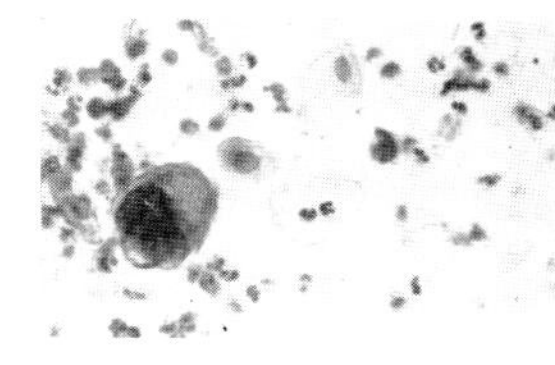

Low Power: Extremely large multinucleated cell in field of leukocytes.

High Power: It is interesting that in this large malignant multinucleated cell from adenocarcinoma that the nuclei appear in such a definite eccentric position in the cell. Compare nuclear position to that of malignant giant cell in squamous carcinoma. The nuclei vary tremendously in size and very little in shape. Cytoplasm is irregular in appearance.

Adenocarcinoma of Endometrium

The diagnosis of adenocarcinoma of the endometrium by vaginal smear is not as satisfactory as that of squamous cell carcinoma. On a percentage basis adenocarcinoma is missed about twice as often as squamous on examination of one vaginal smear. There are two reasons for this. First is the fact that often extreme degeneration takes place in the desquamated cells before they reach the vagina. Second, the identification of certain well differentiated adenocarcinoma cells is difficult cytologically. We do not mean to suggest that the vaginal smear is not of definite value in the diagnosis of carcinoma of the endometrium but we would like to emphasize that it is neither as accurate nor as clear-cut cytologically as that of squamous carcinoma.

Difficulties in Interpretation of Cells from Adenocarcinoma: The greatest difficulty encountered is that of making a distinction between atypical endometrial and endocervical cells and true carcinoma. Except for these two types none of the other normal cells seen in the vaginal secretion should be confusing. Endometrial cells normally have small regular nuclei which are finely granular (see Fig. 24), or pyknotic, depending on the state of preservation of the cells. When these cells desquamate from a hyperplastic endometrium the picture may be somewhat different. Cells may vary in size quite considerably. They occur in tight groups occasionally with vacuolated cytoplasm. In groups such as these, identification of the nuclei as atypical benign depends entirely on chromatin structure. The nuclei of such endometrial cells still retain their vesicular pattern with one or two prominent clumps of chromatin. Endometrial cells with pyknotic nuclei are not confusing, since the nuclei are small and show a smooth denseness of the chromatin.

As in endometrial cells, the basis for distinguishing enlarged endocervical cells as benign, rather than malignant, depends entirely on the chromatin structure. Nuclei must show marked irregularity of chromatin material to be considered malignant, and the nuclei of even enlarged endocervical cells do not present that pattern. The distinction between atypical normal cells and malignant cells may be quite difficult but if the chromatin structure of the nucleus is scrutinized carefully accurate identification can usually be made.

It is not always possible to classify a group of malignant cells as adenocarcinoma but in a fair number of instances it can be done quite accurately. A group of cancer cells showing indistinct cell borders with vacuolization of the cytoplasm can be identified as adenocarcinoma. The vacuolization is a helpful criterion and is quite characteristic. It must be distinguished from the vacuolization which is caused by degeneration, in which the vacuoles are usually small and contained well within the cytoplasm. (See Fig. 39.) The vacuolization of adenocarcinoma presents a "pouching out" appearance of large vacuoles which may contain leukocytes.

Criteria for Identification of Adenocarcinoma Cells: 1. Differentiated cells: Round cell with eccentric nucleus with irregular chromatin arrangement. Cellular border is visible but not distinct. Cytoplasm is inadequate for size of nucleus and is not smoothly stained. It often contains large vacuoles. Cells occur in groups. 2. Undifferentiated cells: Round or oval nuclei with irregular chromatin arrangement. Variation is more in size than in shape. No cellular borders are present and cytoplasm appears as a background which may show vacuolization. These cells occur in tight clusters with nuclei overlapping. 3. The structure of the nuclei identifies the cells as malignant, the vacuolization classifies the cell as adenocarcinoma.

CHAPTER VI

ADENO-ACANTHOMA OF THE UTERUS

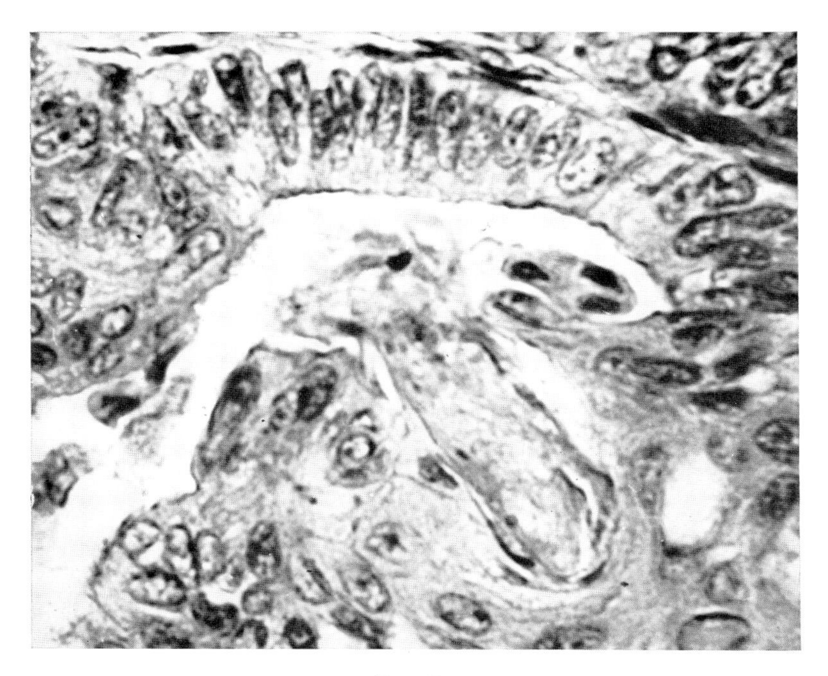

Fig. 70

HISTOLOGIC SECTION: ADENO-ACANTHOMA OF THE UTERUS

The section above illustrates the characteristics of the mixed tumor which contains both squamous and adenocarcinoma elements. As in histologic interpretation of this tumor, the cytologic diagnosis of this mixed tumor depends on finding well differentiated cells of both types of carcinoma. If typical malignant tadpole or fiber cells are seen in a smear which has also well differentiated groups of adenocarcinoma, a diagnosis of *positive* and consistent with adeno-acanthoma may be made.

It should be stated that it is rare to see such a striking example of the combination of the two types of cells as in the following plate. Usually groups of vacuolated differentiated adenocarcinoma cells will be seen and groups of fiber cells or single tadpole cells.

This type of tumor is relatively rare. We merely include it here to explain the occasional presence of cellular elements from both types of tumor. In some cases only one type of cell seems to desquamate. The smear will contain only differentiated squamous carcinoma cells or all cells will show the characteristics of differentiated adenocarcinoma. The reverse is also true; smears may have distinctive cells of each group and the biopsy may show only one type of tumor. In mixed tumors diagnosis depends on both types being present.

DESCRIPTION OF CELLS FOUND IN ADENO-ACANTHOMA

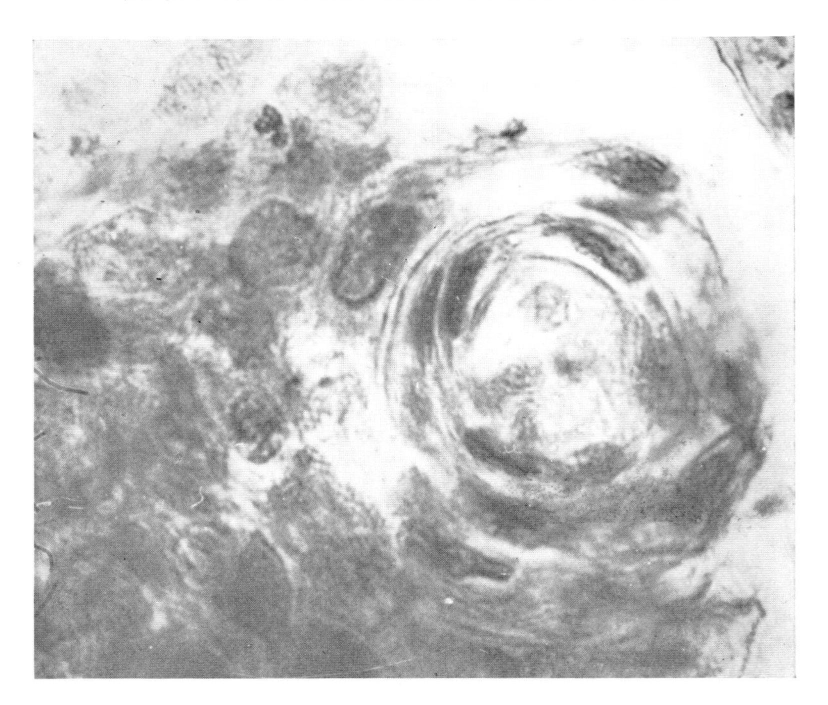

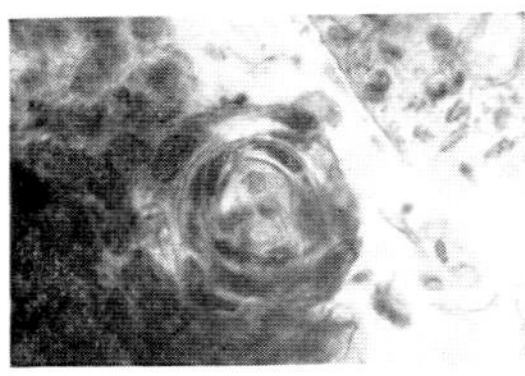

Low Power: Cluster of dark-staining nuclei, some being elongated and in a whorl formation.

High Power:

A. Characteristics of Nucleus: 1. Wrinkled well defined nuclear borders. 2. All of the malignant nuclei show many dispersed clumps of chromatin, groups 1 and 4 being more hyperchromatic than the other groups. 3. The undifferentiated cells in groups 2 and 3 show variation in size and shape, overlapping or piling qualities and vacuolated cytoplasm, which are characteristics indicative of adenocarcinoma. Group 4 is a collection of differentiated fiber cells with elongated wrinkled nuclei in a pearl formation, which is seen only in squamous carcinoma. When these two different types of cells are found in the same smear it is consistent with the mixed tumor of adeno-acanthoma.

B. Characteristics of Cytoplasm: 1. Groups 1 and 2 have no cytoplasm, but the cells in group 3 exhibit cytoplasm with pouched-out vacuoles. Compare to classical adenocarcinoma plate. 2. The nuclei of the malignant pearl, group 4, have adequate cytoplasm at each end, but the cytoplasm is conglomerated and it is difficult to distinguish individual cellular borders. 3. Staining reaction: Nuclei, purplish pink. Cytoplasm in differentiated cells, orange.

C. General Characteristics of Group: Two separate groups of malignant cells, one of overlapping undifferentiated cells varying in size and shape with vacuolated cytoplasm, and the other of elongated differentiated fiber cells in a pearl formation.

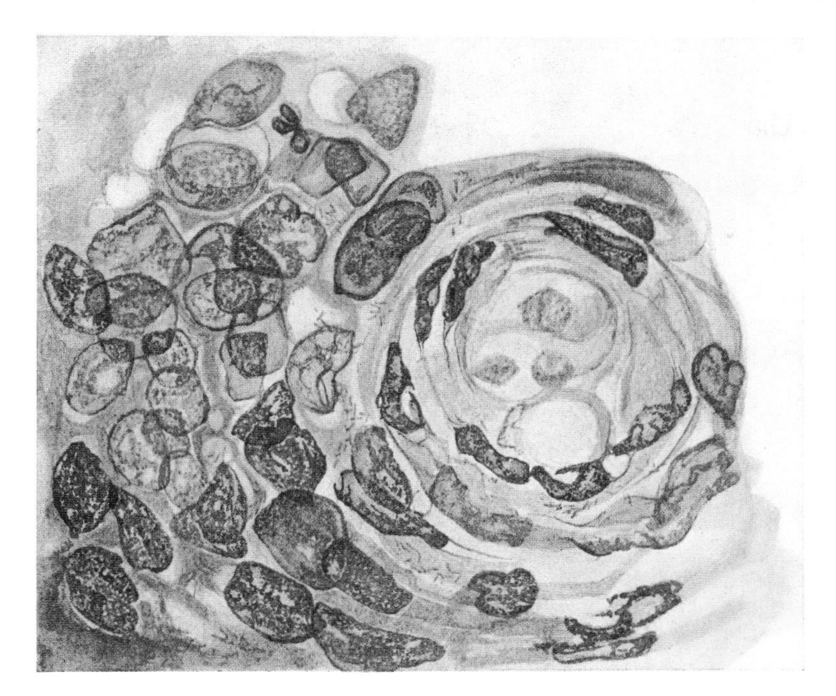

PLATE 12

KEY TO ADENO-ACANTHOMA CELL PLATE

1. Three undifferentiated hyperchromatic nuclei, varying more in shape than in size, with clear nuclear borders and no cytoplasm.
2. Group of undifferentiated cells showing the characteristic piling of adenocarcinoma. They vary in size and shape and have sharp nuclear borders and no evidence of cytoplasm.
3. Group of undifferentiated cells with large vacuoles in the cytoplasm, also indicative of adenocarcinoma.
4. A collection of degenerate differentiated fiber cells in a pearl formation with adequate conglomerated cytoplasm showing indefinite cellular borders.

CHAPTER VII

OTHER TUMORS OF THE FEMALE GENITAL TRACT

We are including here the cells which desquamate from tumors other than the two common malignancies of the female genital tract, squamous cell carcinoma of the cervix and adenocarcinoma of the endometrium.

Leiomyosarcoma of the uterus is a rare entity. It would not be expected to desquamate cells which would accumulate in the vaginal secretion unless it was far advanced and erosion of the mucosa had taken place. We have found cells which could be identified as malignant in only one case of leiomyosarcoma. Two other cases classified as sarcomas, a sarcoma botryoides and a carcinosarcoma, desquamated malignant cells. All three had tight clusters of undifferentiated malignant cells showing nuclear abnormalities and variation in size.

Occasionally malignant cells from carcinoma of the ovary will desquamate, remain preserved and be found in the vaginal secretion. We have had such cases in which the tumor was primary in the ovary and there were no metastases to the endometrium. Cells from adenocarcinoma of the ovary in our experience fulfill all the criteria for classification of adenocarcinoma. (See Fig. 74.) A far advanced case of malignant teratoma of the ovary desquamated undifferentiated malignant cells. The occurrence of cancer cells in the vaginal secretion from extra-uterine tumors is unusual, but if definite undifferentiated malignant cells are found in the smear and no tumor found in either cervix or endometrium, it is a diagnosis which should be considered.

Cells from carcinoma of the vulva may appear in the vaginal secretion as contaminants. As the pipette is withdrawn, cells may be aspirated from the vulva. The cells are either similar to the third type differentiated cells of squamous carcinoma of the cervix or they appear undifferentiated. The third type cell seen in vulva carcinoma is often larger than that encountered in squamous carcinoma.

We have had two cases of bladder carcinoma in which the cells appeared in the vaginal secretion as contaminants. The criteria for identifying bladder carcinoma will be given in the chapter on carcinoma of the urinary tract. Both cases were investigated thoroughly and no evidence of a vesicovaginal fistula could be found. If great numbers of tumor cells are desquamating, cells may occur as true contaminants in the vaginal secretion, apparently carried there in the urine at the time of voiding.

It should be emphasized that the appearance of cells from the tumors described above is extremely rare. They are included here as an interesting group of cases, and for possible clarification of the unusual case where definite malignant cells are found in the smear and no tumor can be discovered in the uterus.

Fig. 71

SARCOMA BOTRYOIDES

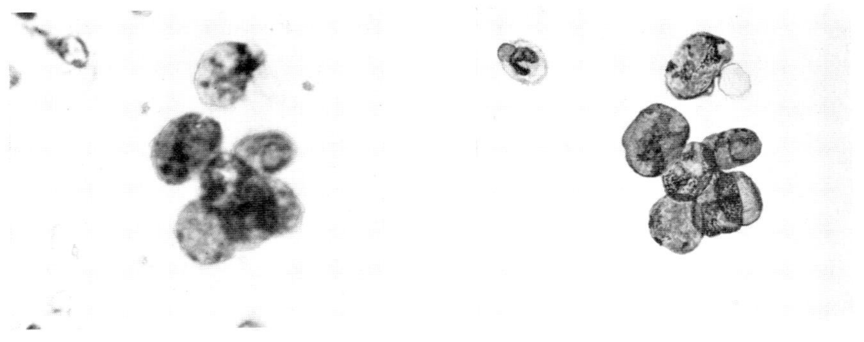

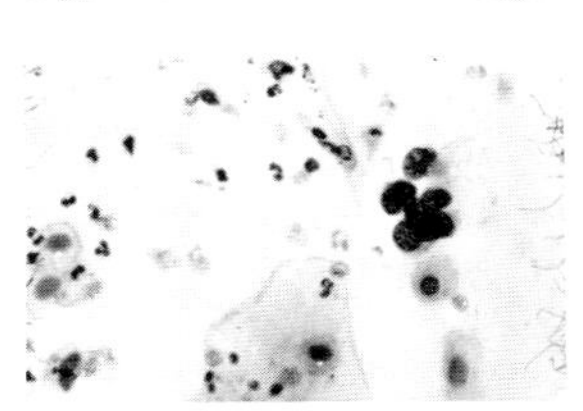

Low Power: Group of hyperchromatic nuclei in field of epithelial cells.
High Power: These densely staining nuclei are tightly clustered and show overlapping. They are typical undifferentiated malignant nuclei with considerable variation in size. The nuclei present marked evidence of chromatin aberration. There is irregularity of nuclear pattern and increase in nuclear material. These nuclei are indistinguishable from those seen in squamous carcinoma.

Fig. 72

LEIOMYOSARCOMA OF THE UTERUS

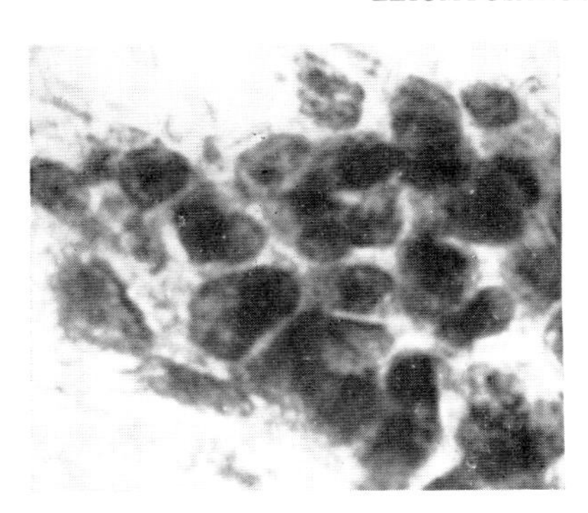

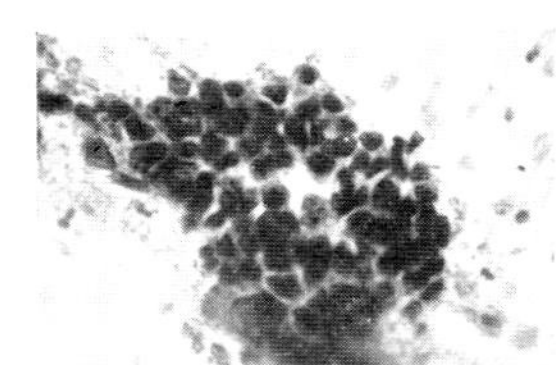

Low Power: Dense group of extremely hyperchromatic nuclei.
High Power: These cells show no differentiation. It would be impossible to distinguish them from a carcinoma. They fulfill all the criteria for identification of malignant cells: distinct aberration, increase in nuclear chromatin, sharp nuclear borders and great variation in size and shape. Cytoplasm is completely absent. The background is blood pigment.

Fig. 73

MALIGNANT TERATOMA OF THE OVARY

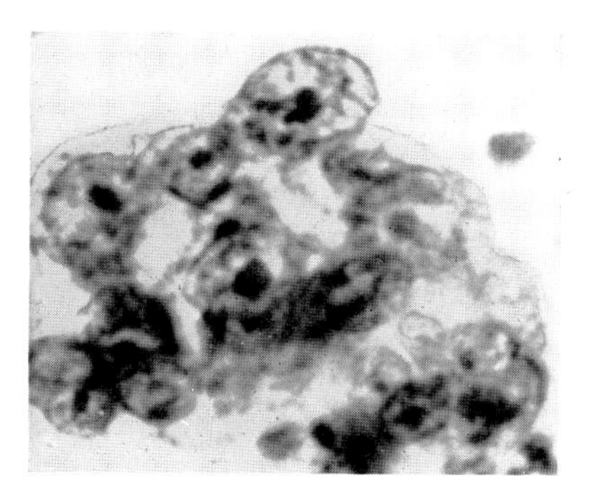

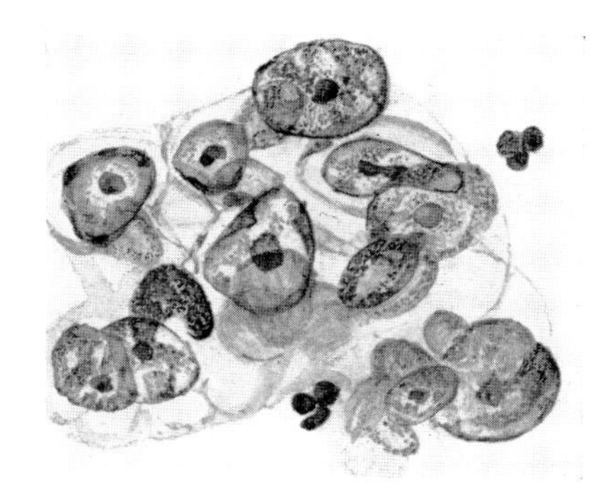

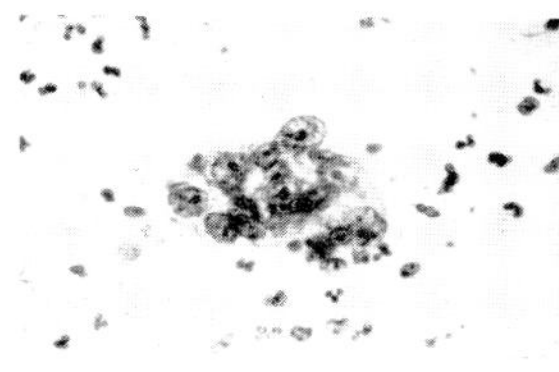

Low Power: Group of abnormally granular nuclei in field of leukocytes.

High Power: The most striking fact about this group of cells is the presence of the large nucleoli. They are exceptionally prominent and stain a reddish purple. They appear lighter than the large clumps of chromatin seen in previous figures and their outline is more regular. The other nuclear changes are characteristic of malignant cells.

Fig. 74

ADENOCARCINOMA OF THE OVARY

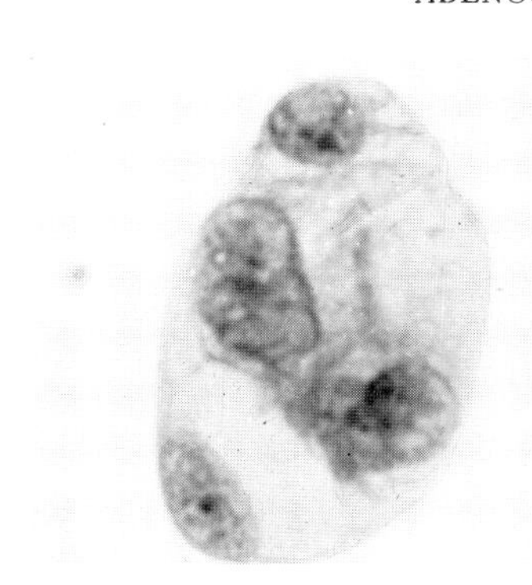

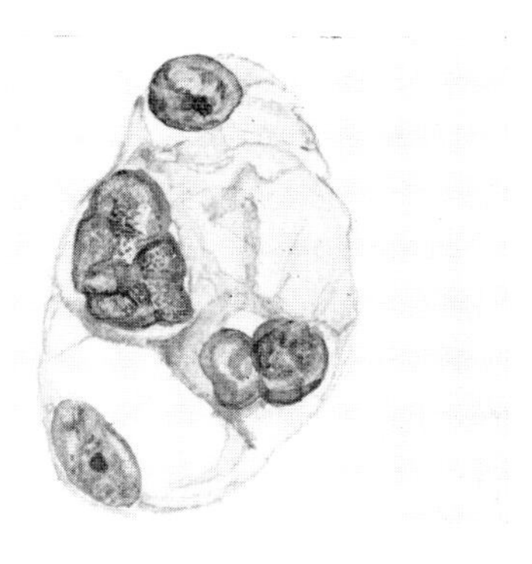

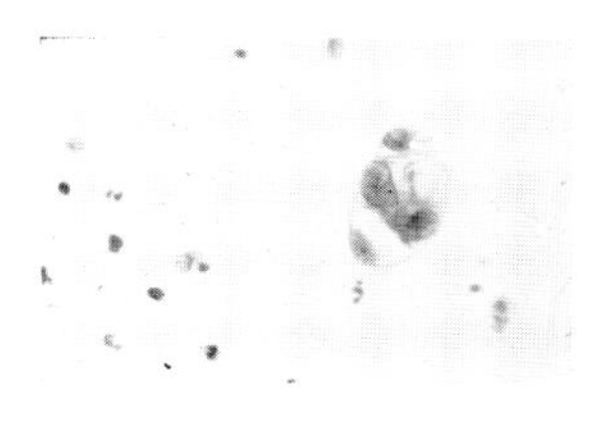

Low Power: Group of vacuolated cells.

High Power: This group of cells fulfills the criteria for adenocarcinoma. The nuclei show changes in the chromatin, the large nucleus presenting extreme abnormalities of nuclear structure. The chromatin is condensed at the periphery of the nuclei, making the nuclear borders very definite. The vacuolization of the uneven cytoplasm is characteristic of adenocarcinoma.

SUMMARY

There are several general characteristics which are common in positive vaginal smears and which serve as indications that the smear should be searched intensively for possible malignant cells. The presence of blood is always a suspicious finding. We consider it especially questionable if there is fibrin and blood pigment present. This indicates bleeding over a period of time and is quite a characteristic finding in positive smears. Originally, we felt that a positive diagnosis should not be made if some evidence of blood was not present, but further study showed that the positive smears from extremely early carcinoma of the cervix were completely free of blood. Thus the presence of blood, since it is seen at the time of the normal menstrual period, or its absence does not determine whether a smear is called positive or negative. But it is the exception to find positive smears showing no signs of blood. Leukocytes occur in great numbers in positive vaginal smears. They often appear in large clumps. Any smear in which the leukocytes appear increased in numbers and in large aggregates should focus attention upon the possible presence of malignant cells.

It is extremely rare to see a positive smear which is atrophic. We designate smears as atrophic which contain only basal cells indicating a very low epithelium. Independent of the age groups, positive smears seldom contain *only* basal cells as the normal cellular elements. It should be emphasized that we are referring to slides that contain no cornified or precornified cells. The typical positive smear has representative groups of all cell types. Occasionally the cells may be all superficial cells. (See Plates 7 and 10.) Whether this association of superficial cells with the presence of malignant cells represents a correlation between the amount of estrin present and carcinoma we are not prepared to say. However, it is certainly true that superficial cells are far more likely to be in the majority than basal cells in a positive vaginal smear.

Because of the rarity of an atrophic vaginal smear containing carcinoma cells, we are reluctant to call such a smear positive. If there are cells which are extremely suspicious, we ask that the patient take 1 mg. of stilbestrol daily for ten days, at which time the smear is repeated. If the cells in question were from carcinoma there will be no question of the diagnosis, since they are obvious in a smear containing only superficial cells. If they were only atypical normal cells caused by atrophy, they will disappear entirely. If this method is followed for suspicious atrophic smears, serious errors will be avoided, especially for the beginner in cytologic diagnosis who is not completely familiar with the variety of forms which atypical basal cells may assume.

The staining characteristics of the cytoplasm of malignant cells in vaginal smears are of little importance. They vary tremendously. For instance, some differentiated cells from squamous carcinoma stain basophilic, others acidophilic. It is the density of the nuclear stain which is important, not the staining reaction of the cytoplasm.

In contrast to the diagnosis of malignancy—histologically, the presence of mitosis in smears is of little significance. Histiocytes are more commonly seen in mitosis than any other cell type. Mitosis in malignant cells is quite rare. There are two possible reasons for this, first that the cells on the surface of a tumor are not in division, and second, if they are dividing they complete the division even after desquamation. At any rate, whatever the explanation of the lack of mitosis in smear preparations, it should be emphasized that the presence of a mitosis does not alone indicate malignancy.

It may be of interest to describe the way in which we report smears. We have three different diagnoses, *i.e.*, negative, doubtful and positive. The doubtful smears are repeated and if a second smear is doubtful repeated again. On the third or at the most the fourth smear all slides are reviewed and the smear called either negative or positive. We dislike leaving smears in the doubtful category since we believe that causes the cytologic method to lose much of its specificity. In the positive smears we try to classify the type of tumor present and the report reads "positive, consistent with adenocarcinoma." If we are unable to distinguish the type of tumor since only undifferentiated cells are present, the report reads "positive, undifferentiated carcinoma cells present."

CHAPTER VIII

RADIATION CHANGES IN NORMAL AND MALIGNANT CELLS OF THE VAGINAL SECRETION

Though the vaginal smear is used primarily for the diagnosis of untreated malignancy, it has a definite application in the follow-up of cases of treated carcinoma for the question of possible recurrence. A fair percentage of the vaginal smears examined in any laboratory will be those from patients treated for malignant disease. Some of these patients will be treated surgically but the majority will have been treated by radiation. Since the cellular changes produced by radiation are pronounced and varied, the student must become familiar with them in order to interpret these postradiation smears correctly.

In the early work in our laboratory we found that occasionally postradiation smears were extremely difficult to interpret. Several times the cellular changes, which we later learned to be aberrations in the normal cells caused by irradiation, were considered erroneously as malignant. In an attempt to understand more thoroughly the changes which radiation produced, we examined the smears of patients with carcinoma of the cervix during their entire course of radiation treatment. For the most part these patients were treated by x-ray, followed by radium implantation.

Examples of these changes in both normal and malignant cells are shown in the following plates. The drawings and photomicrographs are at the same magnification as all other plates. Days indicated in the paragraphs below are counted from the beginning of radiation treatment and are counted whether or not treatment was given on each day. The radiation therapy consisted of an average of 6000 r of x-ray given over a period of twenty-one days followed by two radium implantations of 2000 mg. hrs. four days apart.

NORMAL EPITHELIAL CELLS

Basal Cells: These are the first cells to show the effect of radiation. The earliest change encountered is in the staining reaction of the cytoplasm. This does not occur in all cells but is a frequent enough occurrence to be a distinct effect. The cytoplasm stains neither the basophilic stain of well perserved cells or the acidophilic one of degenerate cells. Instead the cytoplasm appears as a yellowish brown. (See Plate 13.) The cytoplasm may be finely vacuolated in this beginning change. This effect of radiation may be seen as early as the second day after x-ray treatment, but is usually not evident until the ninth or tenth day. At the same time that the staining reaction of the cytoplasm changes, the basal cells begin to assume aberrant forms. The cells appear elongated and often have long tails of cytoplasm. These aberrant forms are similar to those seen in atrophic smears. (See Fig. 5.) The nuclei, too, show signs of degeneration. Pyknosis and karyorrhexis are common findings. These effects of radiation are not as specific as the changes described below, since occasionally a very atrophic smear will contain cells with these same degenerative characteristics, *i.e.*, brownish stain, aberrant forms and degeneration of the nuclei.

Beginning usually around the twelfth day the most striking effect of radiation on the cells of the basal layer is seen. There is a great increase in the size of the cells. They become three to four times their normal size. The cytoplasm of these enlarged cells often shows distinct fibrils throughout. (See Fig. 75.) It is important to realize that these "blown-up" cells show no change in the cytoplasmic nuclear ratio. It remains constant. The nucleus and the cytoplasm have increased in the same proportion. (See Fig. 75.) The nucleus in one of these large cells is even more finely granular than in the normal. It has a very smooth speckled appearance in most instances. These cells might be interpreted as normal precornified cells if care is not taken. The total size of the cell is often that of the precornified cells. Two criteria point to their origin in the basal layer. First, the cytoplasm does not have the true transparency of precornified cells; second, the nucleus is much too big for a normal precornified cell. We have never seen this tremendous ballooning of cells except in smears from a radiated patient.

These "blown-up" basal cells often contain brownish yellow central deposits. They have been interpreted as glycogen. (See Fig. 76.) Cells with these deposits are not specific as a radiation change, but their occurrence is far more frequent in postradiation smears.

A specific change in the basal cells is the abnormal vacuolization of the cytoplasm. (See Plate 13.) This usually occurs in cells which have increased in size and is a late effect. It is usually not marked until the fifteenth day and increases progressively after that. The cells contain tremendous vacuoles which almost completely fill the cell. Some of the cells may have smaller vacuoles containing polymorphonuclear leukocytes, but this is not as common in the basal cell as in the precornified cells.

The occurrence of an abnormal number of nuclei in postradiation smears is common. (See Plate 13.) There may be as many as five nuclei in one basal cell with well defined cytoplasm. The nuclei are usually fairly regular in size, but occasionally they exhibit marked differences. They may be piled one on another in the center of the cell or spread discretely. The presence of abnormal numbers of nuclei is evidence of the abnormal division associated with radiation.

The last change to occur in the cells of the basal layer is the appearance of bizarre forms. The cells may be almost any shape. They are usually of the "blown-up" type so that the shape is even more striking. (See Fig. 77.) This change is much more pronounced than in the aberrant forms mentioned earlier. The cells may assume dumb-bell shape, elongated or tadpole forms. These cells must be distinguished from the differentiated malignant cell. The nucleus is the distinguishing characteristic. If large it is smoothly granular and not hyperchromatic, or it is very small and pyknotic.

DESCRIPTION OF RADIATED BASAL CELLS IN VAGINAL SECRETION

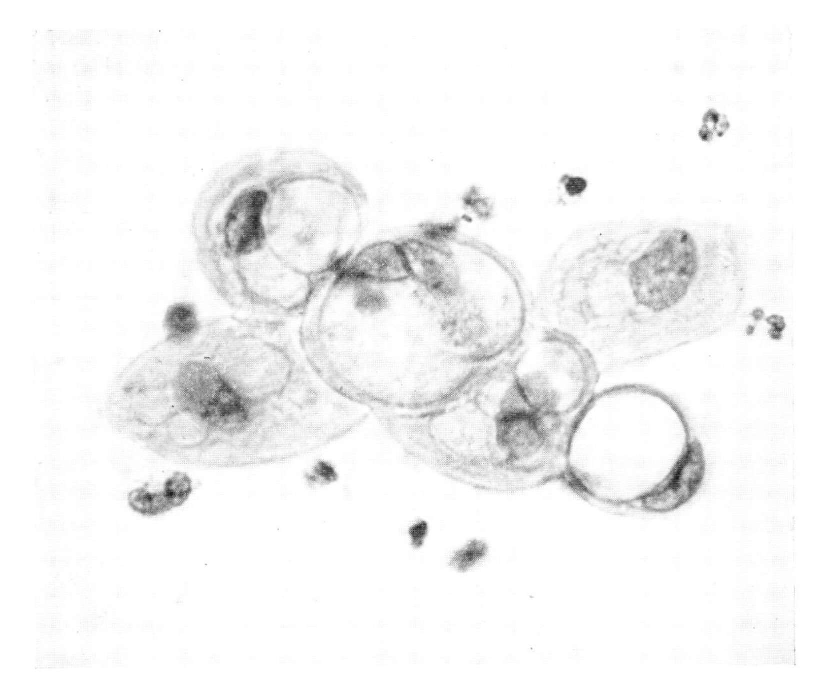

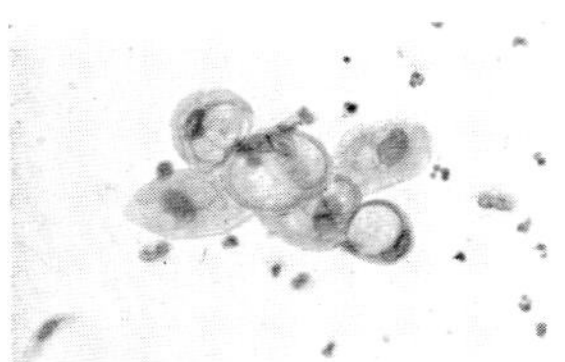

Low Power: Group of inner layer basal cells showing vacuolization.

High Power:

A. Characteristics of Nucleus: 1. Definite nuclear border, but chromatin is not condensed at border. 2. Chromatin content is finely granular. Where there are wrinkles on the surface of the nuclei, cells 2 and 3, the nuclear material is condensed and appears darker. No two nuclei exhibit the same amount of chromatin. Compare nuclei of cell 2. 3. The shape of the nuclei is for the most part oval. However, in cells 3 and 5, there is some change of shape. 4. Nuclei exhibit moderate variation in size. (See nuclei of cell 2.) 5. Cell 2 has three nuclei, evidence of abnormal division stimulated by radiation.

B. Characteristics of Cytoplasm: 1. Distinct cellular borders. 2. the most striking change is the extreme vacuolization of the cytoplasm. Vacuoles may overlie the nuclei (see cells 2 and 5), or be contained entirely within the cytoplasm, as in cells 1 and 6. Vacuoles may be very large (cell 2), or small and scattered throughout the cytoplasm (cell 3). 3. Staining reaction: light brown with shades of green and yellow. Compare to normal basals, Plate 1.

C. General Characteristics of Group: Basal cells showing moderate variation in size and shape. Both nuclear and cellular borders are definite. Marked vacuolization of cytoplasm is present.

PLATE 13

KEY TO RADIATED BASAL CELLS

1. Basal cell with fine vacuolization of cytoplasm, one large vacuole.
2. Triple nucleated basal cell with two large vacuoles. Nuclei vary in size and chromatin content.
3. Basal cell showing wrinkling of nucleus and fine vacuolization of cytoplasm.
4. Basal cell with discrete vacuole overlying the nucleus.
5. Vacuolated basal cell with irregularly shaped nucleus.
6. Basal cell with large vacuole producing signet ring type of cell.

Fig. 75

RADIATION CHANGES IN BASAL CELLS: INCREASE IN SIZE

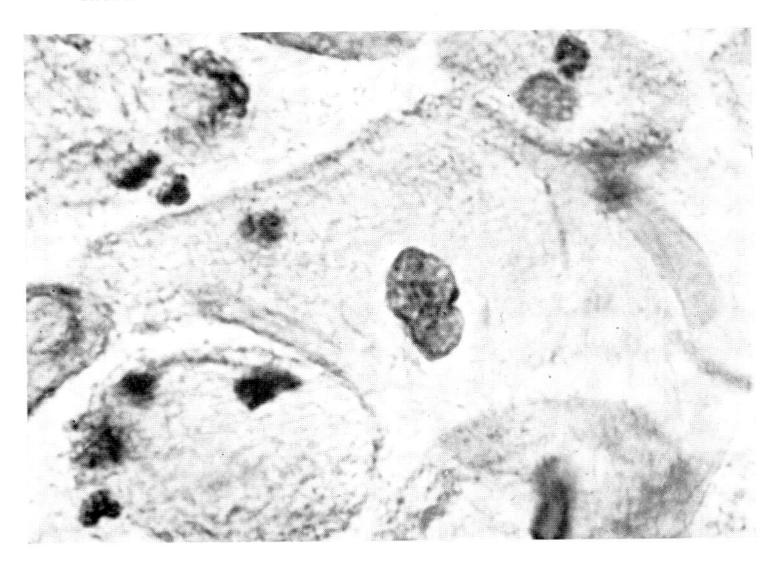

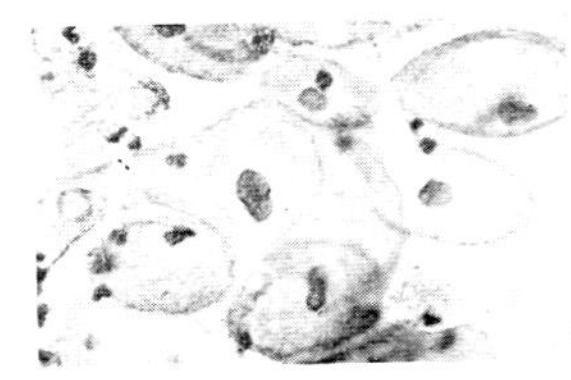

Low Power: Field of enlarged epithelial cells.
High Power: All the basal cells in the field are enlarged except for one in upper right. Nuclei have enlarged in proportion to the cytoplasm. Close inspection shows the chromatin to be evenly divided except where the nucleus is wrinkled. Fine fibrils are seen in the cytoplasm.

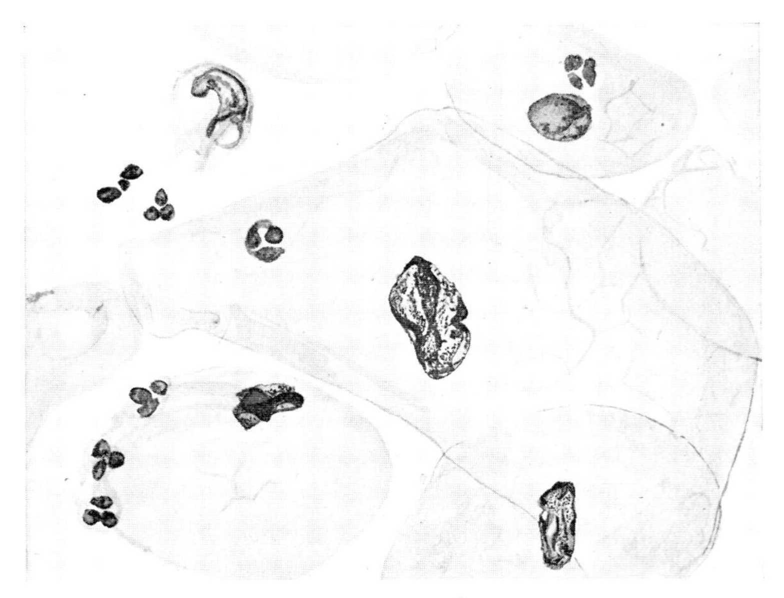

FIG. 76

RADIATION CHANGES IN BASAL CELLS: GLYCOGEN DEPOSITS

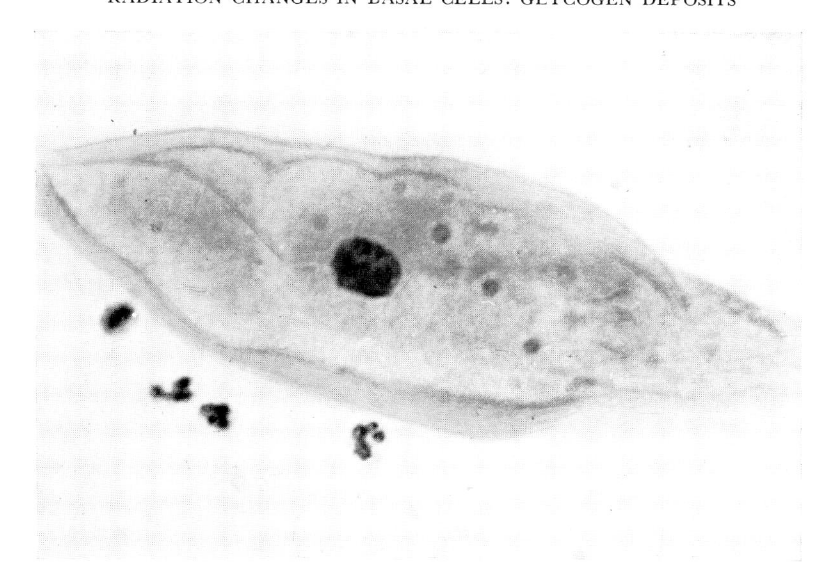

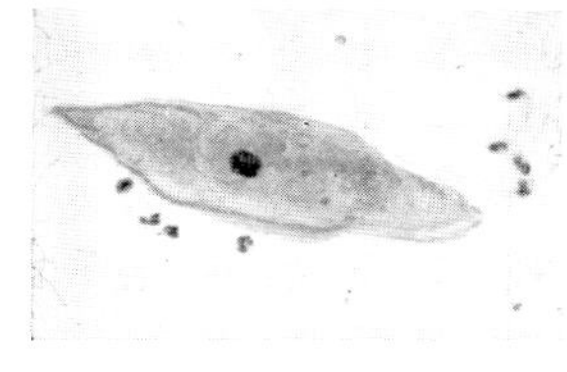

Low Power: Enlarged, elongated epithelial cell. *High Power:* The cytoplasmic border of this elongated outer layer basal is much thicker than usual. The central irregular deposits stain yellow and have been interpreted as glycogen. Compare size to normal basal cells, Fig. 3. Cellular deposits of this kind are common after radiation.

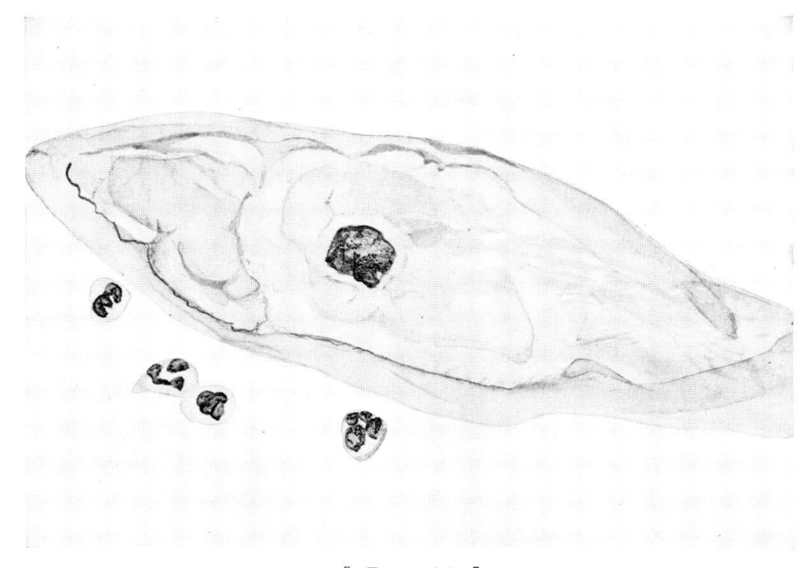

FIG. 77

RADIATION CHANGES IN BASAL CELLS: BIZARRE FORMS

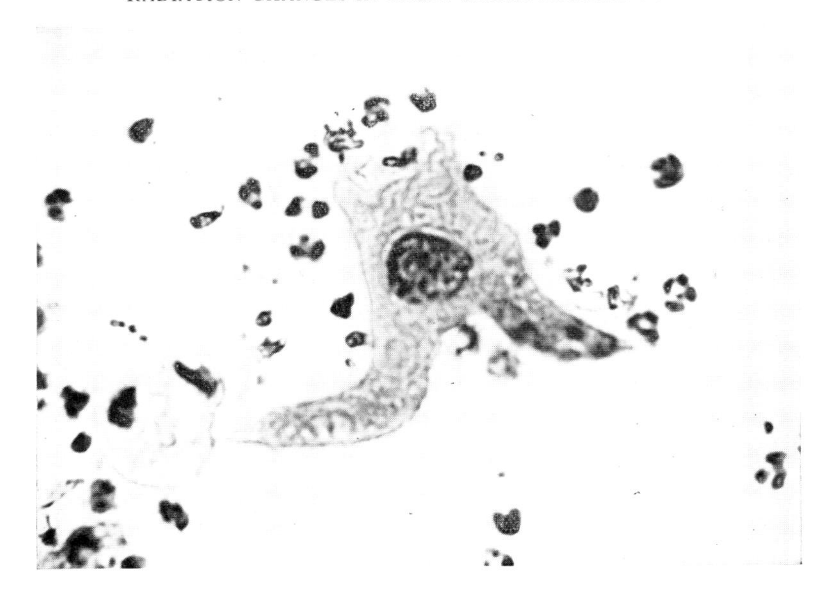

Low Power: Ameboid epithelial cell in field of leukocytes.

High Power: This is an example of the bizarre forms of which there are a great variety. Though nucleus and cytoplasm are increased in size, the cytoplasmic-nuclear ratio is that of a basal cell. This fact plus the lack of transparency of the cytoplasm identifies this cell as a basal.

FIG. 78

RADIATION CHANGES IN BASAL CELLS: POLYMORPHONUCLEARS IN CYTOPLASM

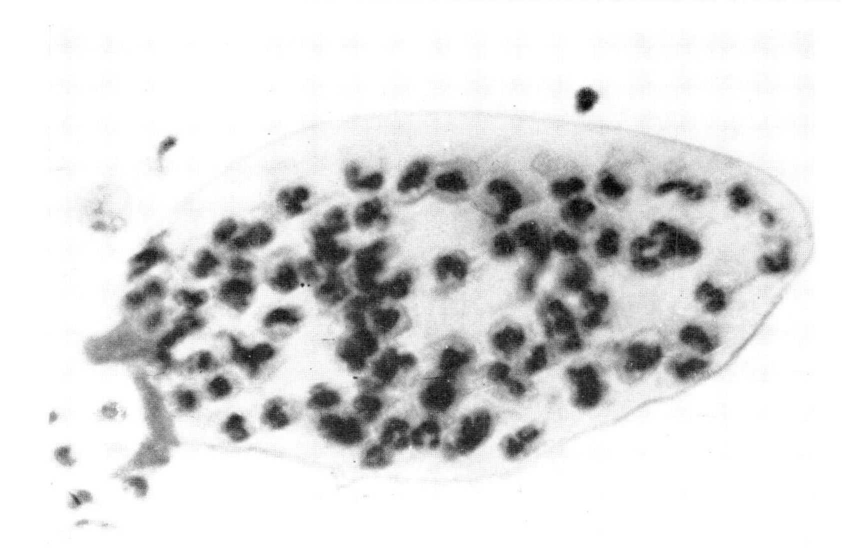

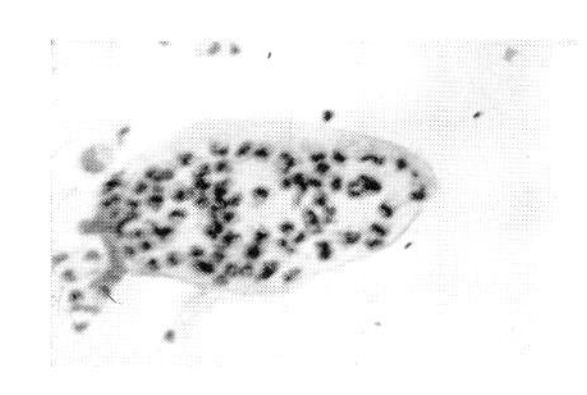

Low Power: Large basal cell containing leukocytes.

High Power: One of the last radiation effects to occur is the invasion of epithelial cells by polymorphonuclear leukocytes. This does not occur with regularity and should be distinguished from the occurrence of leukocytes superimposed on cells, which is common. The nucleus has disappeared. Oval shape of cell identifies it as an outer layer basal.

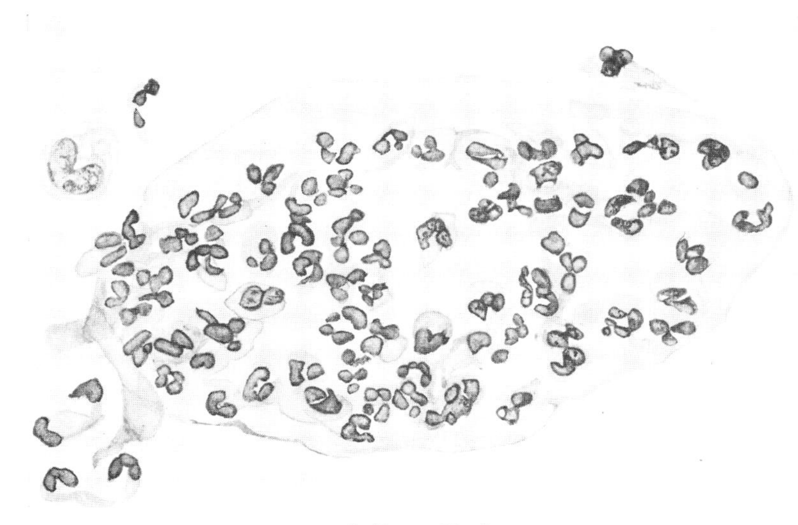

Radiation Changes in the Superficial Cells: The precornified cells also exhibit changes after radiation, though they occur later than the effects seen in the basal cells. The first reaction to occur in this layer of cells is an increase in size. They become from three to six times the size of a normal precornified cell. Often cells are seen which will fill a high dry field. The cytoplasm and the nucleus increase correspondingly, so that the cytoplasmic-nuclear ratio remains undisturbed. (See Plate 14.) On the average this effect is apparent about the fifteenth day of treatment.

At about the same time as this increase in size takes place there are beginning changes in the nuclei of the precornified cells. At first the only change is one of size, and the nuclear material appears finely granular. However, very soon the chromatin structure of the nucleus beings to show degenerative changes. The nucleus becomes dark and wrinkled, with less structural detail. It has an inactive, almost pyknotic look despite its abnormal size. The nuclear borders lose their smooth appearance and are very irregular. The surface of the nucleus appears folded and wrinkled. Often the nucleus has contracted, leaving a perinuclear space. Occasionally, although it is not seen as often as in the cells of the basal series, the nuclei will become fragmented and appear only as small dark particles, the nuclear borders having completely disappeared. Multinucleated cells are not uncommon in these precornified irradiated cells. (See Fig. 80.) These multinucleated cells may have the "blown-up" vesicular type of nuclei or degenerate ones. The presence of more than one nucleus is probably evidence of the abnormal division stimulated by radiation.

When nuclei begin to exhibit evidence of degeneration the cytoplasm also shows signs of radiation effect. The first evidence of degeneration in the cytoplasm is the appearance of very fine fibrils which may occur in concentric circles around the nucleus or may run from the nucleus to the cell border. (See Plate 14.) This change is seen more frequently in the precornified cells than the basal cells. Degeneration is also shown by the striking vacuolization of these enlarged superficial cells. There may be many small vacuoles scattered evenly throughout the cytoplasm or there may be only one large vacuole which almost replaces the cellular substance. (See Fig. 82.) We regard this marked vacuolization as a specific effect of radiation.

An effect which occurs immediately after the increase in size of the cells and the change in the nuclei is the appearance of precornified cells of queer shape. These bizarre cells are seen occasionally as early as the sixth day of treatment, but usually do not appear until later, around the eighteenth day. The cells are even more striking than the bizarre basal cells because of their size. (See Fig. 81.) They may assume any shape. Some become tremendously elongated, in some instances stretching across a high dry field. Others may assume enormous tadpole shapes. Some have long fibers stretching out from the cell almost resembling pseudopodia.

The last change to occur in the precornified layer is the appearance of polymorphonuclear leukocytes in the cells. This effect is also seen in the basal cells but less frequently than in the precornified. (See Fig. 78.) This takes place usually around the twentieth day of treatment. It has been seen as early as the tenth and as late as the thirty-first day. It is important when looking for this change to be sure that the polymorphonuclear leukocytes are in the cells and not merely on top of the cells. The leukocytes are very numerous in postradiation smears, and often clump on a cell, almost obscuring it completely. Clumping of the leukocytes on top of epithelial cells takes place early in the course of treatment, but the actual inclusion of leukocytes is a later change. It is often impossible to identify the nucleus of the epithelial cell in these instances and this change probably represents the actual phagocytosis of a degenerate cell by leukocytes.

The cornified cells exhibit all the evidences of radiation reaction which are seen in the precornified cell except for the changes in the nuclei. The normal maturation of the epithelial cells of the cervix and vagina seems to be undisturbed by radiation. The large vesicular "blown-up" nuclei of the precornified cells contract to the typical dense pyknotic appearance of cornified nuclei. They may be larger than usual but in other respects are indistinguishable from the normal nuclei of cornified cells. As in the unradiated cells we distinguish between cornified and precornified cells on the basis of whether the nucleus is vesicular or pyknotic. The cytoplasmic stain of radiated epithelial cells shows a great variety of colors from bluish green, brown, orange to deep pink, so that differentiation of the cells cannot be dependent on the staining reaction of the cytoplasm. The changes which take place in the cytoplasm of precornified cells are also seen in the cornified, *i.e.*, fibrils, vacuolization and bizarre forms.

It is important to point out that the changes in the basal cells are seen only if the control smear before treatment contained basal cells as part of the cell population. If the woman is premenopausal, few basal cells are found and the radiation changes are seen first in the precornified cells.

Radiation Changes in Cells from Columnar Epithelium: Endocervical Cells: These cells are seen only infrequently in postradiation vaginal smears. When present they exhibit the same changes seen in the squamous cells. The increase in size of both nucleus and cytoplasm is the most striking effect. If well preserved, they retain their palisade formation but appear three to four times larger than normal. The nuclei appear finely vesicular and are often wrinkled both on the surface and at the nuclear border. Extreme vacuolization of the cytoplasm is apparent.

Endometrial Cells: We have not been able to observe any definite differences in these cells which could be attributed to radiation, except for slight increase in size.

A great increase in polymorphonuclear leukocytes is seen often in the immediate postradiation smears. In some cases they increase so markedly that interpretation of the smear is quite difficult, since the leukocytes obscure much of the picture. The increase in leukocytes is variable in the time it appears. It may occur immediately after treatment is instituted or it may not occur until much later. Often toward the end of x-ray treatment the smear will become quite clean, but immediately following radium implantation there is again a great increase. An increase in small histiocytes usually follows the increase in leukocytes. The response of the phagocytic and foreign body giant cell is variable, but marked responses of both types of histiocytes have been seen.

DESCRIPTION OF RADIATED SUPERFICIAL CELLS IN VAGINAL SECRETION

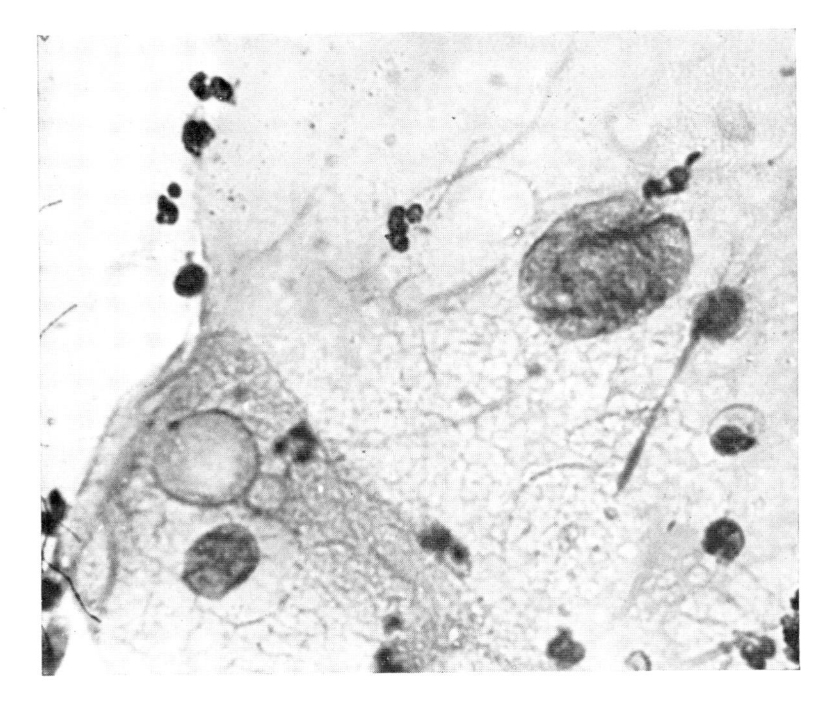

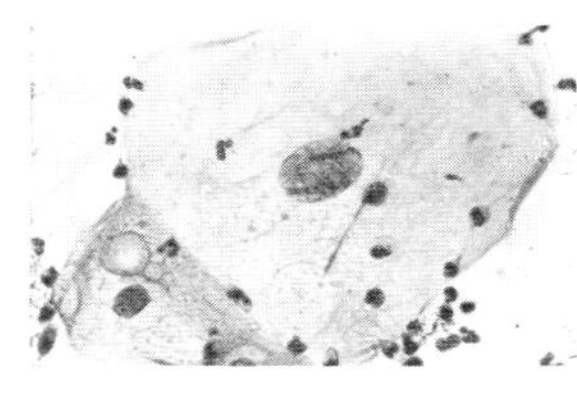

Low Power: Enlarged precornified cell with normal size precornified cell superimposed.

High Power:

A. Characteristics of Nucleus: 1. Definite nuclear borders. 2. Increase in size of nucleus in cell 2. It is approximately six times the size of nucleus in the normal size precornified cell 1. 3. Finely granular chromatin. Despite the increase in size of the nucleus in cell 2 the chromatin appears evenly distributed throughout. The difference in the two nuclei is in size rather than in distribution of nuclear material. Surface of nucleus 2 shows some wrinkling. 4. Though there is great variation in the size of the nuclei there is none in shape.

B. Characteristics of Cytoplasm: 1. Well defined cellular borders. 2. In cell 2, marked increase in amount of cytoplasm. 3. Both cells show changes in the cytoplasm which are due to radiation. The cytoplasm shows definite vacuolization varying from large discrete to very fine vacuoles. Fine fibrils are present throughout the cytoplasm of both cells. 4. Staining reaction: acidophilic.

C. General Characteristics of Group: One large precornified cell, illustrating increase in size which is a characteristic effect of radiation. Nucleus has increased in size proportionately. Thus cytoplasmic-nuclear ratio has remained constant. Normal size precornified cell is present. Cytoplasm of both cells contains vacuoles and fine fibrils.

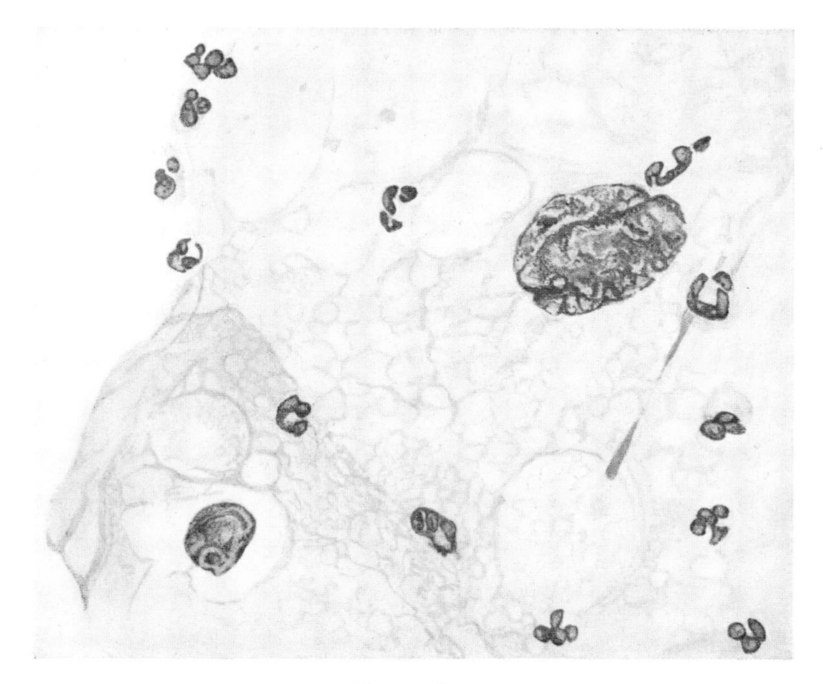

PLATE 14

KEY TO RADIATED SUPERFICIAL CELLS

1. Normal size precornified cell, showing vacuoles and fibrils in cytoplasm. Nucleus is vesicular.
2. "Blown-up" precornified cell with increase in amount of cytoplasm and in the size of the nucleus. Nuclear-cytoplasmic ratio is normal. Vacuoles and fibrils are in the cytoplasm. Large nucleus shows some wrinkling of its surface.

Fig. 79

RADIATION CHANGES IN SUPERFICIAL CELLS: INCREASE IN SIZE

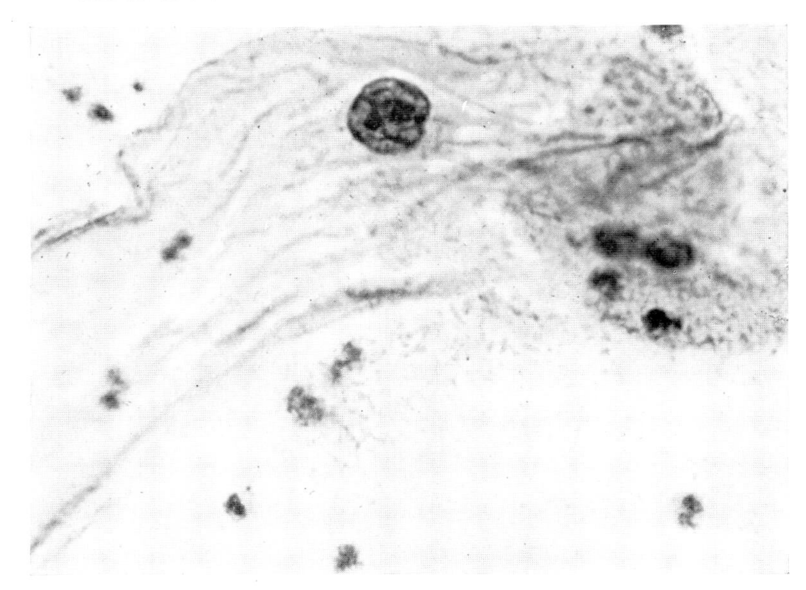

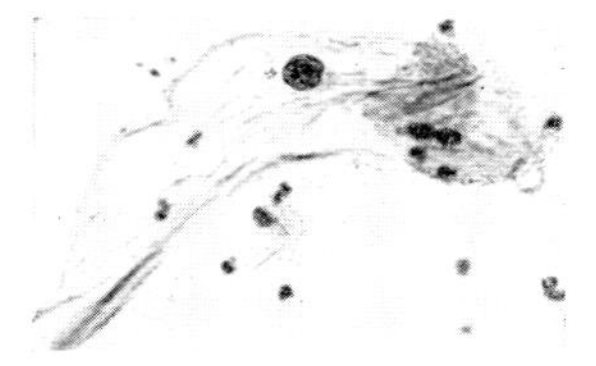

Low Power: Enlarged precornified cell.
High Power: This superficial cell is entirely normal except for the great increase in size. The nucleus is finely granular, the chromatin being distributed evenly. Nuclear-cytoplasmic ratio has remained constant despite increase in size. Enlarged, radiated superficial cells usually stain acidophilic. Compare size to normal superficial cells, Plate 2.

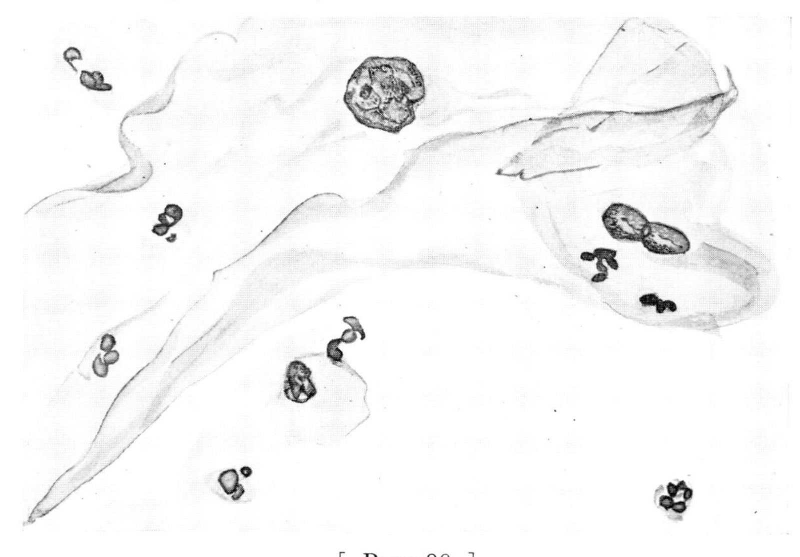

FIG. 80

RADIATION CHANGES IN SUPERFICIAL CELLS: ABNORMAL NUMBER OF NUCLEI

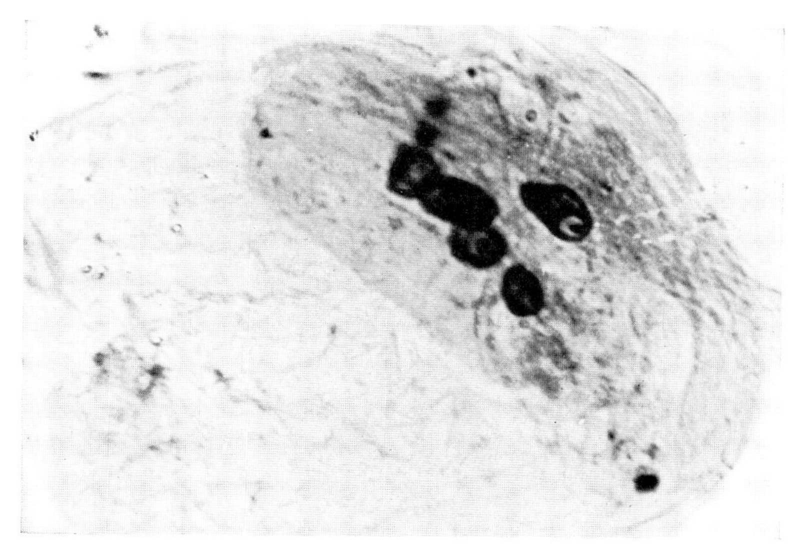

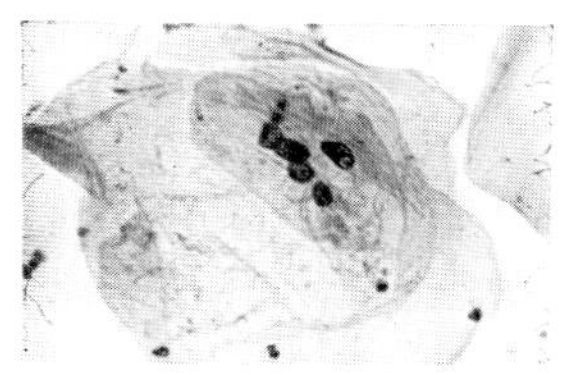

Low Power: Enlarged cell with five nuclei.
High Power: The surface of this large epithelial cell is wrinkled. The abnormal number of nuclei is an indication of the abnormal mitoses stimulated by radiation. As many as nine nuclei are commonly found. The chromatin has condensed and the nuclei have a dead, pyknotic appearance, instead of the fine granularity seen in well preserved cells.

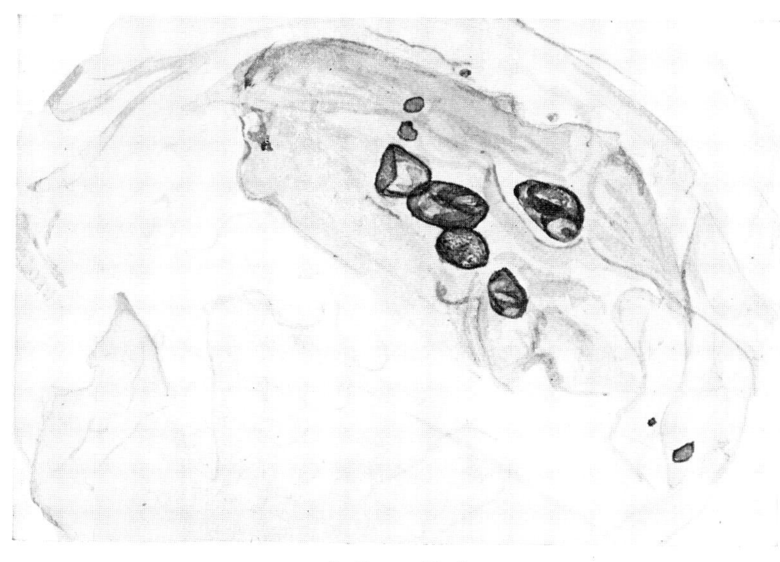

Fig. 81

RADIATION CHANGES IN SUPERFICIAL CELLS: BIZARRE FORMS

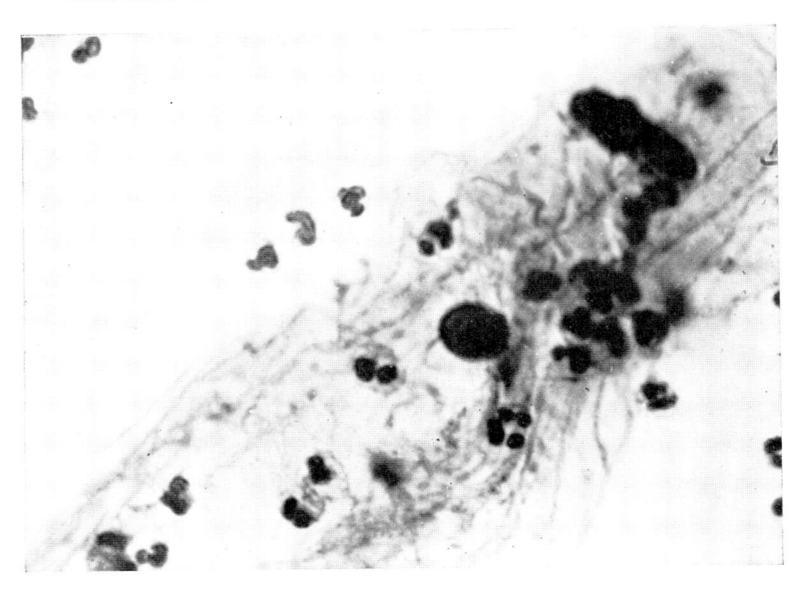

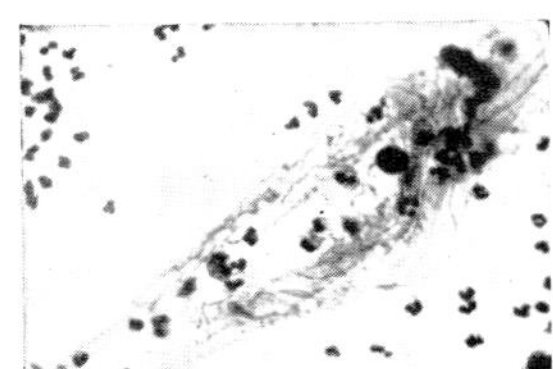

Low Power: Elongated superficial cell.
High Power: The length of this cell is greatly increased, giving it an unusual appearance for a superficial cell. The precornified and cornified cells often assume bizarre forms but, in general, changes in shape are not as common as in the basal cells. The nuclear material is condensed. Fine vacuoles and fibrils are present in the cytoplasm.

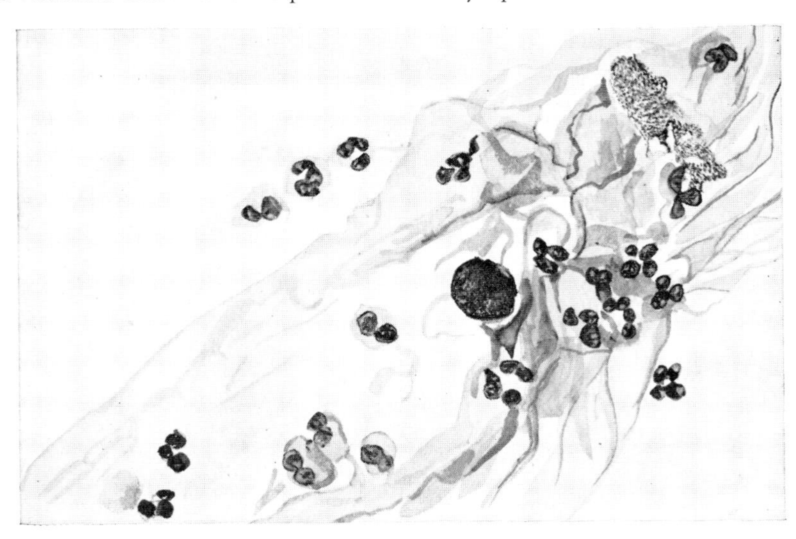

FIG. 82

RADIATION CHANGES IN SUPERFICIAL CELLS: VACUOLIZATION

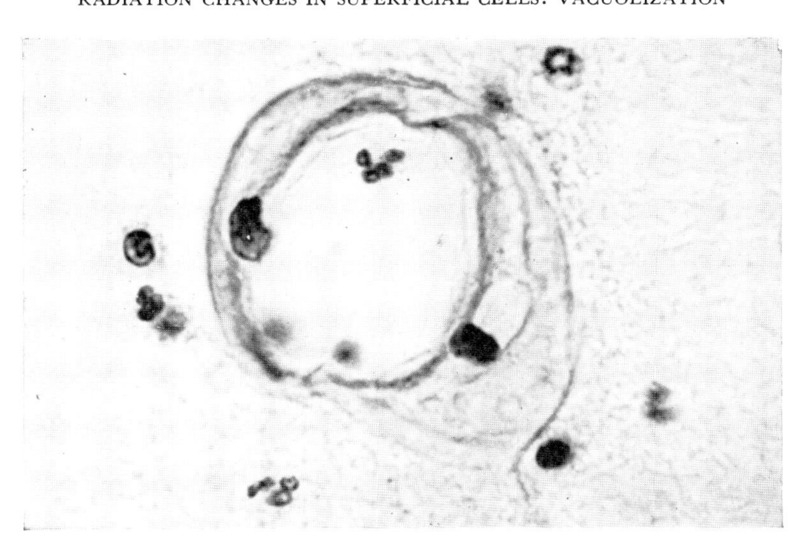

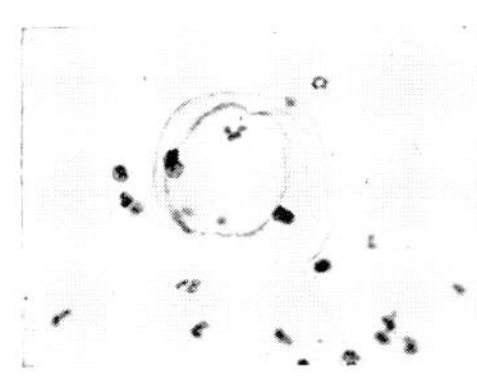

Low Power: Double-nucleated superficial cell with large vacuole.
High Power: This large epithelial cell illustrates the definite vacuolization which is a specific effect of radiation. The entire cytoplasm appears vacuolated and irregular. The large vacuole separates two degenerate nuclei, which appear hyperchromatic owing to the condensation of chromatin.

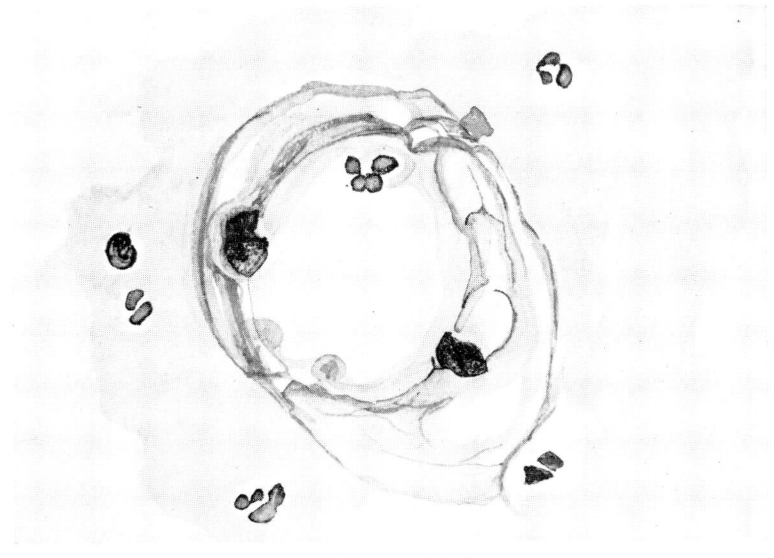

RADIATION CHANGES IN NORMAL CELLS

The changes produced by radiation in the normal cells of the cervical and vaginal mucosa are, as has been shown, both marked and bizarre. Not all of the changes are specific. The fine vacuolization of the cytoplasm of the basal cells, the karyorrhexis and pyknosis of the nuclei are seen in extreme atrophy of the genital tract as well as in radiation. However, the marked increase in size of both nucleus and cytoplasm and the striking vacuolization of the cytoplasm are changes which we regard as specific radiation effects in exfoliated cells. The similar effect of swelling and vacuolization in precornified and cornified cells is also regarded as due to radiation. Bizarre cells, such as dumb-bell shapes, ameboid or tadpole forms, are interpreted as radiation effect. Vacuolization of the nuclei is not seen often, but when present represents an additional change produced by radiation.

The cellular changes caused by radiation are not of long duration. In the cases of cervical carcinoma showing response they reach their peak at the end of x-ray therapy, at which time almost all benign cells show evidence of radiation effect. During the period between x-ray and radium the radiated cells rapidly disappear to reach a smaller peak after radium implantation. Three months after therapy the cells show no distinctive changes due to radiation in most instances. However, occasionally, a few radiated cells may persist for as long as a year and a half, and it is here that difficulty in interpretation arises. When a vaginal smear contains normal epithelial cells and only a rare radiated cell, the cellular changes produced by radiation appear much more suspicious. Especially troublesome are the large bizarre cells. They are apt to be mistaken for the differentiated cells of squamous carcinoma. They have tadpole and fiber forms and their nuclei appear large and granular. It is essential that the cytologist become familiar with the cellular effects produced by radiation in order to avoid misinterpreting these changes as evidence of recurrence.

These cells may be distinguished from those of squamous carcinoma if careful attention is paid to the nuclear structure. It is extremely granular, but is regular in pattern. The chromatin is not condensed at the periphery. The cellular border is apt to be much sharper than that usually present in differentiated carcinoma cells. Still, it will be remembered that fiber cells may have an increase of chromatin until the appearance of the nucleus is pyknotic. This may also occur in radiated cells where the chromatin condenses to form a pyknotic nucleus. For this reason we have found it expedient to base our diagnoses of a recurrent or persistent tumor after radiation therapy on the basis of undifferentiated malignant cells only. This is especially important for the beginner in cytology. Dependence on the presence of undifferentiated malignant cells is a perfectly safe criterion, since local recurrences which desquamate malignant cells show numbers of the undifferentiated type. In addition, it is rare to have any positive vaginal smear contain only differentiated cells, with the possible exception of those smears containing only third type differentiated cells discussed previously. If a positive diagnosis in postradiation smears is made only on the basis of undifferentiated cells, many false positive diagnoses will be avoided.

RADIATION CHANGES IN MALIGNANT CELLS

The changes produced by radiation in the malignant cells of squamous carcinoma are similar to those seen in benign cells. The great majority of the effects are in the undifferentiated cells. The nuclei increase tremendously in size in comparison to those seen in the control smears. Their nuclear characteristics appear accentuated because of their large size. (See Plate 15.) The chromatin appears in heavy clumps and is concentrated at the periphery of the nucleus. Often even though the nucleus is of large size there appears to be a condensation of chromatin so that it appears pyknotic. Variation in size and shape is even more apparent than in unradiated cells. (See Fig. 86.) One of the striking effects in the malignant as well as in the normal cells is the vacuolization of the cytoplasm. Though in control smears the undifferentiated cells may appear to be surrounded by only a thin rim of cytoplasm, when the cells are radiated even this small amount will contain large vacuoles. (See Fig. 85.) As in the benign cells this distinctive vacuolization is one of the most specific effects of radiation.

The differentiated cells from squamous carcinoma do not show the marked effects seen in the undifferentiated cells. In most cases they do not show any change and gradually disappear. Occasionally, however, they will show some change as illustrated in Fig. 84. The nucleus and cytoplasm of fiber cells increase, the nuclear-cytoplasmic ratio remaining unchanged. The cytoplasm shows some evidence of vacuolization. These cells appear as extremely large fiber cells. Rarely evidence of radiation will be seen in the third type differentiated cell. There is an apparent disintegration of nuclear material. (See Fig. 83.) The nuclear border disappears and the heavy aggregates of chromatin spread throughout the cytoplasm. We have not seen distinct radiation effects in tadpole cells but presumably they show changes similar to those seen in the fiber cells.

The malignant cells of adenocarcinoma of the cervix after radiation therapy show similar changes to those seen in the undifferentiated cells. The vacuolization characteristic of adenocarcinoma is much less pronounced than that produced by radiation. Since the undifferentiated cells of squamous carcinoma show such marked vacuolization after radiation, it is almost impossible to classify cells as desquamating from adenocarcinoma in post-radiation smears.

It should be pointed out that the malignant cells do not always show these changes. In many cases they simply gradually disappear from the smear without any evidence of change. Their numbers vary considerably. In some instances there is an apparent increase of malignant cells around the fourth or fifth day after the beginning of therapy. In other cases there is a steady decrease from the initiation of treatment. The greatest number of malignant cells showing radiation response is seen in most instances from the eighth to the twelfth day. The final disappearance of the cancer cells from the smear varies a good deal. They may have a precipitant decrease and be absent entirely by the sixth or seventh day or they may on occasion persist until after the radium implantation.

Examples of the effect of radiation on the malignant cells of squamous carcinoma are presented on the following pages.

DESCRIPTION OF RADIATED MALIGNANT CELLS

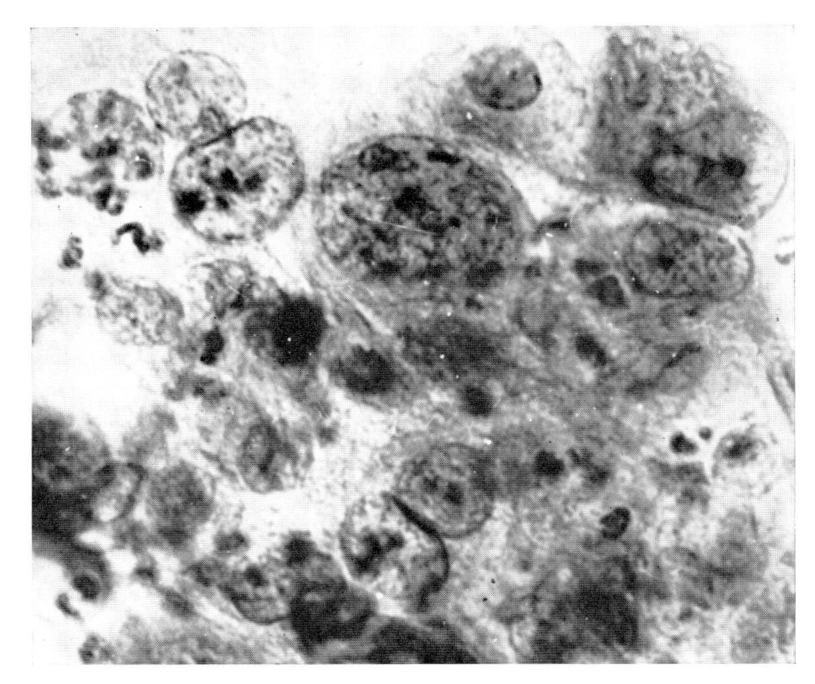

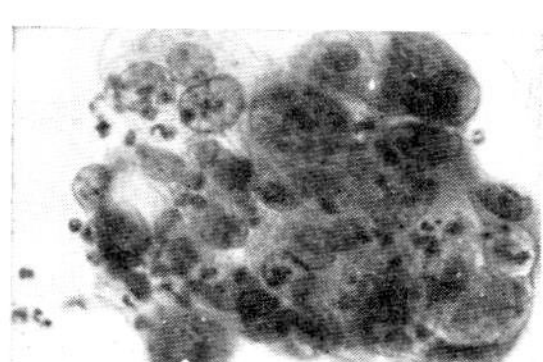

Low Power: Large group of undifferentiated malignant cells showing marked variation in size.

High Power:

A. Characteristics of Nucleus: 1. Sharp nuclear border. Chromatin is condensed at periphery. (See groups 1 and 3.) 2. The nucleus has an uneven appearance, since the aggregations of chromatin vary in size. (See cell 2.) 3. There is extreme variation in size. (Compare cell 2 with the cells in group 4.) 4. Variation in shape is slight. 5. All nuclei show increase in size and appear swollen. (Compare these radiated undifferentiated malignant cells to those in Plate 6.)

B. Characteristics of Cytoplasm. 1. Cytoplasm forms an indistinct background. 2. No cellular borders are present classifying these cells as undifferentiated. 3. Staining reaction bluish purple nuclei with faint acidophilic cytoplasm. 4. Blood pigment and polymorphonuclears form the background. This is a common picture in radiation smears.

C. General Characteristics of Group: The swelling of these radiated malignant cells seems to augment their peculiarities, since they appear more obvious because of the increase in size. They have marked irregularity of nuclear material and condensation of chromatin at border. There is striking variation in size.

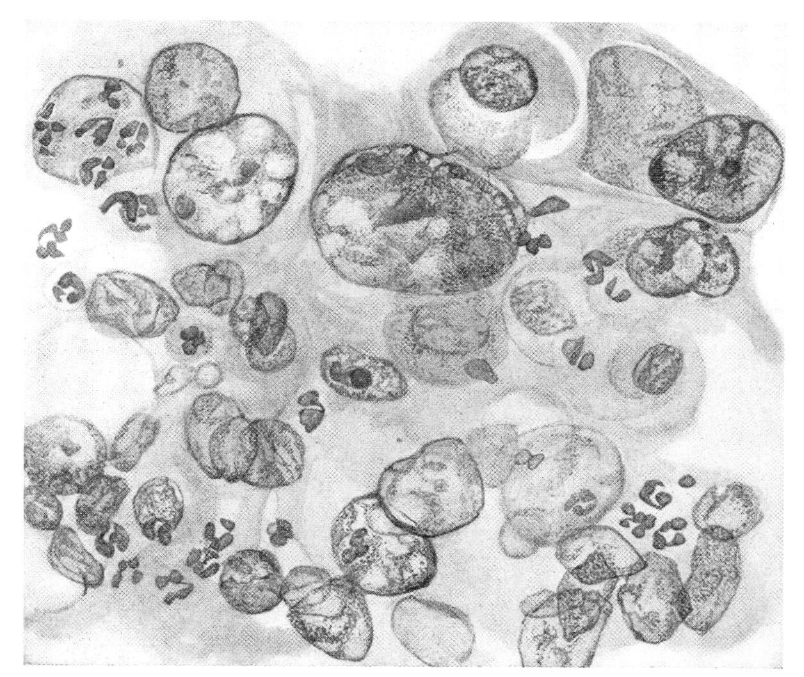

PLATE 15

KEY TO RADIATED MALIGNANT CELLS

1. Three malignant nuclei showing variation in size and sharp borders.
2. Enlarged malignant nucleus with irregular chromatin.
3. Nuclei with chromatin condensed at periphery, large aggregates of chromatin.
4. Group of small overlapping malignant nuclei.
5. Two nuclei exhibiting variation in size.

FIG. 83

RADIATED MALIGNANT CELLS: DISINTEGRATION OF NUCLEUS

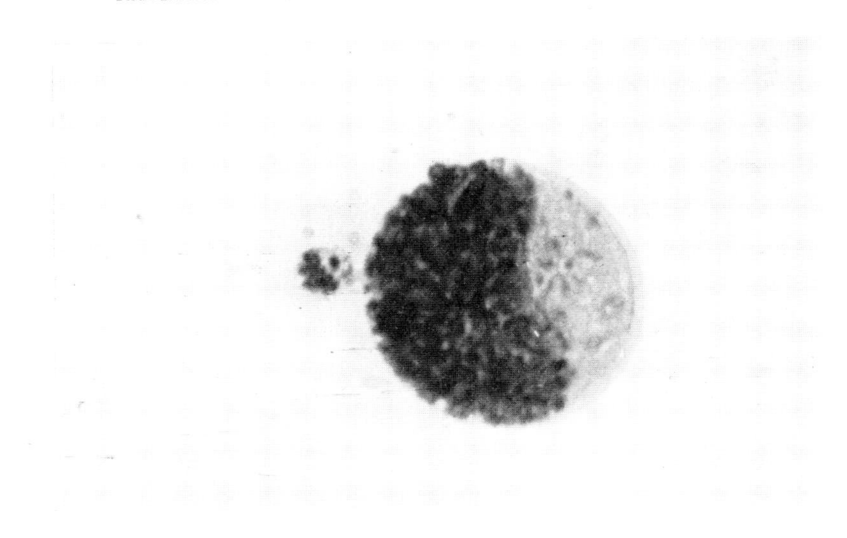

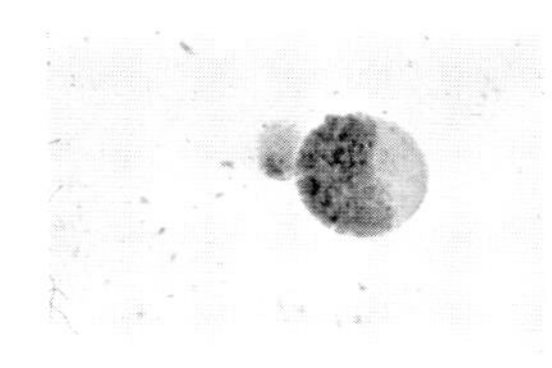

Low Power: Enlarged single malignant cell with granular chromatin.

High Power: The nuclear material is extremely granular and appears increased in amount. Nuclear border is absent and chromatin is not contained within any membrane. This cell is classified as a third type differentiated because of its definite cellular border. Such disintegration of a nucleus probably represents the true destruction of malignant cells by radiation.

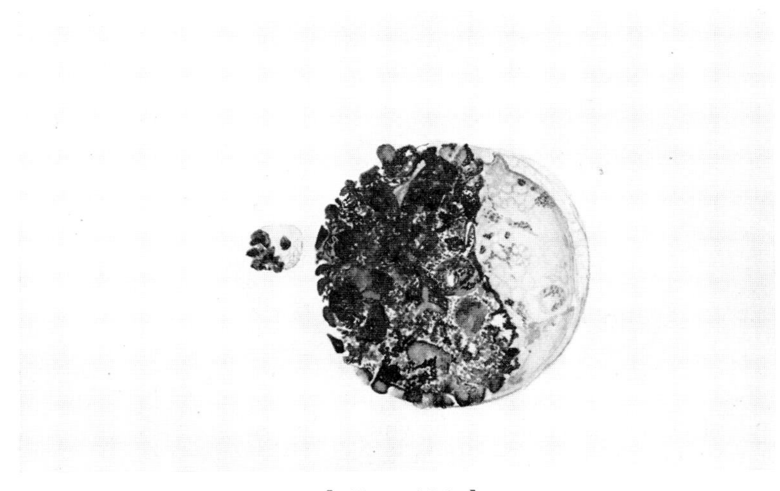

FIG. 84

RADIATED MALIGNANT CELLS: ENLARGED FIBER CELL

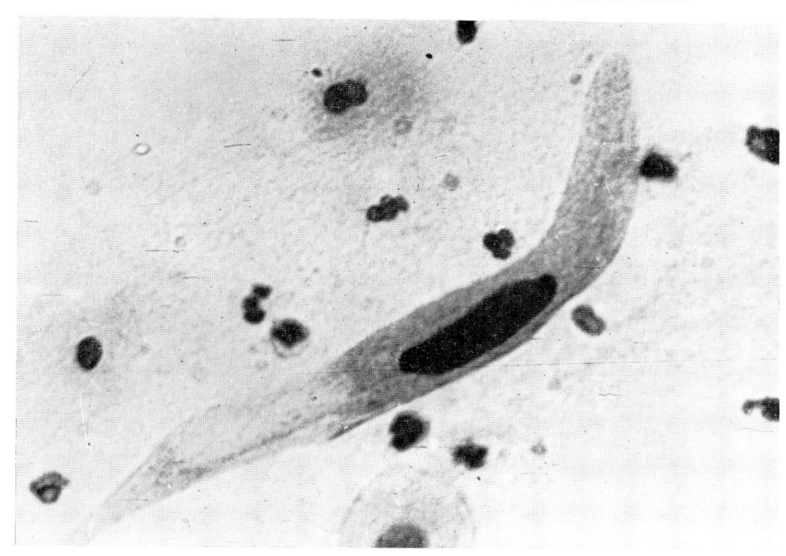

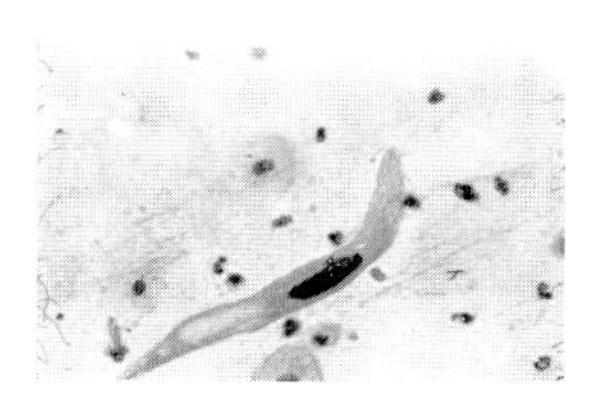

Low Power: "Blown-up" malignant fiber cell in field of basal cells and leukocytes.

High Power: The fiber cells usually disappear from the vaginal smear before any pronounced changes due to radiation are observed. This figure is an unusual example of radiation changes in fiber cells. There is a proportionate increase in size of cytoplasm and nucleus and distinct vacuolization of the cytoplasm.

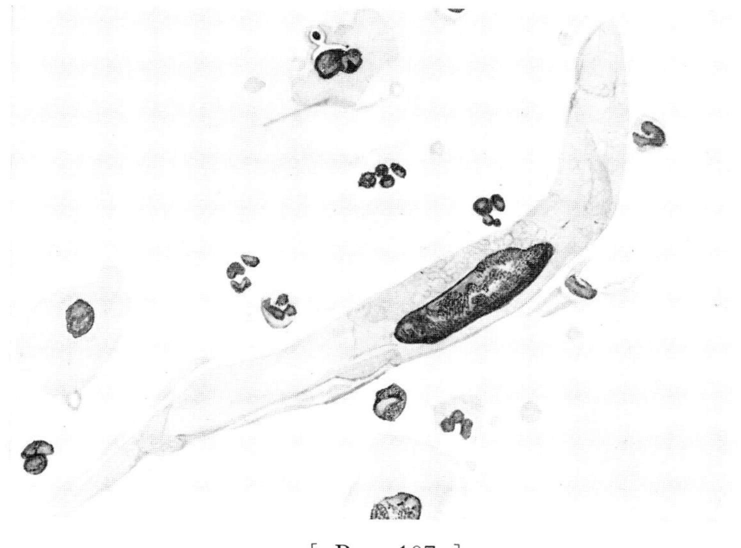

FIG. 85

RADIATED MALIGNANT CELLS: VACUOLIZATION OF CYTOPLASM

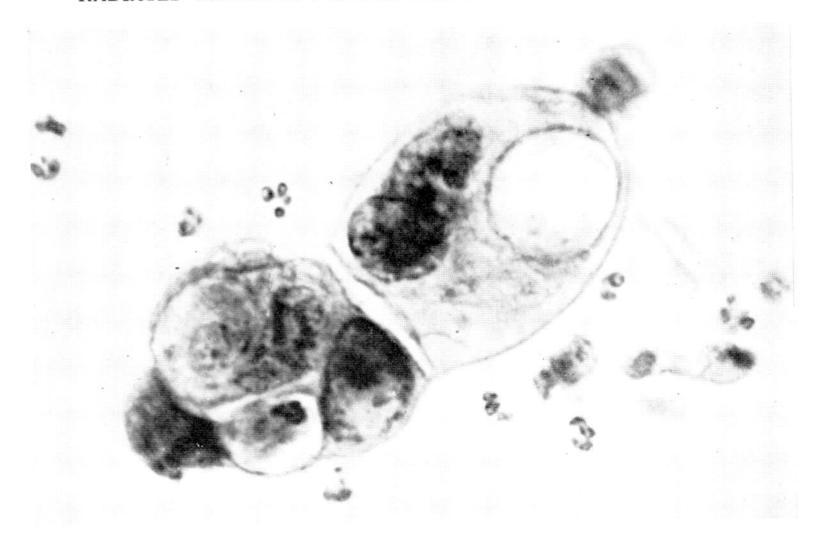

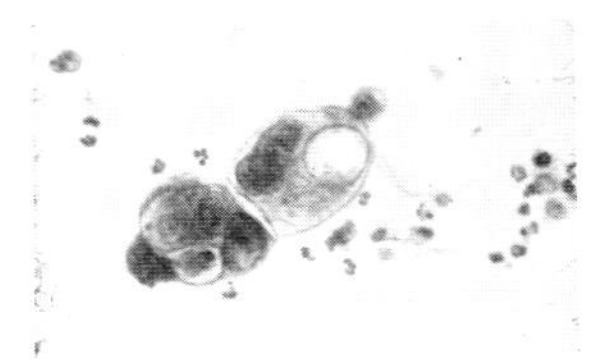

Low Power: Large hyperchromatic nuclei with vacuolated cytoplasm.

High Power: The malignant nuclei are enlarged, but still retain their specific characteristics. The chromatin is irregular and is concentrated at the periphery. The cytoplasm is irregular and contains definite vacuoles. The vacuolization should be differentiated from that of adenocarcinoma. In these radiated cells the vacuoles are contained within the cytoplasm.

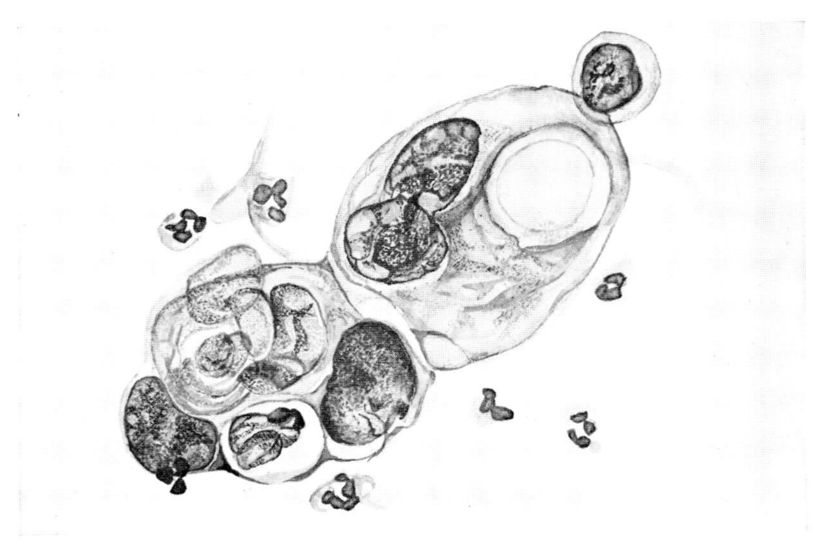

FIG. 86

RADIATED MALIGNANT CELLS: SWELLING OF NUCLEUS

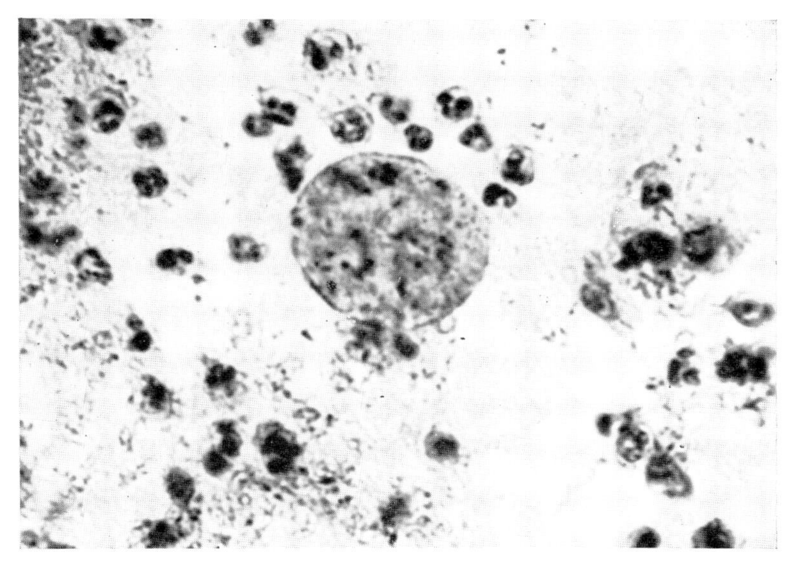

Low Power: Enlarged nucleus in messy field.
High Power: Often single radiated undifferentiated cells present this appearance. The nucleus appears extruded with no visible cytoplasm. The nuclear border is still fairly definite but on the left it appears to be breaking. Chromatin still retains an irregular appearance. Background of leukocytes and blood pigment is typical of radiation smears.

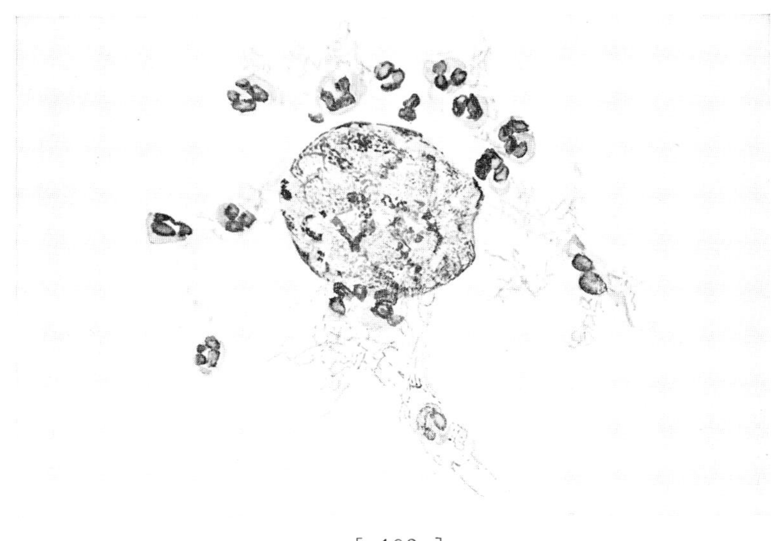

As was indicated in the previous discussion, there is great variability in the time in which malignant cells become absent from a postradiation smear. In some instances the positive cells may disappear and then reappear. Because of this variation, whether or not malignant cells are present in the *immediate* postradiation smear has little correlation with whether tumor is still present. The smear may be negative and yet positive biopsies may be obtained. Because there seemed to be such poor correlation with the presence or absence of malignant cells and the possible success of radiation treatment, we attempted another type of correlation. In some cases of carcinoma of the cervix treated by x-ray and radium both the normal and malignant cells show radiation effect. In others few of either type show changes. At the present time we believe that those patients in whom the great majority of both normal and malignant cells show distinct radiation effect have a good prognosis. Conversely, those patients in whom there is little response in either type of cell have a poor prognosis. Obviously there are other factors to influence the ultimate outcome, such as the clinical extent of the disease and the adequacy of the radiation therapy. However, we have found that this interpretation of depending on radiation response in both normal and malignant cells in the postradiation smear correlates more closely with the ultimate result than the absence or presence of malignant cells.

The vaginal smear can be of extreme value in discovering early postradiation recurrences. Often it is difficult clinically to distinguish between radiation reaction in normal tissue and a recurrence of a malignant growth. Too, the adhesions which may be present after radiation treatment often make direct visual examination difficult. In these instances the vaginal smear is of distinct aid We have had several cases in which typical undifferentiated cells were found in the vaginal secretion, though the patients were thought to be free of disease clinically. In all these cases a small focus of carcinoma was hidden behind the adhesions at the apex of the vagina. These serve as an example of the useful purpose the vaginal smears serve in the gynecologic follow-up tumor clinic.

SUMMARY

The effects of radiation on the normal cells of the vaginal and cervical mucosa are unique. The specific effects are great increase in size of both nucleus and cytoplasm with no disturbance of the cytoplasmic-nuclear ratio, marked vacuolization of the cytoplasm and occasionally of the nucleus, and production of many bizarre forms. The effects on undifferentiated malignant cells are increase in size and vacuolization of the small amount of cytoplasm present. The differentiated cells appear to show the effect of radiation only in rare instances.

It is essential that the cytologist become familiar with the changes produced by radiation, since they represent one of the most difficult differentiations. The study of postradiation smears will prove profitable both from the point of view of ultimate prognosis of cervical malignancy and the detection of early recurrence.

CHAPTER IX

CELLS OF SQUAMOUS EPITHELIUM OF RESPIRATORY TRACT

Sputum acts as a reservoir for cells shed not only from the mucosa lining the tracheobronchial tree, but from the epithelia lining the mouth, nose and nasopharynx. Since sputum specimens always contain some saliva, many cells from the stratified squamous epithelium of the mouth are present in the smears. Other sources of the squamous cells may be the portions of the nasal passage and nasopharynx which are lined by this epithelium. Obviously it is impossible to determine the anatomical source of the superficial epithelial cells, since they have no distinguishing characteristics.

The superficial cells as they appear in sputum specimens are similar to those seen in the vaginal smear. They are large, thin, transparent cells with either a small dark pyknotic nucleus or a round vesicular one. Their cytoplasm may either stain acidophilic or basophilic. They undergo the same types of degeneration as described for the superficial cells in vaginal secretion. Wrinkling and folding of the thin cytoplasm is common and many cells present a ragged cellular outline. The cells do not appear as nicely preserved as the superficial cells of the vaginal secretion and are often in large clusters surrounded by mucus. The superficial squamous cells are by far the most frequent cell seen in sputum smears.

The deeper layers of the stratified squamous epithelium of the mouth and upper respiratory tract do not desquamate cells as readily as the tissue lining the female genital tract. Cells similar to the outer layer basal are seen occasionally but certainly not with the frequency that such cells are seen in the vaginal secretions. These cells are similar in cellular characteristics to the outer layer basal of the vaginal and cervical epithelium. They are round or oval cells, the cytoplasm staining deeper than that of the superficial cells. The nucleus is round or oval, finely granular in appearance and surrounded by a good margin of cytoplasm.

There is one type of benign cell encountered in sputum which has no counterpart in vaginal secretion. These are small "deep" cells. We have interpreted them as desquamating from squamous epithelium, though occasionally their appearance may approach a cuboidal form. These deep cells are small cells, the size of an inner layer basal. They have a centrally located nucleus which, in most instances, is not finely granular, but has condensation of chromatin. The nuclear border is sharp. Cytoplasm is dense in appearance and is adequate for the size of the nucleus. They appear to be extremely active cells. (See Fig. 90.) We do not know the source of these cells. We suspect that they are desquamated from areas of squamous metaplasia rather than from normal epithelium.

Examples of these various types of cells desquamated from squamous epithelium are given on the following pages.

DESCRIPTION OF SUPERFICIAL SQUAMOUS CELLS IN SPUTUM

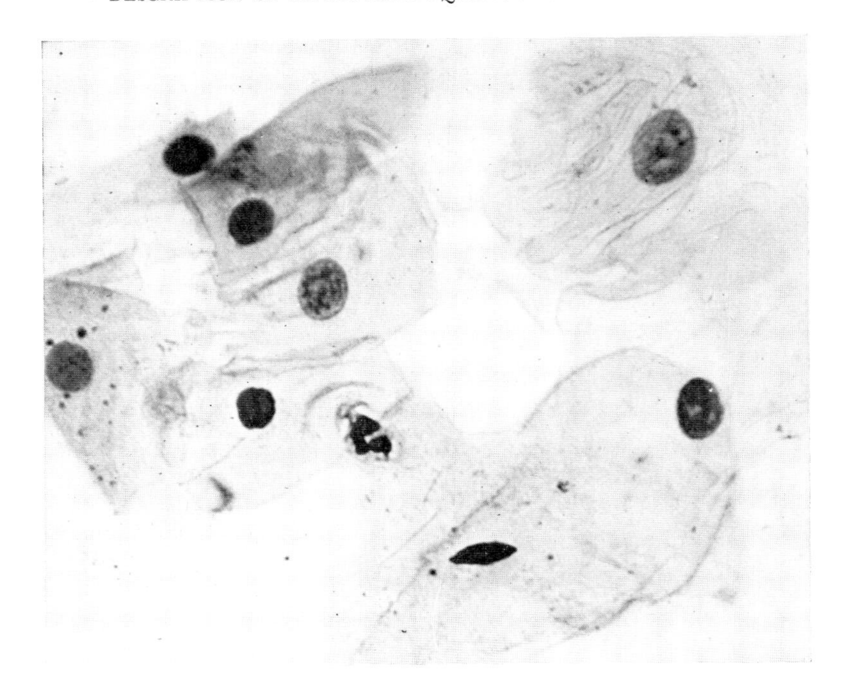

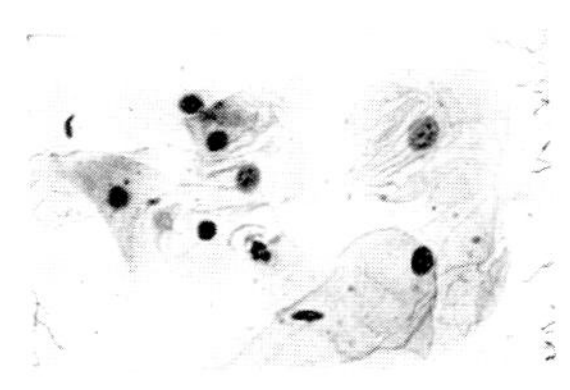

Low Power: Clean field of individual cells with good nuclear cytoplasmic ratios.

High Power:

A. Characteristics of Nucleus: 1. Clear nuclear borders. 2. The nuclei vary slightly in chromatin structure from cell 7, which is dense and pyknotic, to cells 1, 2 and 5, which are finely granulated, to cells 3, 4 and 8, which show some clumps of chromatin. Cell 6 has a dark-staining wrinkled nucleus, but notice that the chromatin particles are evenly distributed. 3. There is little variation in size and shape, except for cell 7, which is elongated, but small. In general, vesicular nuclei are the size of leukocytes or larger.

B. Characteristics of Cytoplasm: 1. Indistinct, irregular cellular borders. 2. All of the cells have a large amount of transparent cytoplasm, which is often wrinkled (as in cell 4), or folded (as in cell 7). 3. Cell 5 shows granules in the cytoplasm which are of no known significance. 4. Staining reaction: pinkish purple nuclei with acidophilic or basophilic cytoplasm.

C. General Characteristics of Group: Group of overlapping cells with pyknotic or vesicular nuclei, surrounded by a large amount of transparent cytoplasm which is frequently wrinkled or folded.

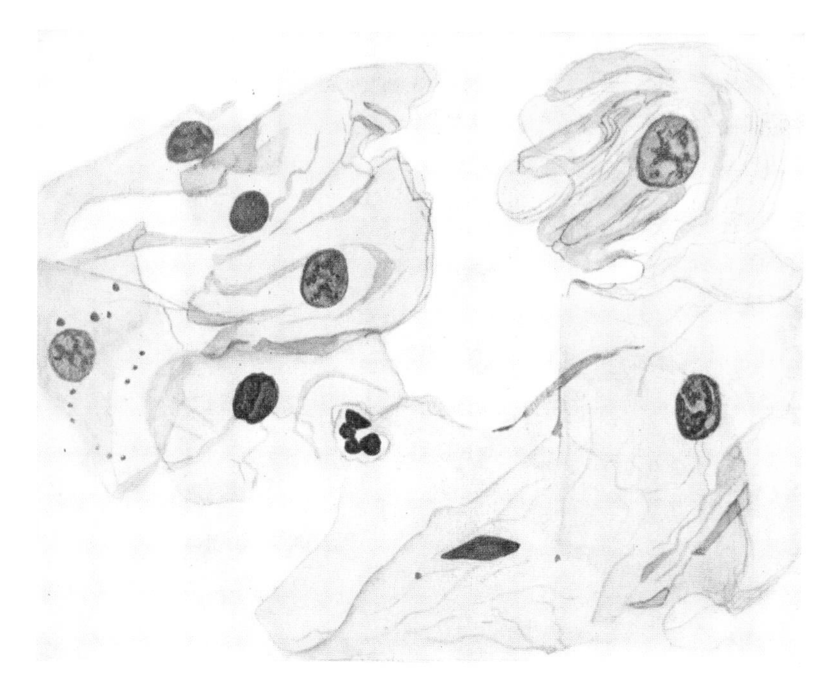

PLATE 16

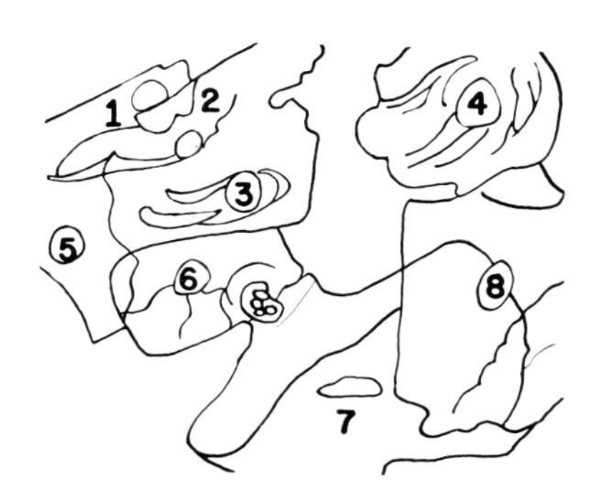

KEY TO SUPERFICIAL SQUAMOUS CELLS IN SPUTUM

1. Degenerate superficial cell, vesicular nucleus.
2. Degenerate superficial cell, vesicular nucleus.
3. Preserved superficial cell with wrinkling.
4. Nucleus of coar ely granular chromatin, and marked wrinkling of cytoplasm.
5. Superficial cell with granulation in the cytoplasm.
6. Superficial cell with wrinkling in the nucleus.
7. Pyknotic nucleus in a superficial cell.
8. Coarse granulation of nucleus.

DESCRIPTION OF "DEEP" SQUAMOUS CELLS IN SPUTUM

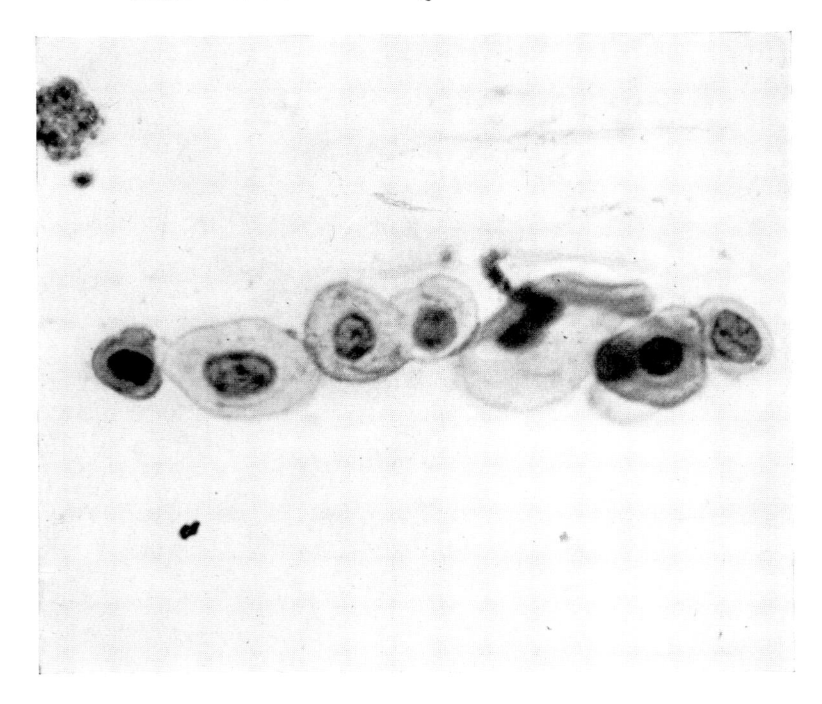

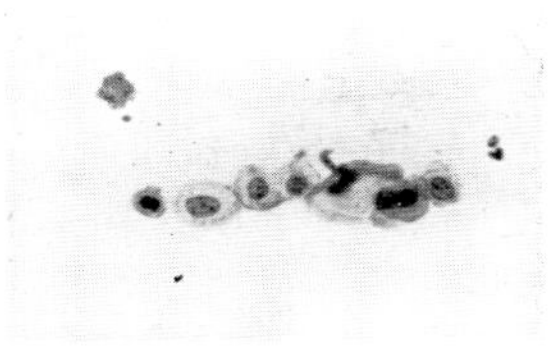

Low Power: Single cells showing good nuclear-cytoplasmic ratio.

High Power:

A. Characteristics of Nucleus: 1. Sharp well defined nuclear borders. 2. In general, all of the nuclei except those of the double nucleated superficial cell, 5, appear active. There is variation in the chromatin content from a condensation of chromatin (as in cell 4) to a granular appearance (as in cell 7). 3. The nuclear size varies according to the size of the cell. Cell 1 has a smaller nucleus than cell 2 but the entire cell is also smaller. Cytoplasmic-nuclear ratio is fairly constant. 4. Shape of the nucleus is oval or round and there is wrinkling of the nuclear borders in cells 1, 2 and 3. Compare these nuclei to those of the superficial cell 5, which shows little chromatin structure.

B. Characteristics of Cytoplasm: 1. Clear cellular borders. 2. Cytoplasm appears dense. In the smaller cells the density is greater than in the larger. As the cells increase in size the cytoplasm become more transparent. (Compare cytoplasm of cells 2 and 7.) 3. Staining reaction: bluish purple nuclei with either acidophilic or basophilic cytoplasm.

C. General Characteristics of Group: Discrete cells, each with a sharply defined active nucleus surrounded by an adequate amount of cytoplasm with definite cellular border.

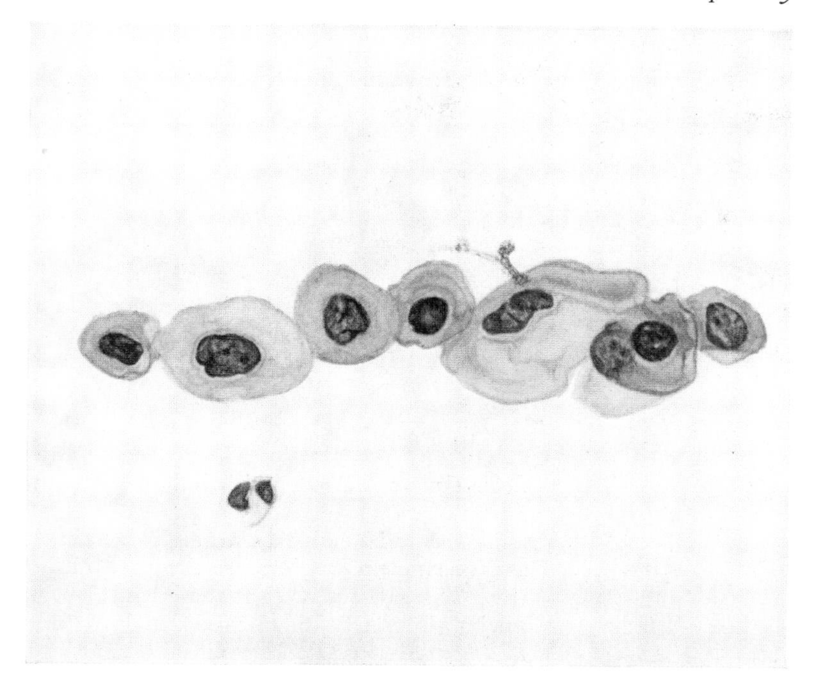

PLATE 17

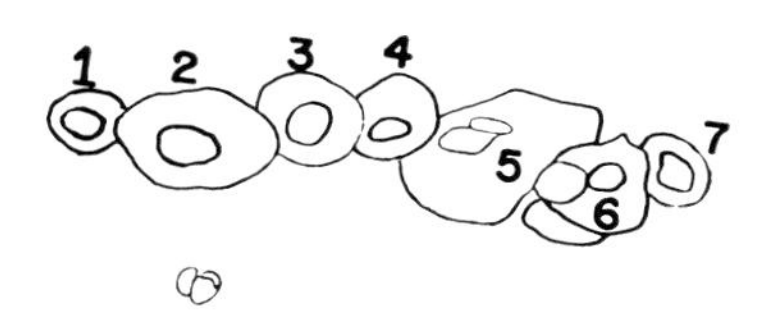

KEY TO "DEEP" SQUAMOUS CELLS IN SPUTUM

1. Small "deep" cell with dark nucleus, active chromatin.
2. Large cell, irregular nuclear border, clumped chromatin.
3. Wrinkled nucleus, irregularity in cytoplasmic stain.
4. Pyknotic nucleus, dense cytoplasm.
5. Double-nucleated superficial cell.
6. Smoothly granular nucleus, wrinkled cellular border.
7. Large granular nucleus, dense cytoplasm.

FIG. 87

DEEP CELLS: WELL PRESERVED

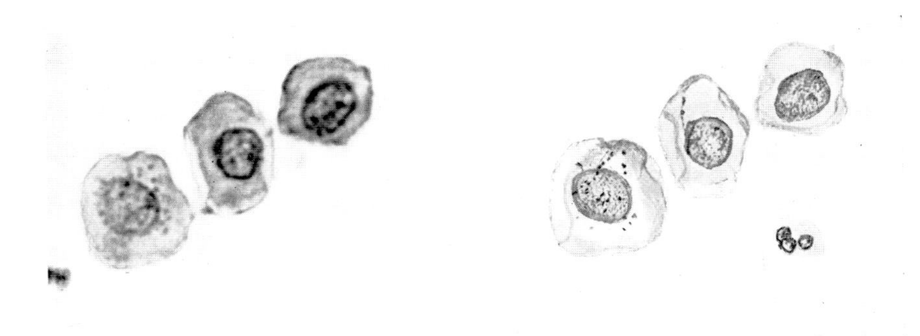

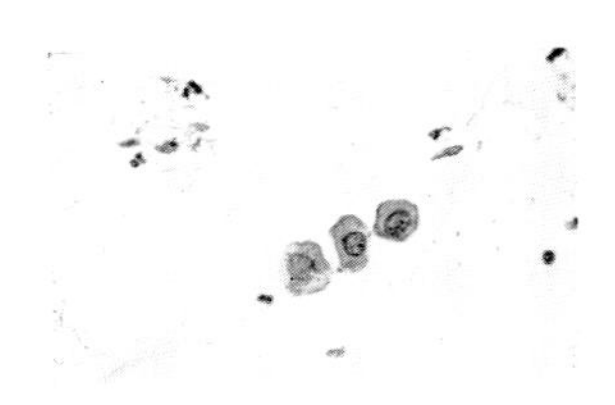

Low Power: Three epithelial cells with even nuclei.
High Power: Nuclei are extremely even in size and shape. Nuclear borders are definite, but chromatin is not concentrated at the edge. Chromatin is finely granular. Cellular border is definite and the wrinkling of the cytoplasmic edge is typical of squamous cells. Cytoplasm is adequate for size of nuclei. Bacteria are superimposed on cell on the left.

FIG. 88

DEEP CELLS: CYTOPLASMIC DEGENERATION

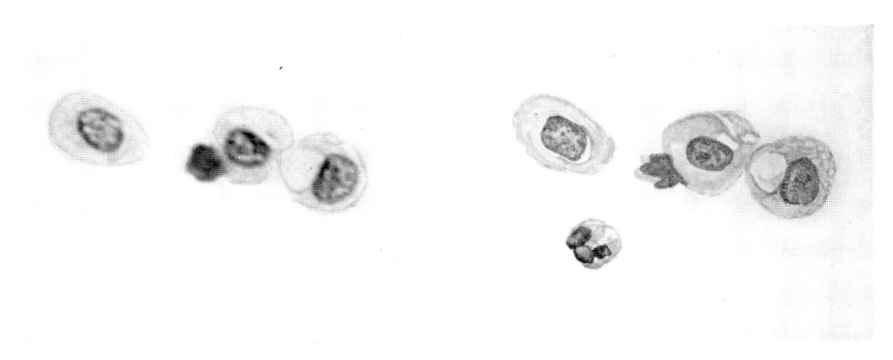

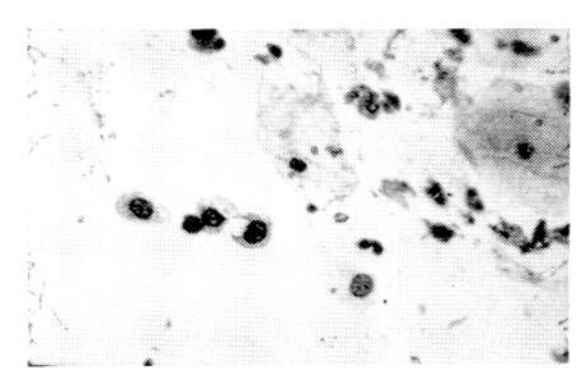

Low Power: Four small epithelial cells showing no variation in cellular or nuclear size.
High Power: We interpret these cells as epithelial in origin, rather than histiocytes. The finely granular nuclei are central in position rather than eccentric as in histiocytes. The vacuolization is irregular, being perinuclear in cell on the left, and distributed unevenly in other two cells. We believe this type of vacuolization is evidence of degeneration.

FIG. 89

DEEP CELLS: ENLARGEMENT OF THE NUCLEUS

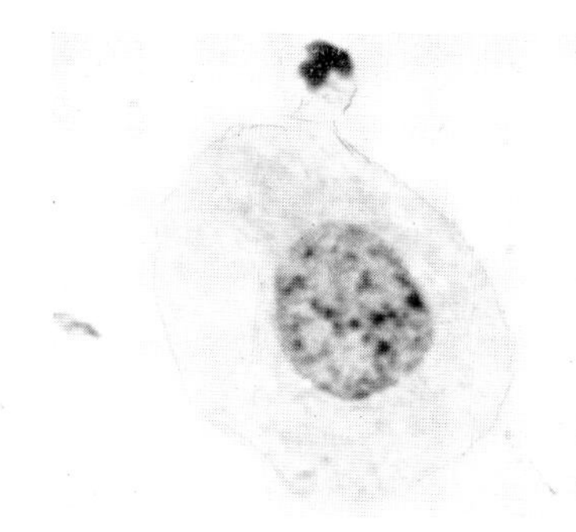

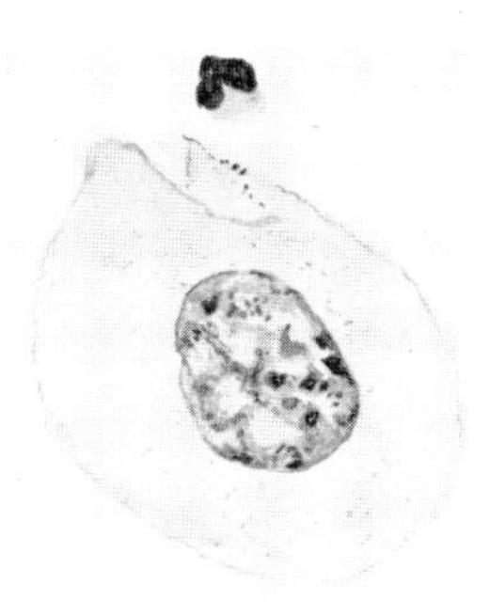

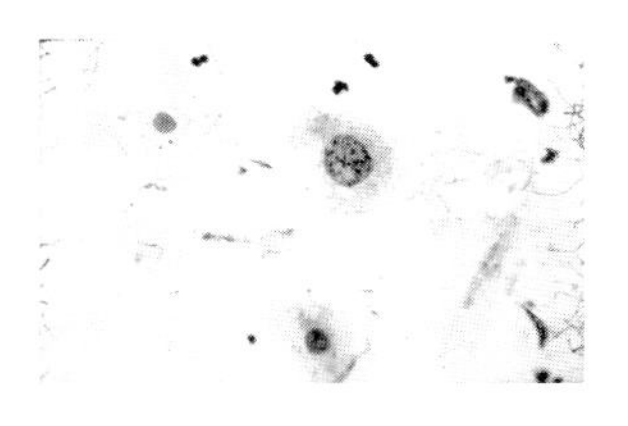

Low Power: Field of epithelial cells; one cell with enlarged nucleus.
High Power: This cell is an example of the atypical benign epithelial cell which occurs in sputum. The size of the nucleus is abnormally large and the chromatin content of the cell is irregular. However, there is no increase in chromatin content and nucleus has a washed-out appearance, its border is not sharp and is smooth rather than irregular. Cytoplasmic-nuclear ratio is still adequate.

FIG. 90

DEEP CELLS: ACTIVE NUCLEAR PATTERN

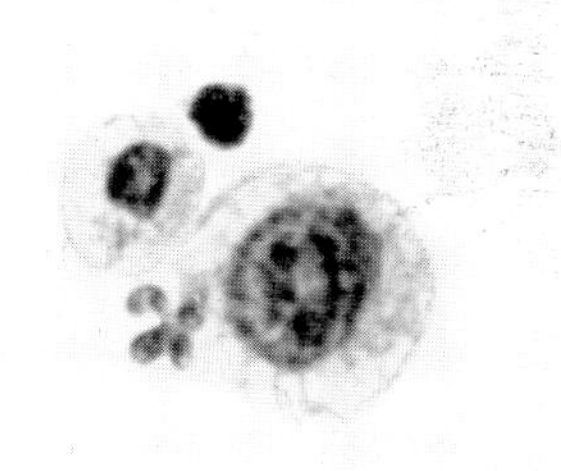

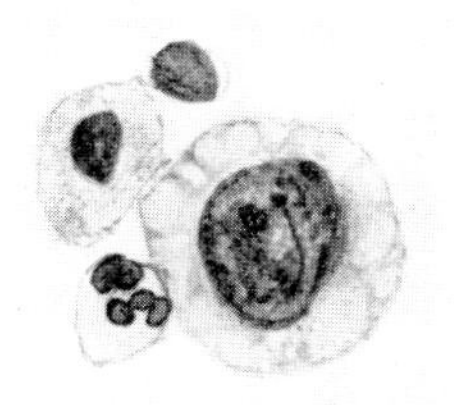

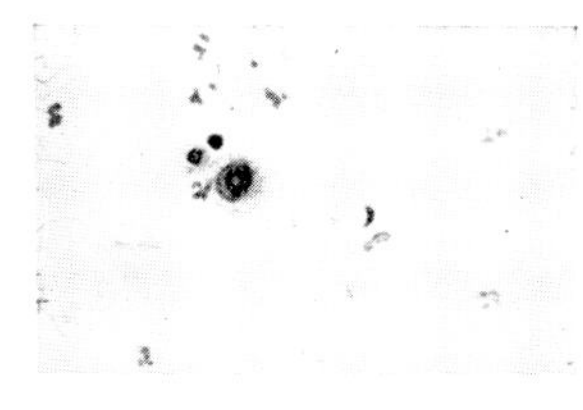

Low Power: Single cells with dark nucleus.
High Power: This cell shows the active nuclear pattern which the deep cells may occasionally have. Though the chromatin is somewhat irregular the nuclear border does not have condensed chromatin and is smooth. Cytoplasm presents some degeneration but cellular border is distinct and cytoplasmic-nuclear ratio is within normal limits. Field contains a lymphocyte and small degenerate epithelial cell.

Squamous Epithelium of Respiratory Tract

Superficial cells from squamous epithelium are the most common cell encountered in sputum smears. In specimens of bronchial secretions they are far less frequent, often being entirely absent. Fortunately they are easily recognized and identified. The "deep" cell from squamous epithelium is seen infrequently. These vary from perfectly normal appearing cells, much like the inner layer basal cell (see Fig. 87), to others which are atypical benign (see Fig. 89). This latter type may occur in smears in which frank carcinoma cells are present. On the other hand, we have seen such cells in sputum from patients where no malignancy was present. Since they do occur in positive smears, their presence indicates extensive search for definite carcinoma cells, but these cells alone do not indicate a positive diagnosis. We have interpreted them as arising from an atypical epithelium, perhaps squamous metaplasia.

Difficulties in Interpretation: The only cell in this group which causes any difficulty in identification is the small "deep" cell. Most of these cells present a smoothly granular nucleus with occasional small clumps of chromatin, adequate cytoplasm and good cellular border. However, an occasional cell will show irregularity of the chromatin structure. (See Fig. 90.) These may be mistaken for third type differentiated malignant cells if care is not taken. The chromatin is not concentrated at the periphery of the nucleus and the nuclear border is smooth. Cytoplasm is adequate. Chromatin does not appear increased, which is important, since in the greatest number of differentiated malignant cells from squamous carcinoma of the lung a definite increase in chromatin is seen. The cells have a regular appearance when seen in groups.

General Criteria for Identification of Squamous Cells in Sputum: 1. Superficial cells: large amount of thin transparent cytoplasm, either round or oval vesicular nucleus on a small, dark, pyknotic one. Cytoplasm may stain either basophilic or acidophilic. 2. Basal or intermediate cells: oval or round with smoothly granular nucleus, adequate amount of dense cytoplasm. 3. "Deep" cells: small round or cuboidal cells, having dense cytoplasm which is adequate, and a round or oval nucleus with smooth border and occasional moderate irregularity of chromatin structure. Cytoplasm may stain either basophilic or acidophilic.

CHAPTER X

THE COLUMNAR EPITHELIUM OF RESPIRATORY TRACT

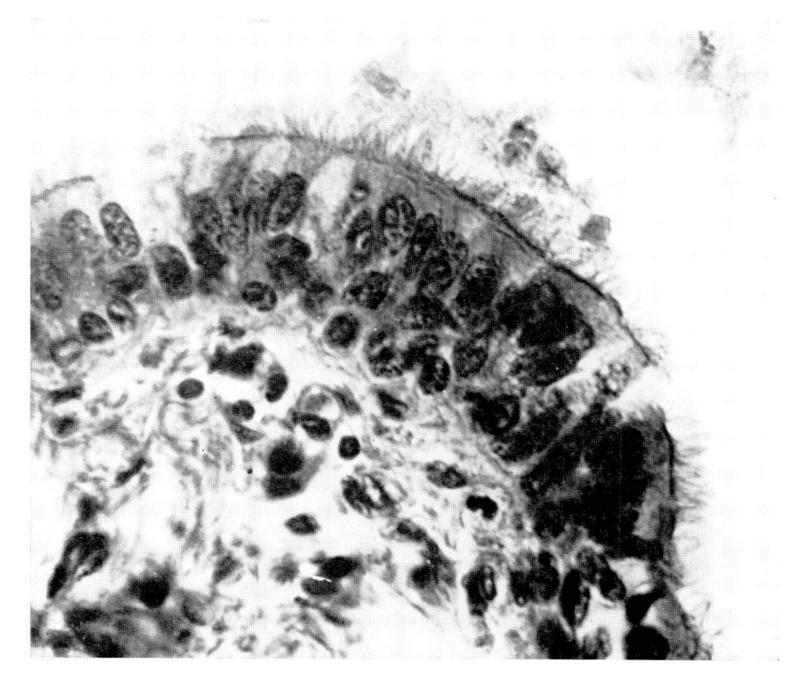

Fig. 91

HISTOLOGIC SECTION: NORMAL BRONCHIAL EPITHELIUM

The lower respiratory tract, large and small bronchi, is lined with ciliated columnar epithelium as shown in Fig. 91 above. These cells desquamate fairly frequently and are seen in smears of sputum. If specimens are made of secretions obtained by bronchial aspiration, the majority of the cells present will be of the columnar type, in various stages of preservation.

As in the endocervical cells seen in vaginal secretions, the columnar cells of the bronchi degenerate quite rapidly. It is rare even in groups where the cilia remain preserved to find distinct cellular borders. The cytoplasm is usually present as a long indistinct background in which a nucleus is present in an eccentric position. The cells may be lined up in a palisade formation and then identification is simple even though the individual cells may not be well preserved. It is when the cells are piled in fairly dense clusters that identification is more difficult. Since the cytoplasm is often indistinct, the nuclei appear as a fairly tight group which often shows variation in size. If it is remembered that the columnar cells of the respiratory tract may degenerate rapidly, and if careful attention is given to the chromatin structure, which is finely granular and not irregular, this type of cell should be identified quite readily.

DESCRIPTION OF COLUMNAR CELLS IN SPUTUM

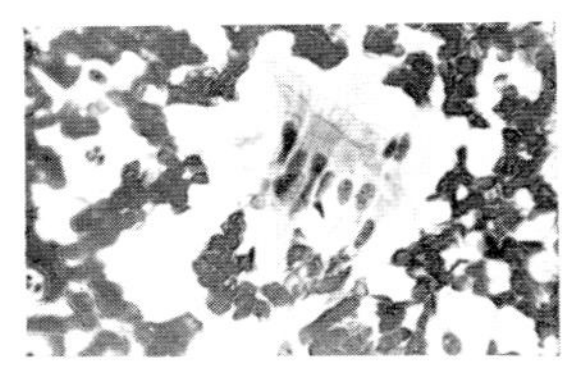

Low Power: Group of elongated cells with nuclei arranged in a unipolar position.

High Power:

A. Characteristics of Nucleus: 1. Clear nuclear borders. 2. All of the nuclei are similar in their chromatin structure, which consists of many small clumps of chromatin evenly distributed throughout the nuclei. 3. The arrangement of the nuclei is typically eccentric (see cells 4, 5 and 6), and in some cases there may appear to be more than one nucleus to a cell. (See cells 2 and 3.) 4. The general shape of the nucleus is oval (see cells 1 and 5), but there is more variation in the size of the nuclei than in the shape. Compare the small nuclei of cells 1 and 7 with those of cells 2 and 3.

B. Characteristics of Cytoplasm: 1. Indistinct cellular borders. 2. In well preserved columnar cells the cilia are clearly visible at broad end of the cytoplasm. 3. In cells 2 to 6 individual cellular borders cannot be seen. However, each cell exhibits a thin tail of cytoplasm at the lower end of the nucleus. In cells 7 and 8 the cytoplasm has almost disappeared because of degeneration. 4. Staining reaction: bluish purple nuclei with lightly staining acidophilic or basophilic cytoplasm.

C. General Characteristics of Group: Elongated cells, each with one or two vesicular nuclei eccentrically located in elongated cytoplasm with definite ciliated border.

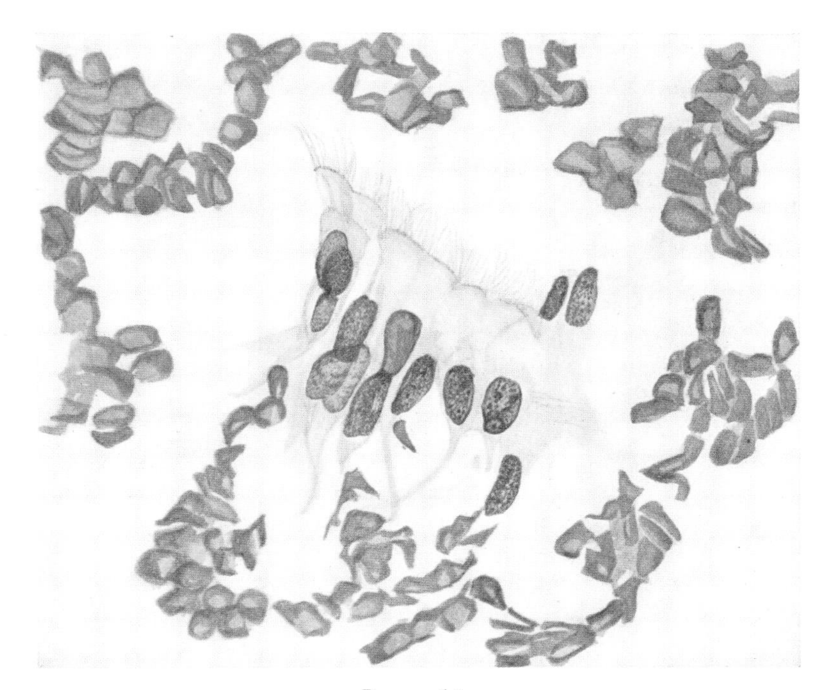

PLATE 18

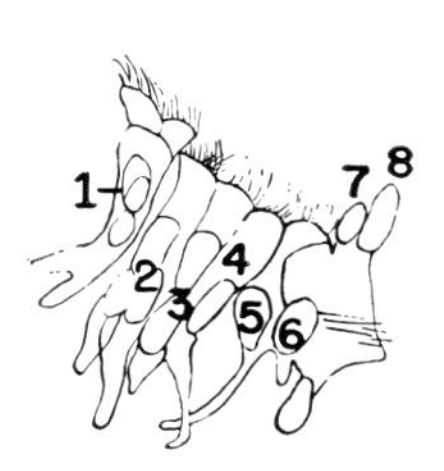

KEY TO COLUMNAR CELLS IN SPUTUM

1 and 2. Columnar cells with indistinct cellular borders showing overlapping nuclei and cilia at the broad end of cytoplasm.
3. Columnar cell with an indefinite cellular border containing two nuclei exhibiting some variation in chromatin content.
4 and 5. Eccentric oval-shaped nuclei in columnar cells with a short tail of cytoplasm.
6. Vesicular round nucleus in cell revealing only a short tail of cytoplasm.
7 and 8. Oval nuclei of degenerate columnar cells which have no cellular borders and no visible cilia. Cluster is surrounded by red blood cells.

Fig. 92

COLUMNAR CELLS: WELL PRESERVED

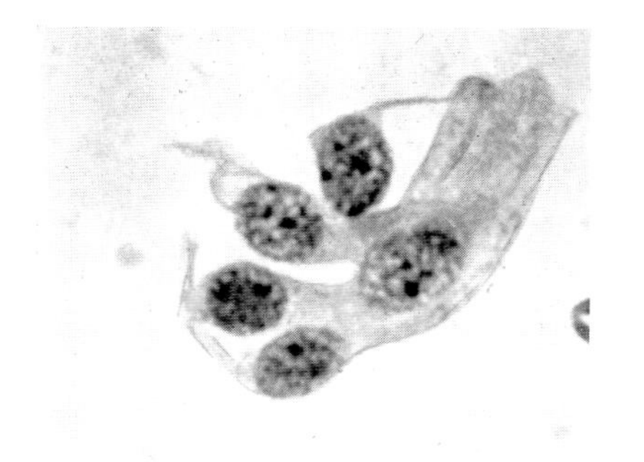

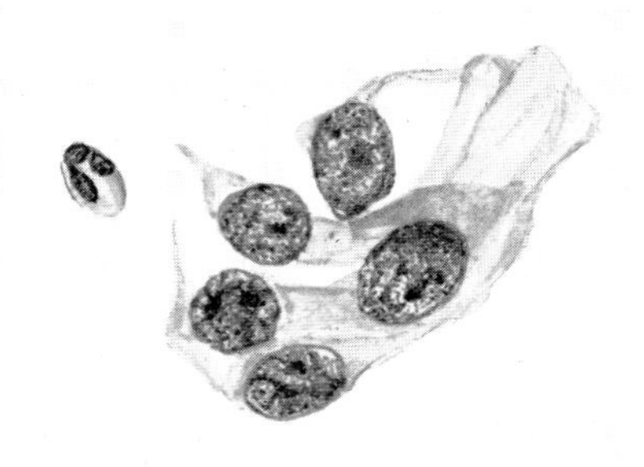

Low Power: Five even nuclei with tails of cytoplasm.

High Power: These columnar cells from bronchial epithelium are fairly well preserved. Cilia are absent but the elongated cytoplasm with eccentric nuclei easily identifies the cells as columnar. The nuclei are even in size and shape and the chromatin is finely granular with occasional prominent clumps of chromatin. Nuclear borders are distinct, but cellular borders are absent.

Fig. 93

COLUMNAR CELLS: DEGENERATION OF NUCLEI

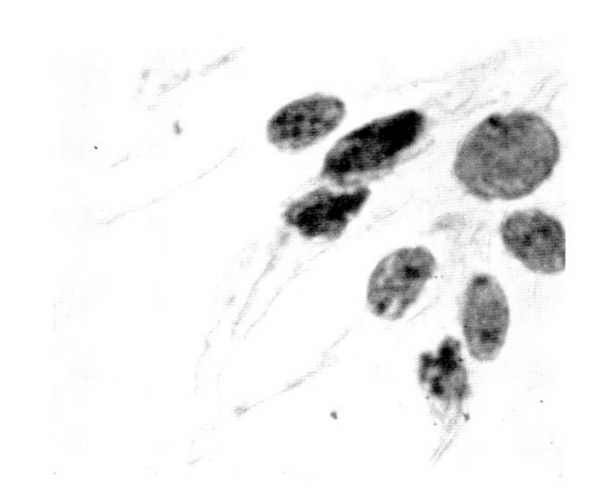

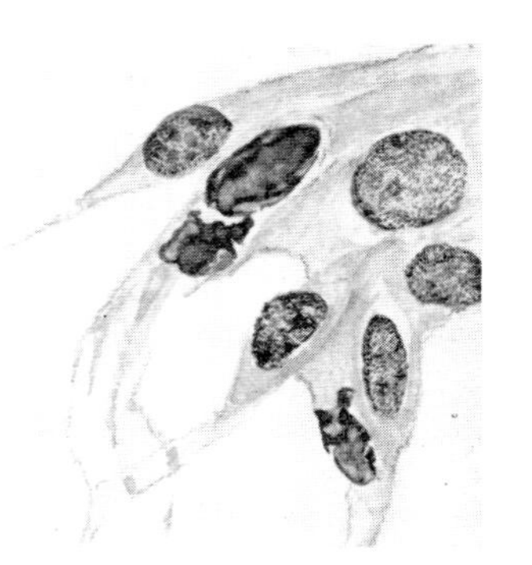

Low Power: Group of nuclei varying in size in background of cytoplasm.

High Power: This group shows more degeneration than those in figure above. Three of the nuclei show marked changes with nuclear borders breaking and chromatin assuming a pyknotic appearance. The other nuclei in the field have finely granular nuclei typically showing more variation in size than in shape.

Fig. 94

COLUMNAR CELLS: OVERLAPPING OF NUCLEI

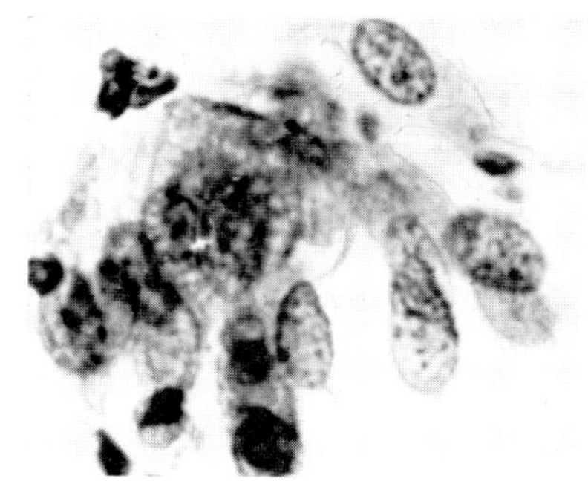

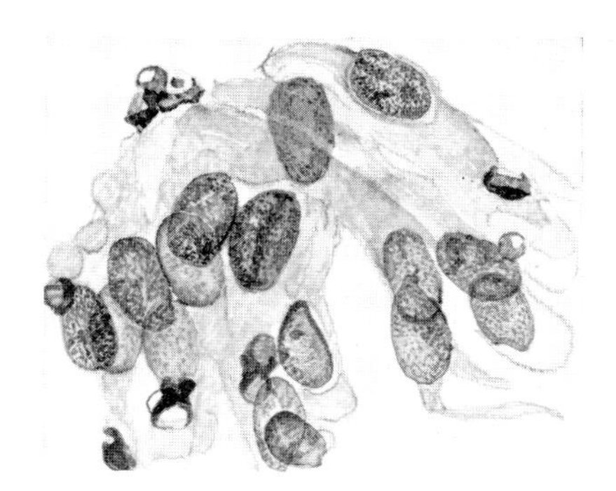

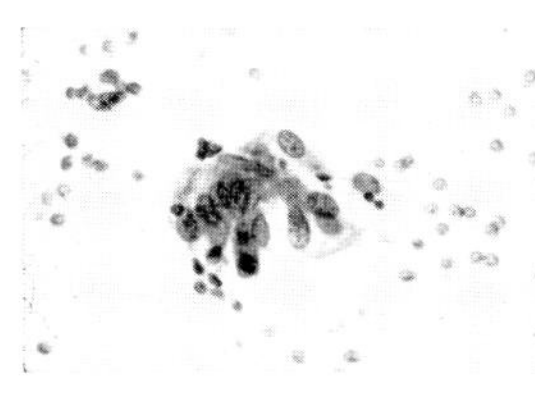

Low Power: Dense group of overlapping nuclei. *High Power:* The cells in this group illustrate the overlapping which is so characteristic of cells desquamated from the columnar epithelium. Careful examination reveals a typical columnar pattern in cells to the right of the field. Examination of single cells at the periphery of such a group often aids in identification of cells where overlapping somewhat obscures the detail.

Fig. 95

COLUMNAR CELLS: CYTOPLASMIC DEGENERATION

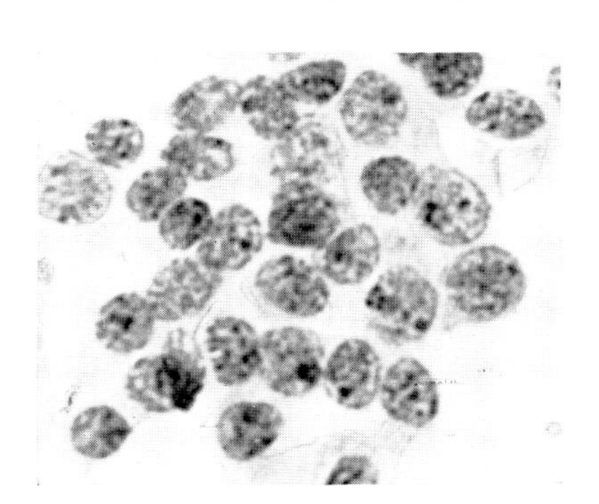

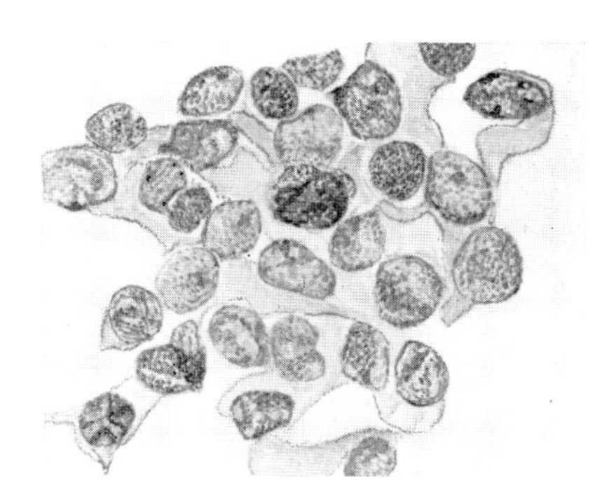

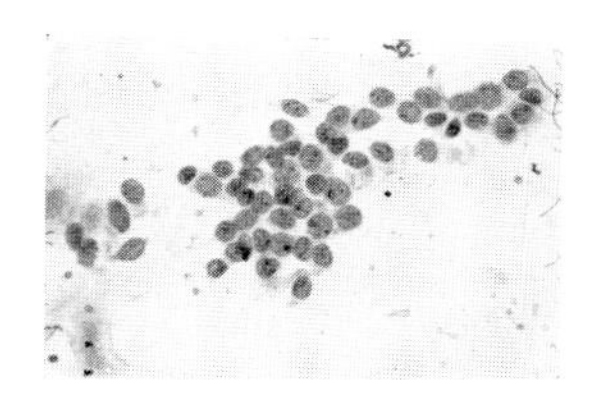

Low Power: Group of even granular nuclei.
High Power: Often columnar cells degenerate to the point where only nuclei remain in the background of cytoplasm. The nuclei above have all the characteristics of those of columnar cells. They are finely granular with occasional prominent clumps of chromatin, and vary more in size than in shape. The cells may degenerate even more, leaving free nuclei as in the endocervical cells in Fig. 22.

Bronchial Epithelium

Columnar cells in various states of preservation are common in both sputum and bronchial secretions. Obviously secretions aspirated directly from the bronchi contain many more columnar cells than sputum. It is important to remember that even in this instance where cells are aspirated directly many show signs of degeneration. The cytoplasm of columnar type cells appears to be extremely fragile and disappears very rapidly.

Difficulties in Interpretation of the Columnar Cells: Because of the degenerative changes in these cells they occasionally may be difficult to identify. They are most confusing when they appear in large groups. In this instance the nuclei show variation in size, overlapping and absence of cellular borders. Such groups as this are occasionally mistaken for a group of undifferentiated malignant cells. The nuclear structure of columnar cells is finely granular with occasional small clumps of chromatin. It does not show the irregular pattern so characteristic of malignant cells. The normal pattern of the nucleus is the most reliable criterion for identification of these cells as benign. The size variation may be extreme, a variation of ten times not being unusual. As indicated before, size variation is the least reliable criterion on which to base a diagnosis of malignancy. In such dense groups of columnar cells the border of the group should be examined carefully. In most instances there will be present one or two cells which have retained a typical columnar pattern, *i.e.*, a cell with elongated cytoplasm with a finely granular eccentric nucleus. When cells at the periphery of a group can be definitely identified as columnar, then comparison of the nuclei of those cells with the nuclei in the group will show the same type of nuclear structure and it may be assumed with some certainty that the cells in the group are of columnar origin also. Thus, if these three points are kept in mind, structure of the nucleus, size variation and comparison of known columnar nuclei with those in the group, such dense clusters of overlapping nuclei may be identified as benign.

General Criteria for Identification of Columnar Cells in Secretions from the Respiratory Tract: Cells with elongated, thin cytoplasm which stains faintly basophilic or acidophilic. Cytoplasm is broad at one end and in well preserved cells cilia may be seen at this end. Opposite end of cells has an extremely narrow projection of cytoplasm. Nuclei are placed eccentrically in the cell, being at the narrow end. Nuclear structure is even, being finely granular with occasional clumps of chromatin. The shape of nuclei may be oval or round. The nuclei are regular in shape, but considerable size variation is common.

CHAPTER XI

CELLS OF NON-EPITHELIAL ORIGIN IN SPUTUM

There are a number of cells which occur both in sputum and bronchial secretion which are not epithelial in origin. The most common is the leukocyte. These are present in every specimen in varying numbers. As in the vaginal smear, the polymorphonuclears are usually not well enough preserved to permit accurate identification of the type. Lymphocytes are seen frequently and can be easily identified. Plasma cells are seen fairly often and retain their usual characteristics. We have not found that the occurrence of leukocytes in dense clusters is especially suggestive of malignancy, as is often true in vaginal smears. Fresh red blood cells, blood pigment and fibrin may be present and are regarded as extremely suspicious, even more so than in the vaginal secretion, since a physiological bleeding occurs in the genital tract, and no such simple explanation can be offered for the presence of blood in the secretion from the respiratory tract.

Histiocytes or macrophages are present in all *sputum* specimens. They are round cells with an oval or round eccentric nucleus, and a foamy, vacuolated cytoplasm. The histiocytes found in sputum are unique in that they contain ingested carbon particles. The carbon appears as small brown or black dots scattered throughout the cytoplasm. It is difficult to distinguish the carbon particles from ingested hemosiderin. Occasionally a histiocyte is seen in which the phagocytosed material is orange brown. These cells probably represent the type known as "heart failure" cells. But by far the greatest number of histiocytes contain carbon rather than blood pigment. These phagocytic cells are the most frequent type of histiocyte encountered. The small histiocyte which shows no phagocytosis is in the minority. These cells often contain more than one nucleus. Double and triple nucleated cells are common. They may occasionally have ten or twelve nuclei arranged peripherally in the vacuolated cytoplasm, and in this instance we have interpreted them as the foreign body giant cell type of histiocyte. They are similar to the same type of cell seen in vaginal secretions with the exception that those in sputum have a more definite cellular border. We have not attempted any correlation between these cells and the presence of tuberculosis of the lung, but we have seen them fairly frequently in sputum from patients with that diagnosis. However, since they may be present in other conditions, their presence is not diagnostic in our experience.

As mentioned before, sputum is a mixture of cells from the mouth, nose and nasopharynx, as well as from the lower respiratory tract. It is important to know whether the specimen examined is actually a specimen of sputum or only a specimen of saliva. If a specimen contains only squamous cells and no histiocytes, we regard the smears as being only saliva and report the "sputum" as unsatisfactory. To make a correct diagnosis for the presence or absence of carcinoma in the respiratory tract, it is essential that adequate samples of actual sputum be examined. Early morning specimens coughed up from "down deep" are the most satisfactory. At any rate, the presence or absence of histiocytes is a suitable criterion for determining the source of the material.

DESCRIPTION OF HISTIOCYTES IN SPUTUM

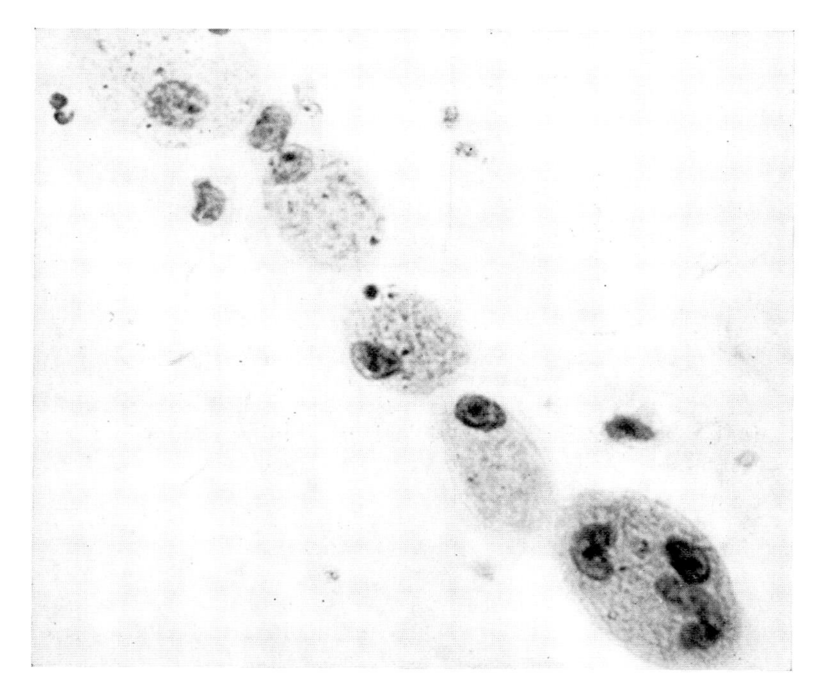

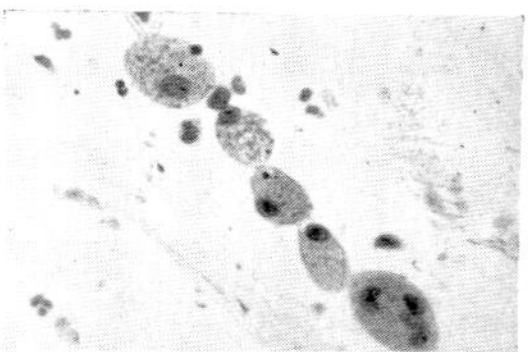

Low Power: Group of individual cells with eccentric nuclei and light-staining cytoplasm.

High Power:

A. Characteristics of Nucleus: 1. Distinct regular nuclear borders. 2. All of the nuclei are smoothly granular and no definite clumps of chromatin are visible. 3. The size of the histiocyte nucleus is normally twice that of a leukocyte or slightly larger. Compare the small nuclei in cell 6 to the size of the nucleus in cell 1. The shape of the nuclei is either round (cell 1), oval (cell 6), or bean-shaped (cell 2). 4. Histiocyte nuclei are usually eccentrically located in the cytoplasm (see cells 3, 4 and 5), unless there are a collection of nuclei in the same cell, in which case they will be arranged peripherally. (See cell 6.)

B. Characteristics of Cytoplasm: 1. Clear cellular borders. 2. In sputum the majority of the histiocytes have a large amount of finely vacuolated cytoplasm often phagocytosing minute particles of carbon. (See cells 1, 4, 5 and 6.) Compare these to cell 2, which has a small amount of cytoplasm showing no phagocytic properties. 3. Staining reaction: purplish blue nuclei with green or brown cytoplasm.

C. General Characteristics of Group: Various shaped bland nuclei eccentrically located in either a small amount of finely vacuolated cytoplasm or a larger amount with phagocytosed carbon particles.

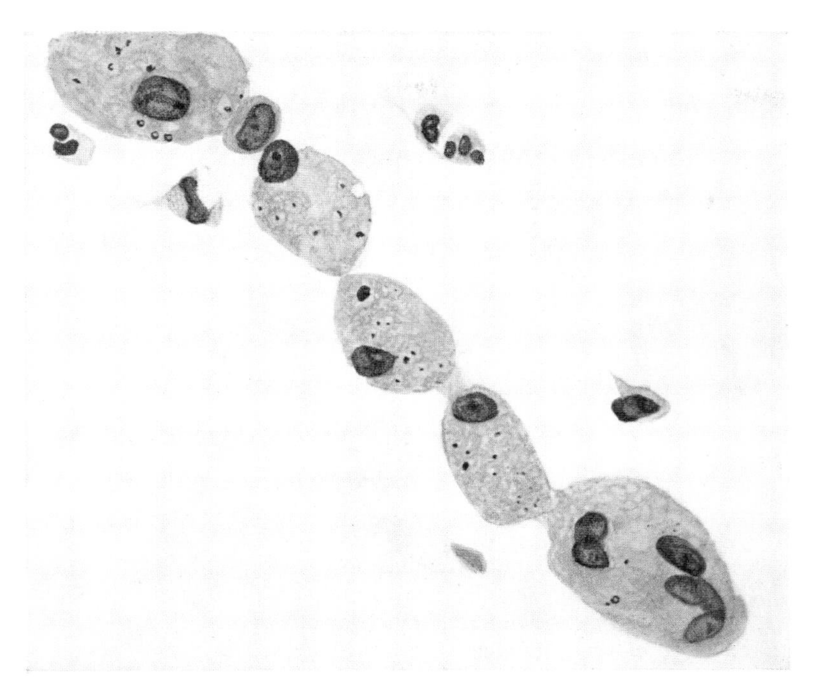

PLATE 19

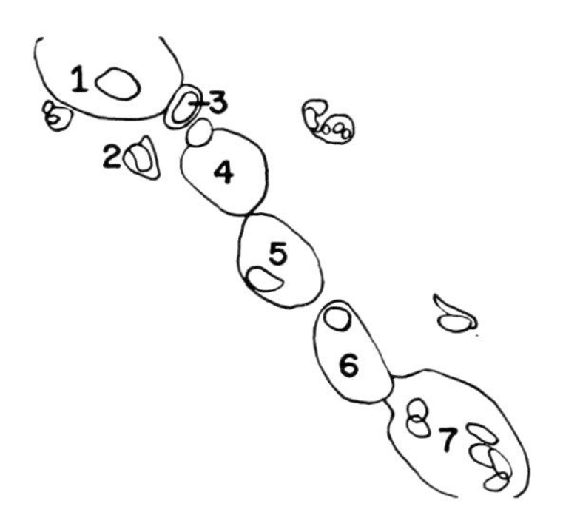

KEY TO HISTIOCYTES IN SPUTUM

1. Histiocyte exhibiting a round nucleus and a large amount of cytoplasm phagocytosing carbon particles.
2 and 3. Bean-shaped nucleus with a small amount of cytoplasm.
4, 5 and 6. Histiocytes with round or oval eccentric nuclei, each with a large amount of vacuolated cytoplasm containing carbon particles.
7. Vacuolated cytoplasm with a collection of small oval-shaped nuclei arranged peripherally. The background is clear except for a few leukocytes.

FIG. 96

HISTIOCYTES: WELL PRESERVED

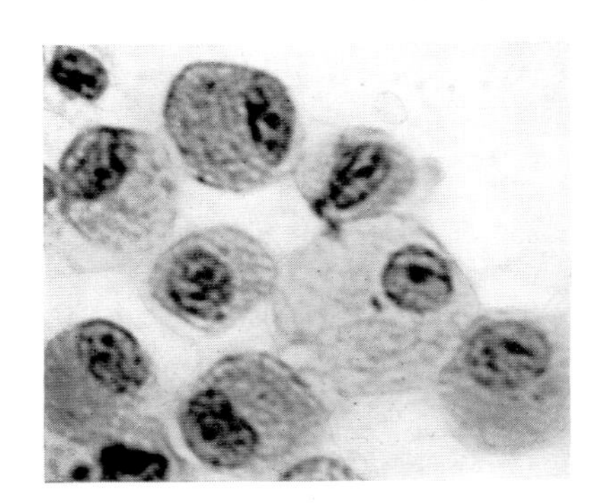

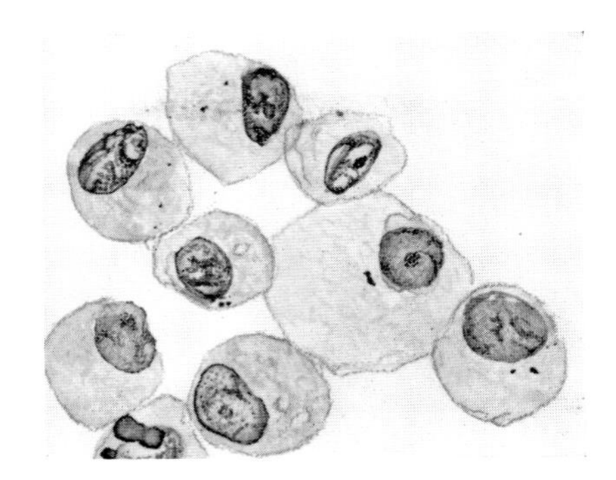

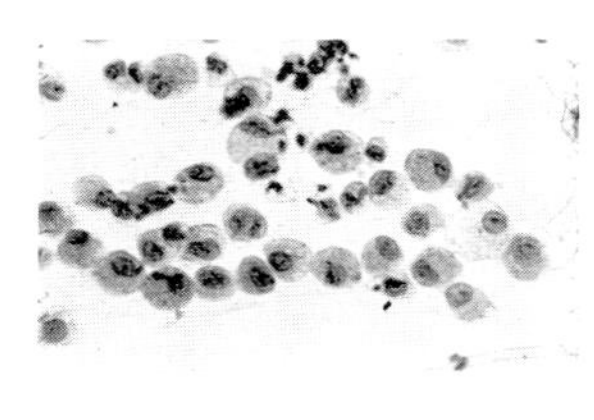

Low Power: Individual cells with well preserved nuclei and cytoplasm.

High Power: All of the nuclei have similar chromatin structure, *i.e.*, clumps and strands of chromatin which do not appear hyperchromatic. The nuclei are also eccentric in relation to the well defined cytoplasm, which shows many small vacuoles. Some of the vacuoles have phagocytosed tiny particles of carbon. The nuclei are round, oval or bean-shaped.

FIG. 97

HISTIOCYTES: FOREIGN BODY GIANT CELL

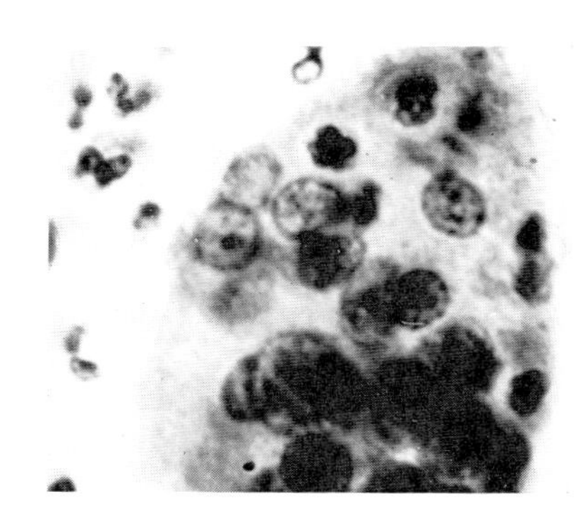

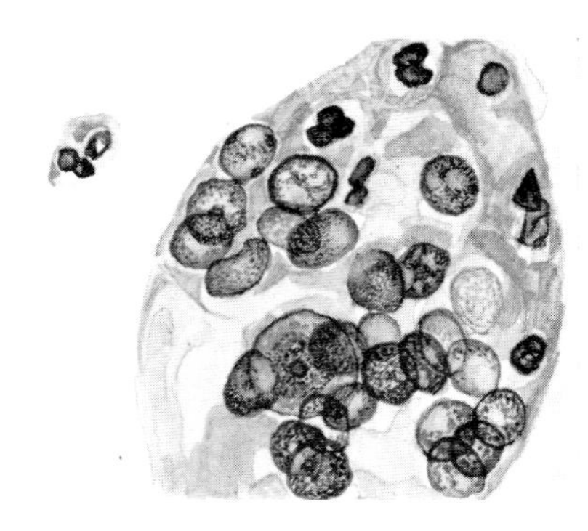

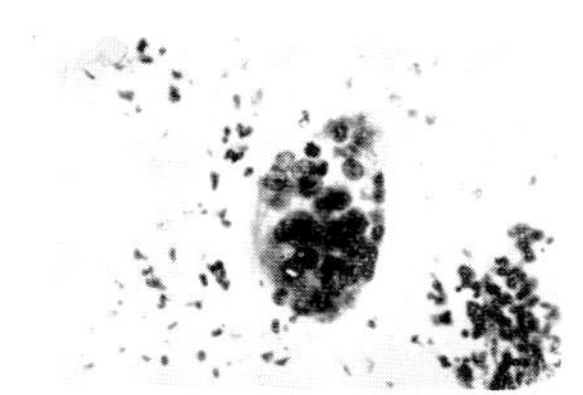

Low Power: Large multinucleated cell in a field of leukocytes.

High Power: The foreign body giant cells of the histiocyte group often are quite large and contain numerous nuclei. In this example there are so many nuclei that the peripheral arrangement of the nuclei is not apparent. Compare nuclear structure to that of single histiocytes in figure above. Nuclei are round or oval.

FIG. 98

HISTIOCYTES: ABSENCE OF INGESTED MATERIAL

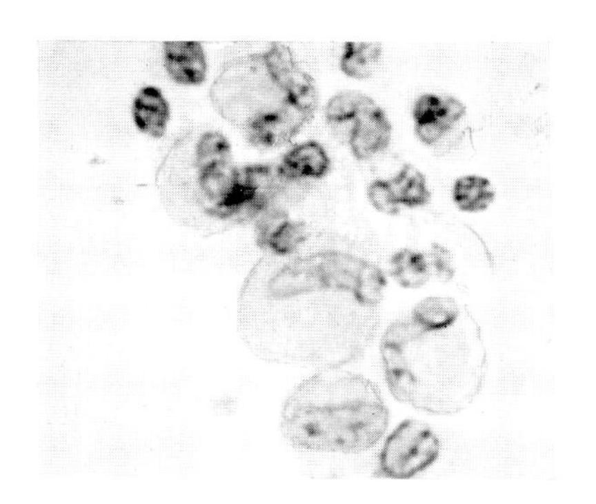

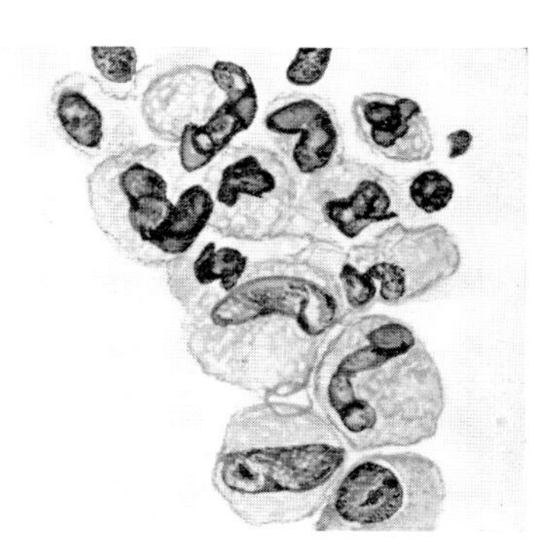

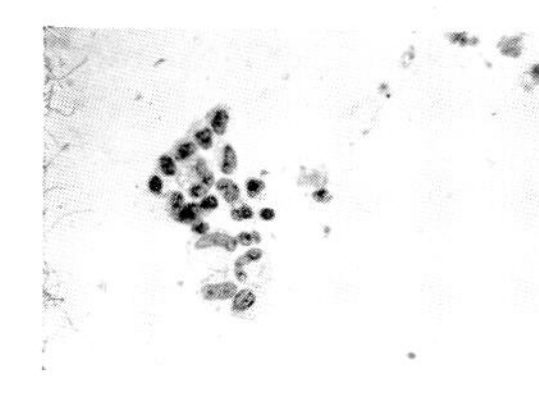

Low Power: Group of nuclei varying in size and shape with adequate cytoplasm.
High Power: Some of the nuclei show prominent clumps of chromatin and others pyknotic qualities. There is also a marked difference in size and shape, but the recognition of several crescent nuclei and the characteristically finely vacuolated cytoplasm around each nucleus identify these cells as histiocytes. Compare with vaginal histiocytes, Fig. 30.

FIG. 99

HISTIOCYTES: ENLARGEMENT OF NUCLEUS

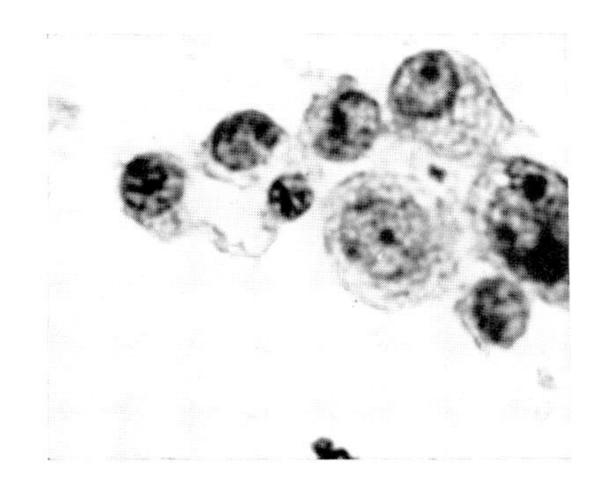

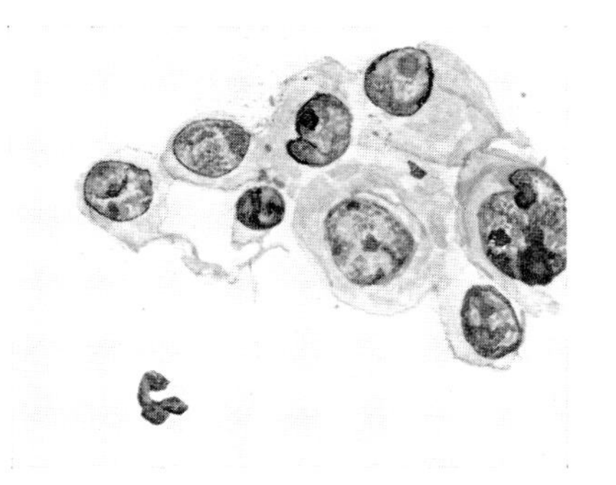

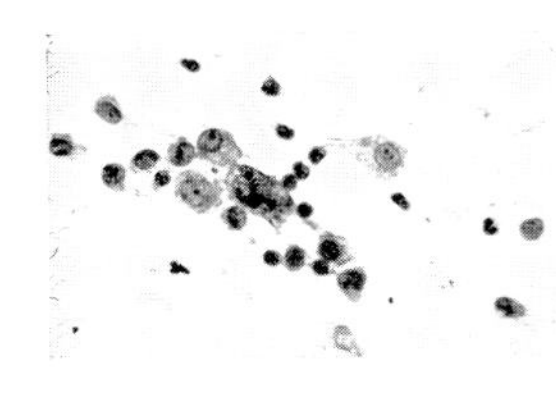

Low Power: Field of granular nuclei with evidence of cytoplasm.
High Power: The nuclei vary in size, but chromatin is evenly distributed throughout except for occasional clumps which do not appear active. The smaller nuclei are eccentric in the foamy cytoplasm, identifying the cells as histiocytes. Nuclear and cytoplasmic structure of the enlarged nuclei is similar to that of the smaller characteristic cells.

Histiocytes in Sputum

In the examination of sputum specimens the observer is immediately impressed by the great amount of mucus present. There are thick strands of mucus throughout the smears. Attempts to be rid of this tenacious substance, which makes preparation of the smears rather difficult, have not been particularly successful. The staining reaction of the cells is changed considerably by any extensive chemical treatment. However, if care is taken in preparation of the smear (see chapter on technic), the presence of mucus is not a real problem. Because of the adhesive properties of the mucus, cells are apt to collect along the edge of the thick strands. This is a common location for the histiocytes and in making a quick appraisal of whether a specimen is actually sputum, search along the borders of strands of mucus often reveals numbers of histiocytes.

Difficulties in Interpretation of Histiocytes: Actually these cells cause little difficulty. The phagocytic properties of the cells are generally so obvious that identification is readily made. The only type which is at all confusing is that shown in Fig. 99. These cells have a centrally located nucleus which is unusually large for a histiocyte. However, the nucleus is finely granular, with only small clumps of chromatin. The cytoplasm exhibits the foamy vacuolated appearance so characteristic of histiocytes. Careful attention to the nuclear chromatin and the characteristics of the cytoplasm will identify the cells.

General Characteristics: Round cells with round or oval vesicular nuclei, eccentrically placed. Many nuclei may be present in one cell. Cellular border is distinct. Cytoplasm is finely vacuolated and may stain acidophilic or basophilic. Phagocytosis is usual, particles of carbon being the common ingested material.

CHAPTER XII

SQUAMOUS CELL CARCINOMA OF THE LUNG

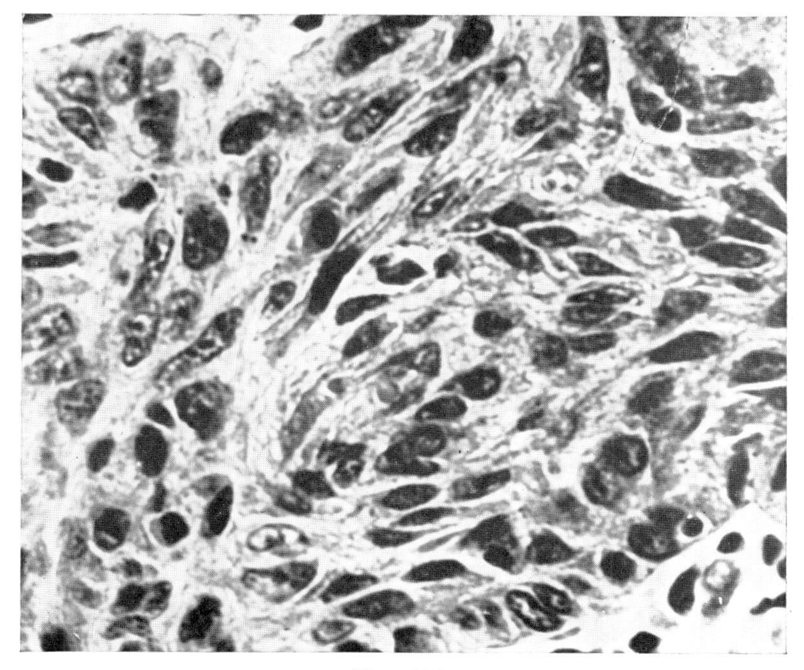

Fig. 100.

HISTOLOGIC SECTION: SQUAMOUS CELL CARCINOMA OF THE LUNG

As shown in the photomicrograph above, squamous cell carcinoma of the lung presents a picture of variation in nuclear size, shape and especially in chromatin content. These same characteristics are seen cytologically. The malignant cells of squamous carcinoma may vary from typical, undifferentiated malignant cells to extremely well differentiated malignant cells.

We have preferred to classify the malignant cells in much the same way as those of squamous cell carcinoma of the cervix. The undifferentiated cells have no cellular border and cytoplasm is indistinct. The differentiated cells show well defined cytoplasm and fairly distinct cellular borders.

The undifferentiated malignant cells have nuclei which have the common characteristics: irregularity of chromatin structure, condensation of chromatin at nuclear border, increase in chromatin and variation in size and shape. The differentiated cells are much the same as seen in vaginal preparations though the cells appear to exhibit greater pleomorphism in sputum than in vaginal secretion. Fiber and tadpole cells may be present but their occurrence is much less frequent than in cells desquamated from carcinoma of the cervix. The most common differentiated cell is the third type differentiated cell. It may be round, or often has a definite irregular cytoplasmic outline, with a centrally located nucleus which is extremely active and hyperchromatic.

DESCRIPTION OF SQUAMOUS CELL CARCINOMA IN SPUTUM

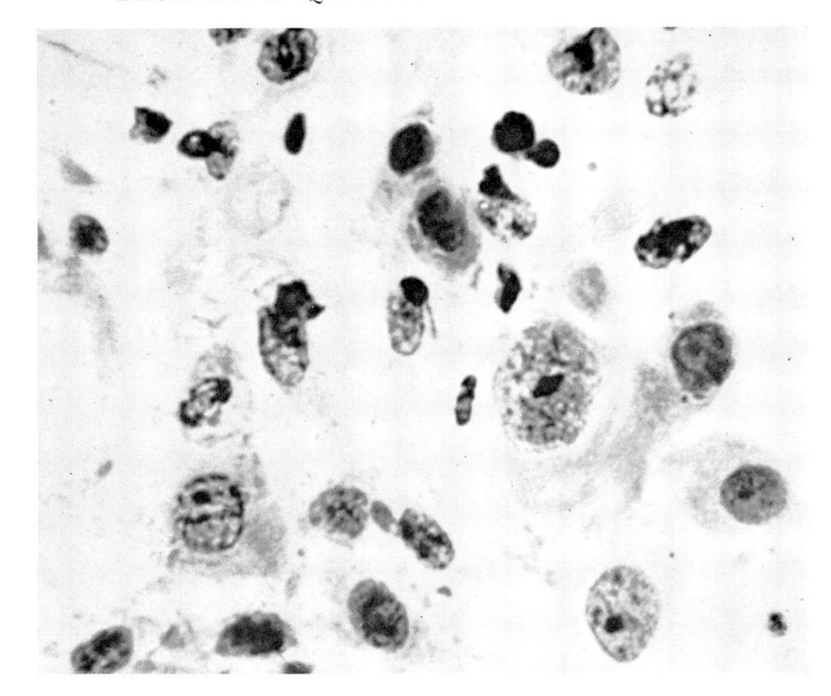

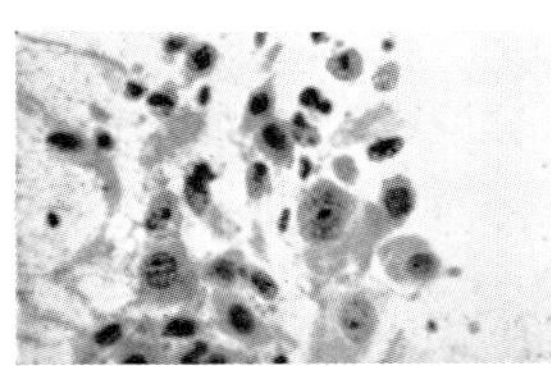

Low Power: Field of nuclei with and without cytoplasm, showing great variation in size and shape.

High Power:

A. Characteristics of Nucleus: 1. Sharp, irregular nuclear borders. 2. There is marked variation in the chromatin content of the nuclei. Cells 1 and 5 have many fine chromatin particles scattered throughout the nucleus plus one prominent nucleolus (note acidophilic stain), whereas cells 2 and 4 are more degenerate and show wrinkled nuclear borders and strands of chromatin. Cell 6 is even more degenerate and individual particles of chromatin are not clearly visible. 3. The nuclei vary greatly in size and shape. Cell 5 is about five times larger than cell 2 and cell 3 is elongated as compared to cell 1, which is round.

B. Characteristics of Cytoplasm: 1. Indistinct cellular borders. 2. Cells 1, 2, 5 and 6 have visible, irregular-shaped cytoplasm with relatively clear cellular borders which identify them as differentiated malignant cells, consistent with squamous cell carcinoma. However, cells 3 and 4 do not have clearly visible cytoplasmic borders and are therefore undifferentiated malignant cells. 3. Staining reaction: bluish purple nuclei with acidophilic or orange-staining cytoplasm.

C. General Characteristics of Group: An irregular pattern of differentiated, malignant cells with cytoplasm and undifferentiated malignant cells without cytoplasm, showing variation in size and shape and chromatin content.

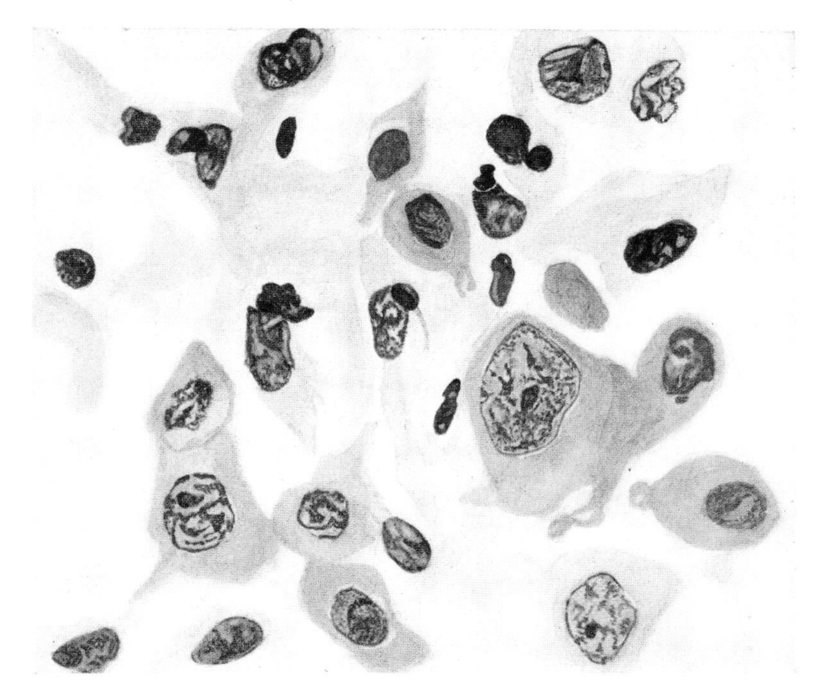

PLATE 20

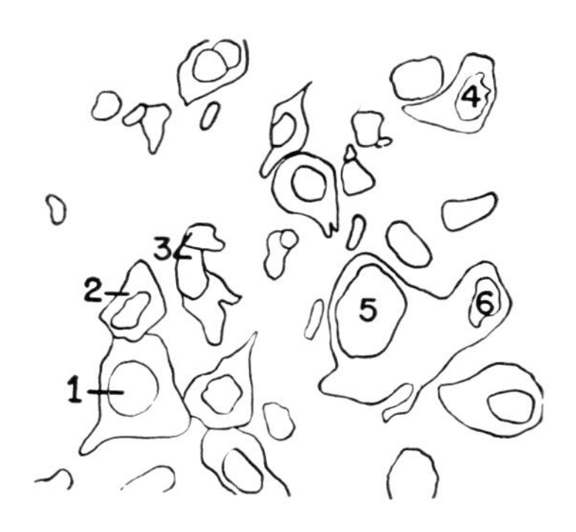

KEY TO SQUAMOUS CELL CARCINOMA IN SPUTUM

1. Differentiated malignant cell with irregular strands of chromatin and prominent nucleolus.
2. Irregular malignant nucleus with well defined cytoplasmic border.
3. Elongated undifferentiated malignant cell without cellular borders.
4. Wrinkled undifferentiated nucleus with strands of chromatin, indistinct cytoplasm.
5. Large irregular differentiated cell with abnormal chromatin pattern and large nucleolus.
6. Nucleus with irregular nuclear pattern but no distinct chromatin.

The field is filled with many other differentiated and undifferentiated malignant cells varying in size and shape.

Fig. 101

SQUAMOUS CARCINOMA OF LUNG: THIRD TYPE DIFFERENTIATED CELL

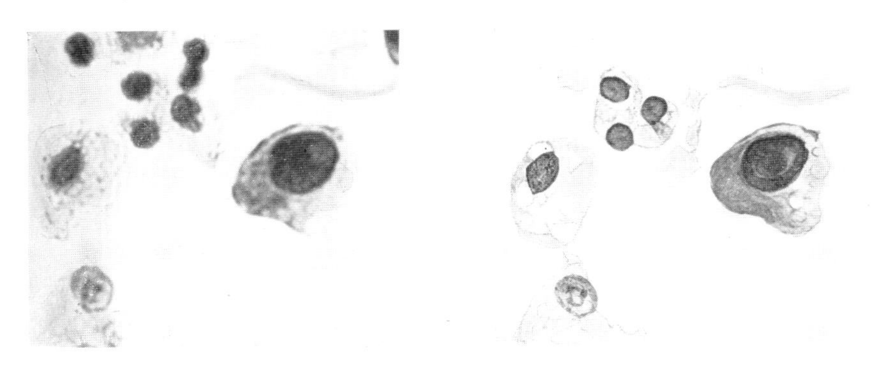

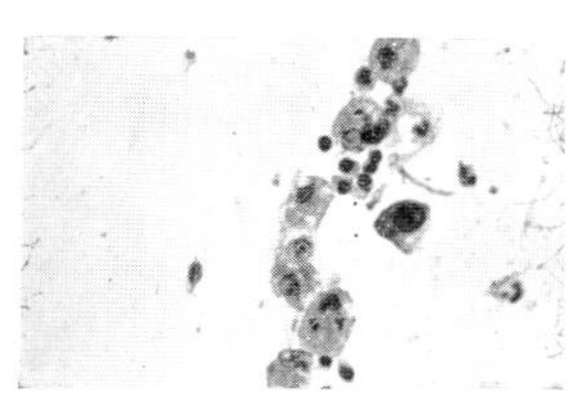

Low Power: One large hyperchromatic cell in a field of histiocytes.
High Power: This is a malignant third type differentiated cell because most of the cellular border is clearly defined and the nucleus is large and dark owing to the many particles of chromatin. The right-hand edge of the cytoplasm shows vacuoles, as it is partially degenerated. The cytoplasm usually stains acidophilic.

Fig. 102

SQUAMOUS CARCINOMA OF LUNG: THIRD TYPE DIFFERENTIATED CELL

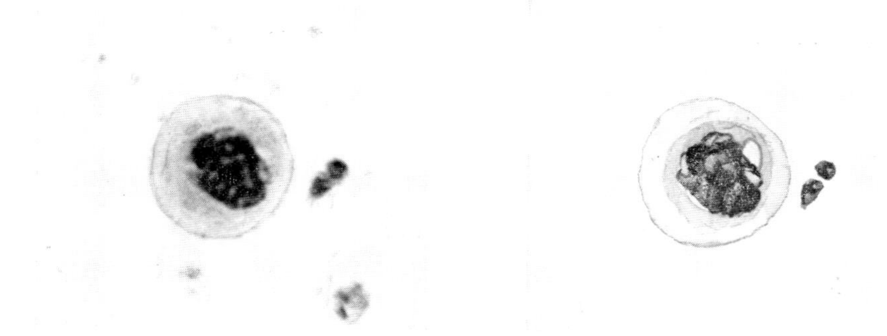

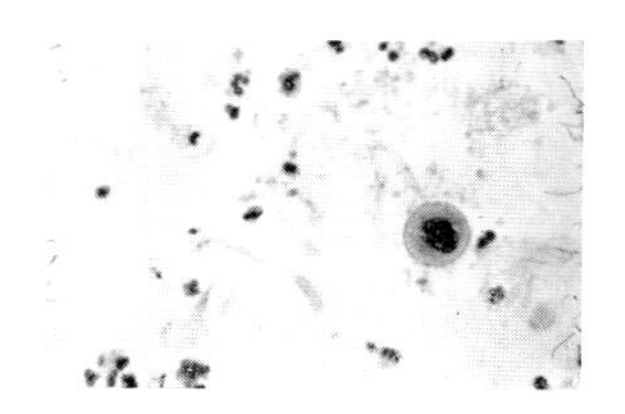

Low Power: Large discrete cell in a relatively clear field.
High Power: A typical third type differentiated cell exhibiting all the diagnostic criteria, *i.e.*, sharp cytoplasmic border, dark-staining nucleus with coarse clumps of chromatin and a wrinkled irregular nuclear border, and a nucleus which is too large for the amount of cytoplasm surrounding it. Compare this third type cell to those found in vagina. (See Plate 10.)

Fig. 103

SQUAMOUS CARCINOMA OF LUNG: SINGLE UNDIFFERENTIATED CELL

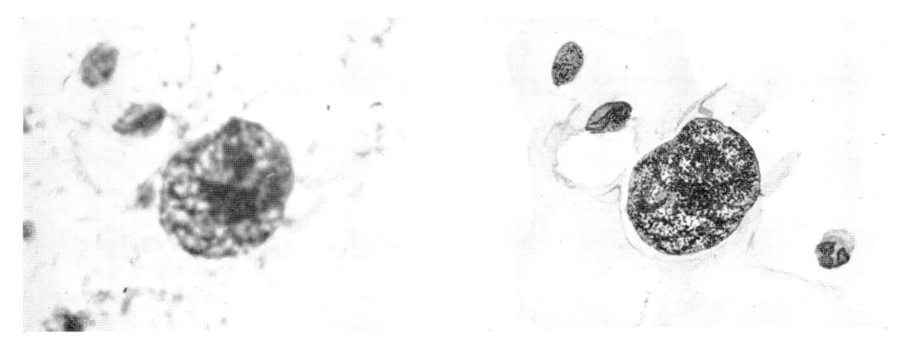

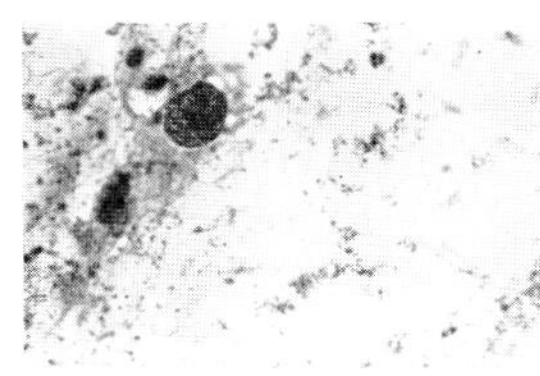

Low Power: Extremely large nucleus in field of leukocytes and blood.
High Power: Large undifferentiated malignant cells such as this occur occasionally in squamous carcinoma. No cytoplasm is visible. The material in the background is blood pigment. The chromatin of the nucleus is markedly increased and appears in heavy strands and large granules. The border of the nucleus shows slight irregularity.

Fig. 104

SQUAMOUS CARCINOMA OF LUNG: GROUP OF UNDIFFERENTIATED CELLS

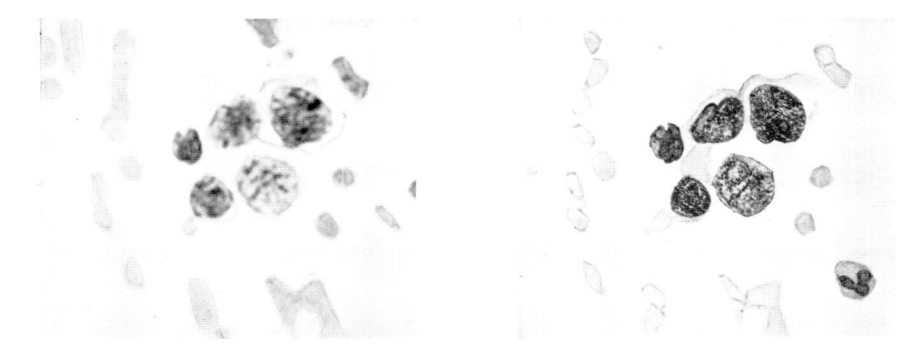

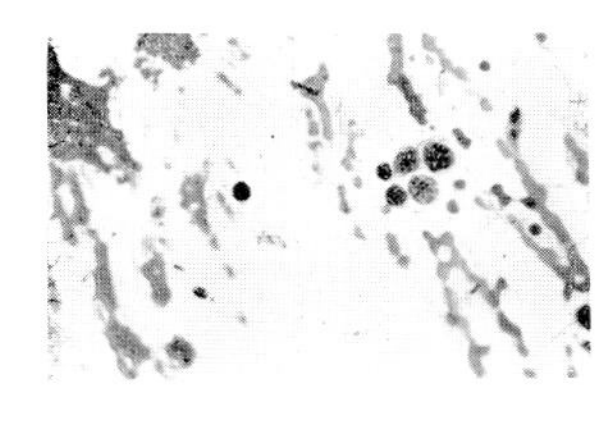

Low Power: Free nuclei varying in size in field of blood.
High Power: These five nuclei fulfill all the criteria for undifferentiated malignant cells. They have a faint background of cytoplasm, but no cellular borders. They show marked variation in size and shape of the nucleus. Nuclear borders are irregular. Chromatin content and irregularity in nuclear appearance are increased.

FIG. 105

SQUAMOUS CARCINOMA OF LUNG: FIBER CELLS

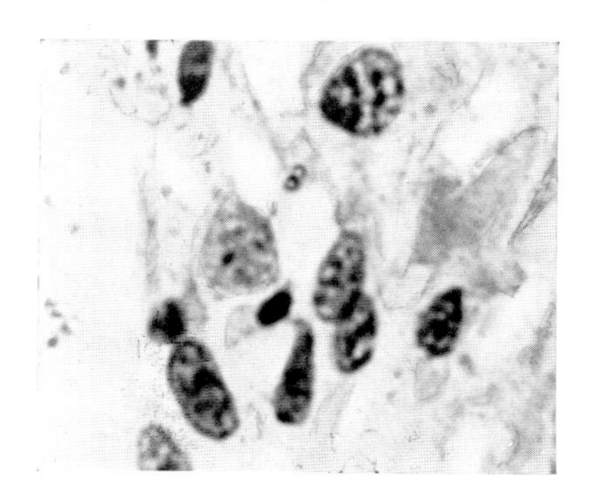

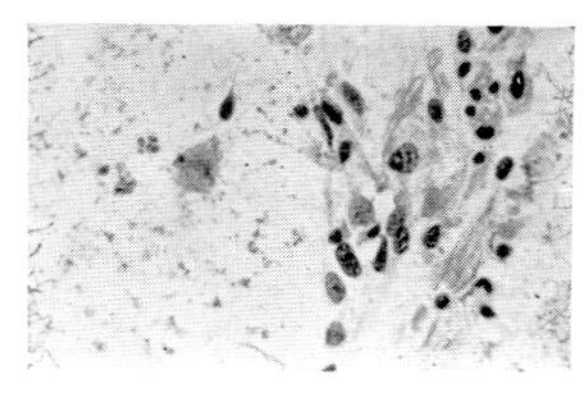

Low Power: Messy field with active nuclei arranged in an irregular pattern.
High Power: The nuclei in the lower portion of this field are quite similar to the fiber cells seen in squamous carcinoma of the cervix. Their nuclear shape is elongated and the chromatin irregularity and hyperchromasia is typical of malignant cells. Though cytoplasmic borders are absent, the shape of these nuclei suggests differentiated cells from squamous carcinoma.

FIG. 106

SQUAMOUS CARCINOMA OF LUNG: BIRD'S-EYE CELL

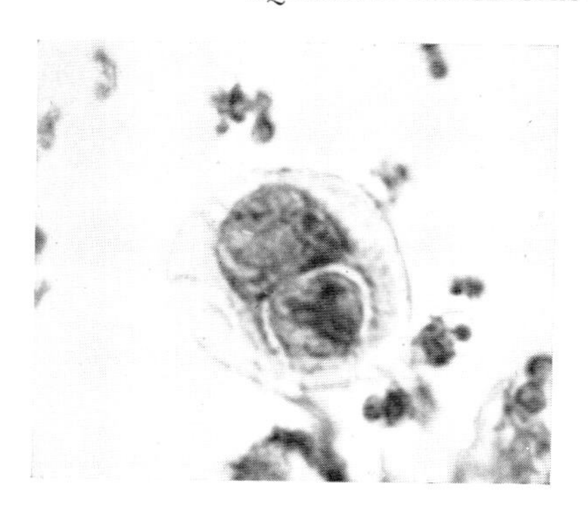

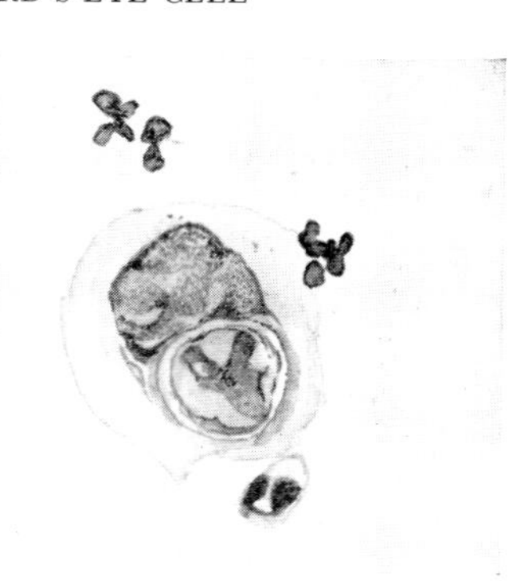

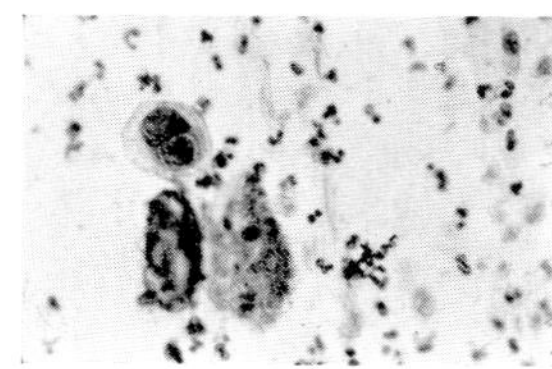

Low Power: Large single cell with two hyperchromatic nuclei in a field of leukocytes.
High Power: Differentiated malignant cells which have phagocytosed another malignant cell are called bird's-eye cells. Careful examination reveals that this cell is not double nucleated. The cytoplasm has a phagocytosed degenerate malignant nucleus, as is shown by the rim of space around the lower of the two nuclei.

Fig. 107

SQUAMOUS CARCINOMA OF LUNG: CONDENSATION OF CHROMATIN

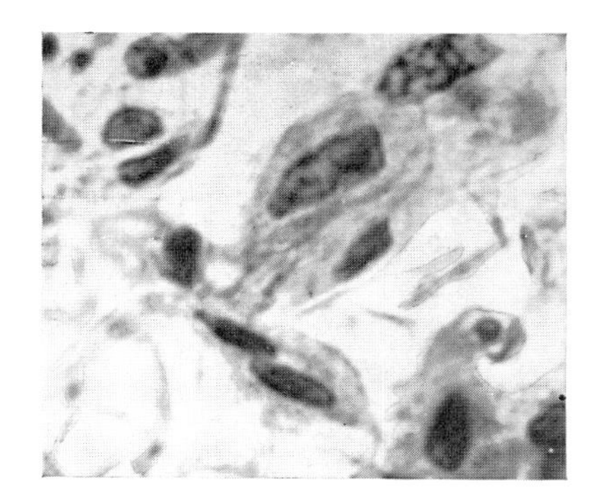

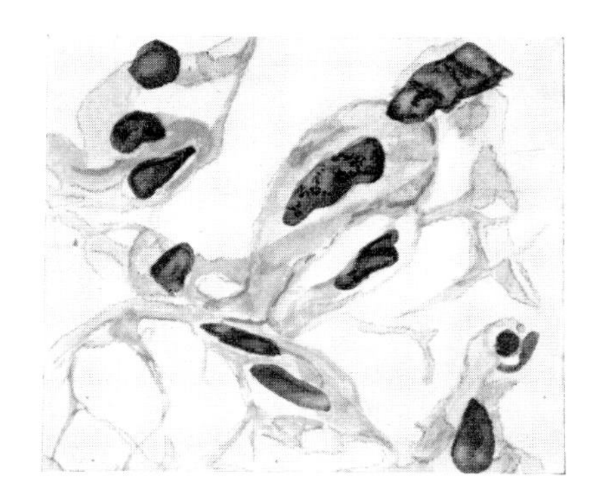

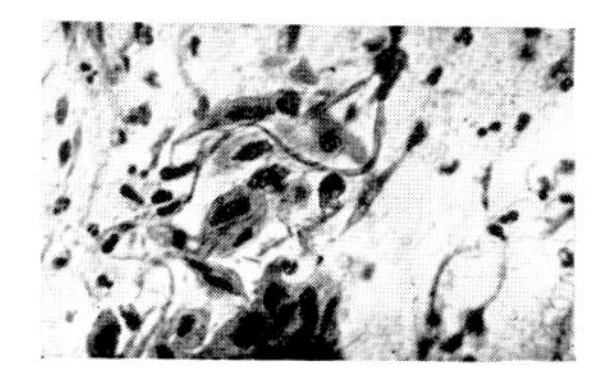

Low Power: Various shaped nuclei arranged in an irregular pattern.

High Power: The nuclei not only show marked variation in size and shape, some resembling fiber cell nuclei, but also have a smooth or glazed appearance which is due to the condensation of the chromatin particles. Cells of this quality are seen quite frequently in squamous cell carcinoma of the lung.

Fig. 108

SQUAMOUS CARCINOMA OF LUNG: MALIGNANT PEARL FORMATION

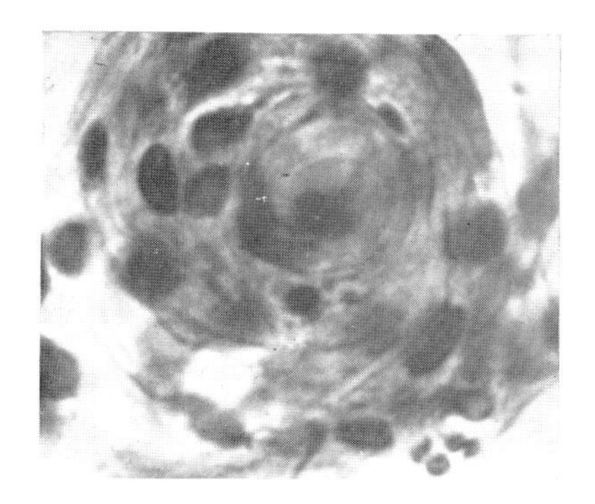

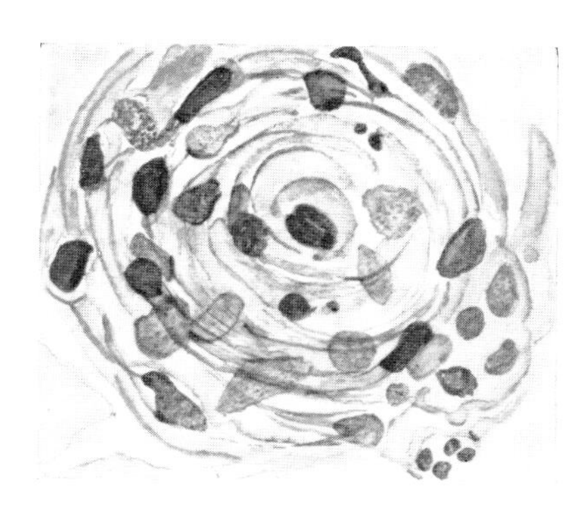

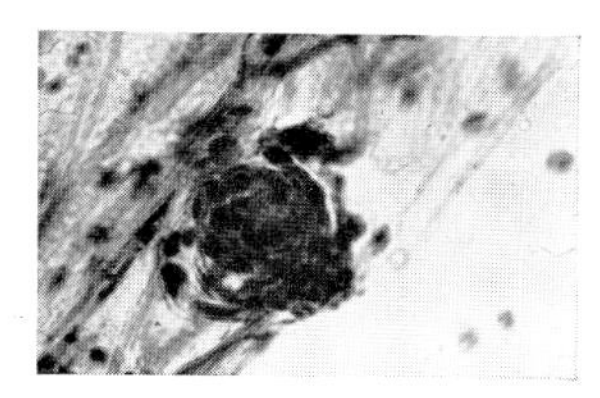

Low Power: Thick, dense group of nuclei in a circular formation.

High Power: Individual cellular borders are not visible, instead the cytoplasm is seen as a whorl formation with many malignant nuclei irregularly scattered throughout. Notice the varying degrees of chromatin density and size and shape of the nuclei as compared to the normal pearl formation seen in the vaginal secretion, Fig. 15.

The nuclei of malignant cells from squamous carcinoma of the lung are unusually hyperchromatic. They often show such an increase in chromatin content that they appear almost pyknotic. Under low power the density of the nucleus is apparent. The nuclei of squamous cancer cells may have true nucleoli. (See Plate 20). As previously stated, we have interpreted dense dots which stain deep purple as prominent and enlarged clumps of chromatin rather than nucleoli. We believe that true nucleoli will stain acidophilic and have a smooth border. This type of nuclear body is more frequent in neoplastic cells seen in sputum than in those in vaginal secretion.

The search for a specific stain for malignant cells has been extensive and arduous and has met with little success. We do not see any specific staining qualities in smears prepared by Papanicolaou's stain. However, the nearest approach to a specific staining property is seen in the differentiated squamous cancer cells found in smears of sputum and bronchial secretion. More often than not their cytoplasm will stain a dense pinkish orange. This should not be interpreted as meaning that all orange cells in sputum are carcinoma. Normal epithelial cells may have this same color. The nucleus must still have definite abnormalities before any such interpretation can be made. On the other hand, the pinkish orange stain is so common in differentiated malignant cells that it is well to examine all cells of that color under high power.

Difficulties in Interpretation: The cells which may be confusing are the "deep" cells from squamous epithelium. They may be mistaken for the differentiated cancer cells. Again the nuclear arrangement, content and shape should be the deciding factors. The nuclei of differentiated malignant cells are extremely hyperchromatic, show definite abnormalities in arrangement, and have borders which are irregular. The deep cells do not show great increase in chromatin, have smooth, nuclear borders, and chromatin is only slightly irregular.

True fiber cells are rare in sputum smears, especially ones in which the cytoplasm is well preserved. This is mentioned because occasionally inexperienced observers have misinterpreted unusually elongated, columnar cells as malignant. When columnar cells become elongated the nucleus may appear elongated also. However, the nucleus retains the fairly granular appearance so characteristic of columnar cells.

General Criteria for Identification of Malignant Cells from Squamous Carcinoma: 1. Undifferentiated cells have a background of cytoplasm and lack cellular borders. Nuclei vary in size and shape. Chromatin is irregular and increased and condensed at the nuclear border. 2. Differentiated cells show a multiplicity of forms. Cytoplasm is present and cellular borders are apparent. Tadpole and fiber cells are not common, but the third type differentiated cells are. Nuclei show increase in chromatin, often to point of an appearance of pyknosis. Chromatin is abnormal in appearance. Nuclear border is irregular and wrinkled.

CHAPTER XIII

OTHER TYPES OF PULMONARY CARCINOMA

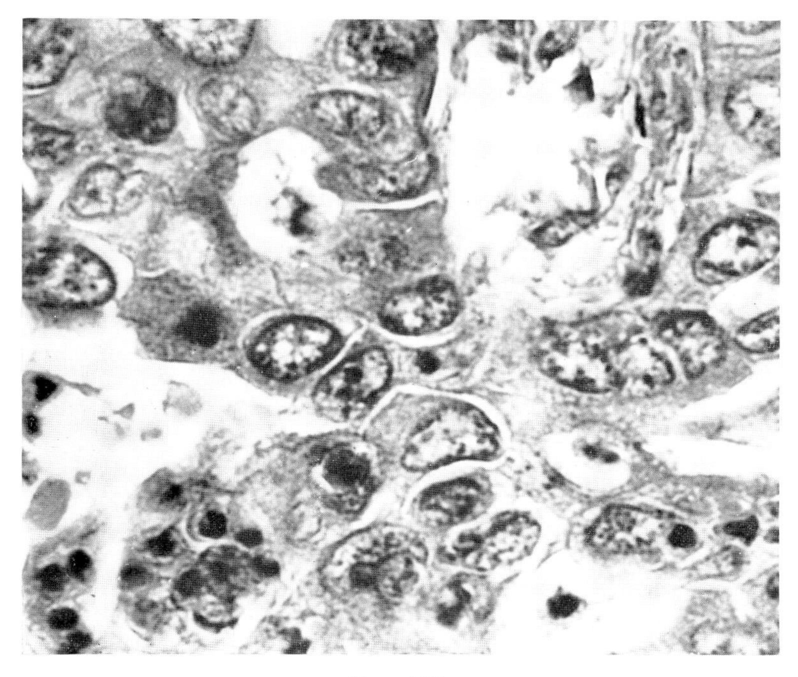

Fig. 109

HISTOLOGIC SECTION: ADENOCARCINOMA OF THE LUNG

The section above shows the characteristic appearance of adenocarcinoma of the lung, *i.e.*, the definite acini, the vacuolization and irregular large nuclei. Much the same picture is seen in the sputum smear from patients with this type of tumor, with the exception, of course, that definite architecture is not present.

The cells from an adenocarcinoma of the lung usually appear in groups, though single ones are occasionally encountered. We have not been able to see any distinguishing characteristic between adenocarcinoma cells which desquamate from a tumor in the lung and those which are shed from a tumor of the endometrium. The cells may appear as undifferentiated ones with only a faint background of cytoplasm and no characteristic vacuolization. More commonly they occur in groups with distinct vacuolization, as illustrated on the following pages. The vacuolization has the same pouched-out effect that is seen in adenocarcinoma of the endometrium. Cellular borders may be present. If so, the nuclei appear in an eccentric position in the cell. The nuclear appearance is characteristic of any malignant cell. Nucleoli may occasionally be present.

DESCRIPTION OF ADENOCARCINOMA CELLS IN SPUTUM

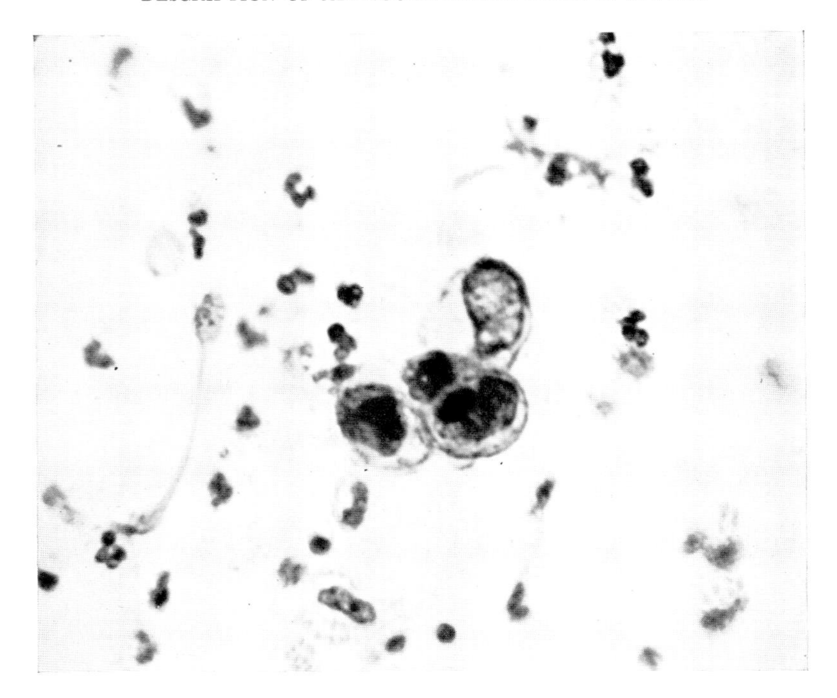

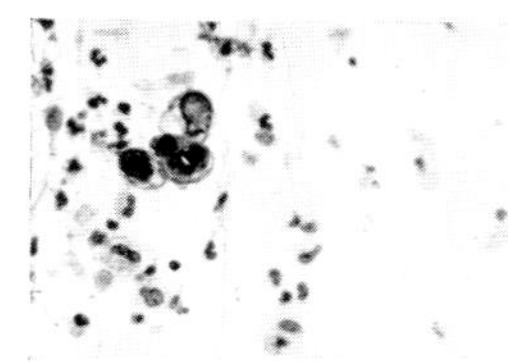

Low Power: Cluster of cells with large hyperchromatic nuclei.

High Power:

A. Characteristics of Nucleus: 1. Irregular, well defined nuclear borders. 2. All of the nuclei are hyperchromatic but there is variation in the density of the chromatin particles. Cell 1 shows strands of chromatin, which gives a wrinkled or folded appearance to the nucleus, whereas cell 3 shows individual clumps of chromatin throughout the nucleus and thick condensation of chromatin particles at the nuclear border. In cell 4 the smallest of the three nuclei is almost black, because of the many thick clumps of chromatin. 3. There is noticeable variation in both size and shape of these nuclei. Compare cell 2, which is fairly round and not much larger than a leukocyte, with cell 3, which is oblong and about three and a half times larger. Nuclei are eccentric in position with the exception of 4, which is multinucleated.

B. Characteristics of Cytoplasm: 1. Clearly visible cellular borders. 2. There are varying amounts of cytoplasm around the nuclei, from a small rim in cell 2 to an ample amount in cell 3. The large "pouched-out" vacuoles in cells 1 and 3 characterize these cells as being adenocarcinoma cells. 3. Staining reaction: bluish purple nuclei with acidophilic cytoplasm.

C. General Characteristics of Group: A group of vacuolated cells, each varying from the other in their amount of cytoplasm and in the size and shape of their individual hyperchromatic nuclei.

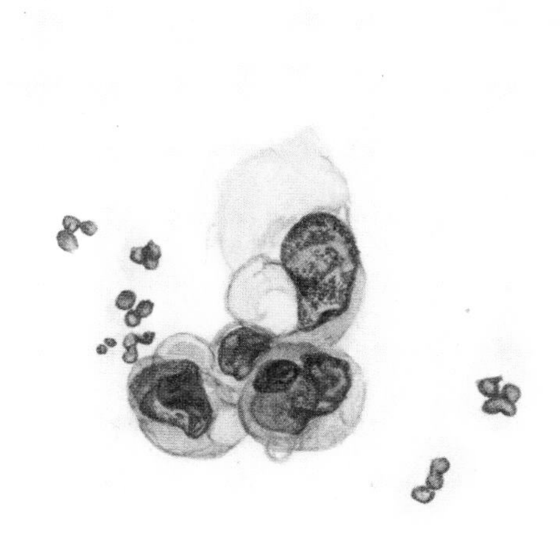

PLATE 21

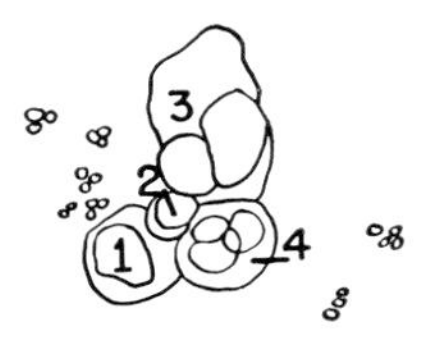

KEY TO ADENOCARCINOMA CELLS IN SPUTUM

1. Vacuolated differentiated cell with a large wrinkled nucleus.
2. Small differentiated cell with very little cytoplasm in comparison to the size of the nucleus.
3. Large coarsely granulated nucleus with a sharp nuclear border and an ample amount of cytoplasm exhibiting a vacuole.
4. Differentiated cell with three hyperchromatic nuclei showing marked variation in size, shape and chromatin content. The background is clean, except for scattered leukocytes.

FIG. 110

ADENOCARCINOMA OF LUNG: OVERLAPPING OF NUCLEI

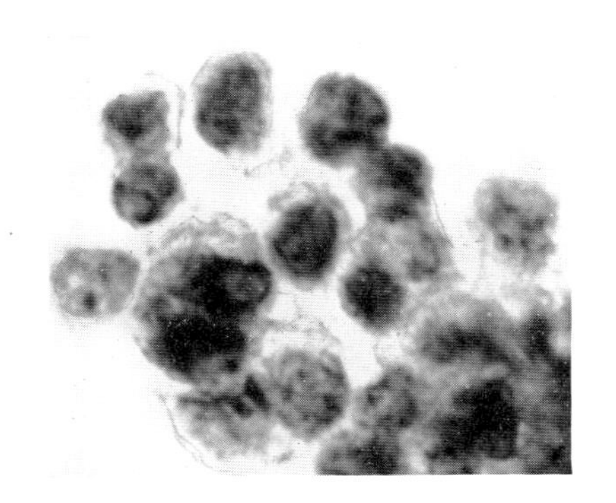

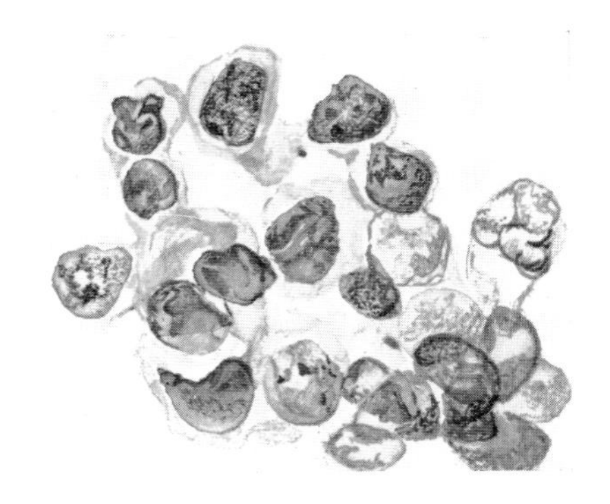

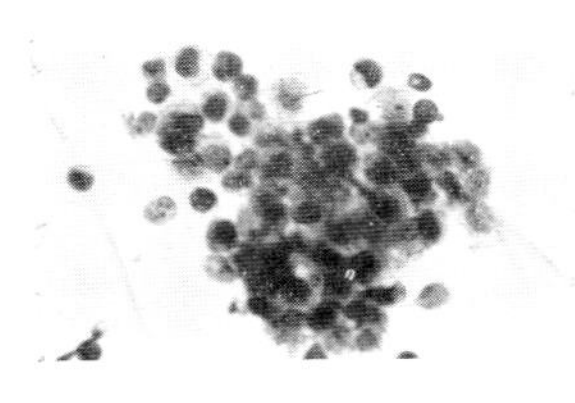

Low Power: Dense cluster of nuclei varying in size and shape.
High Power: The nuclei have sharp but uneven nuclear borders and no definite cellular border. The shapes are not uniform and there is evidence of size variation throughout the group. In the lower right-hand corner, the nuclei are overlapping or piled, which is seen more often in adenocarcinoma than in squamous carcinoma.

FIG. 111

ADENOCARCINOMA OF LUNG: METASTASES FROM PROSTATE

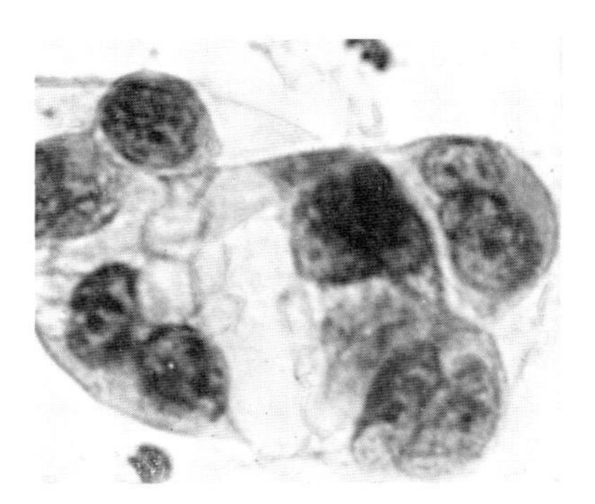

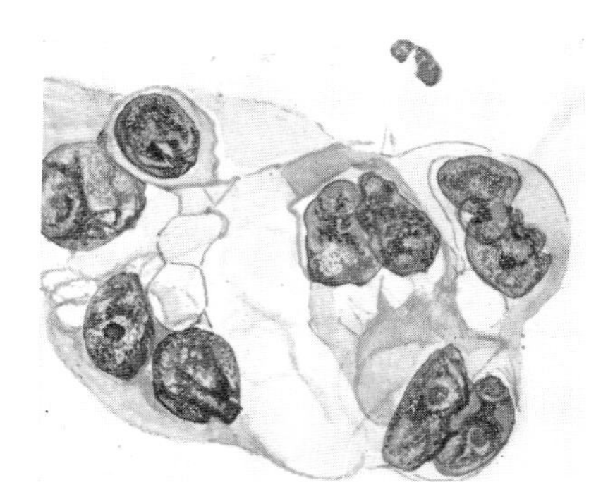

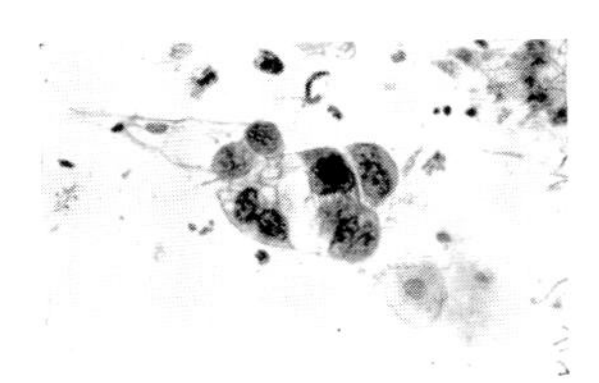

Low Power: Five pairs of hyperchromatic nuclei seeming to have a common cell border.
High Power: The noticeable feature of this group is the enormous vacuoles which have destroyed the individual cellular outlines, leaving a cell wall enclosing all of the malignant nuclei. This abnormal vacuolization plus the irregular chromatin structure of the nuclei distinguishes this group as being consistent with adenocarcinoma. Compare with vaginal Fig. 65.

OAT CELL CARCINOMA OF THE LUNG

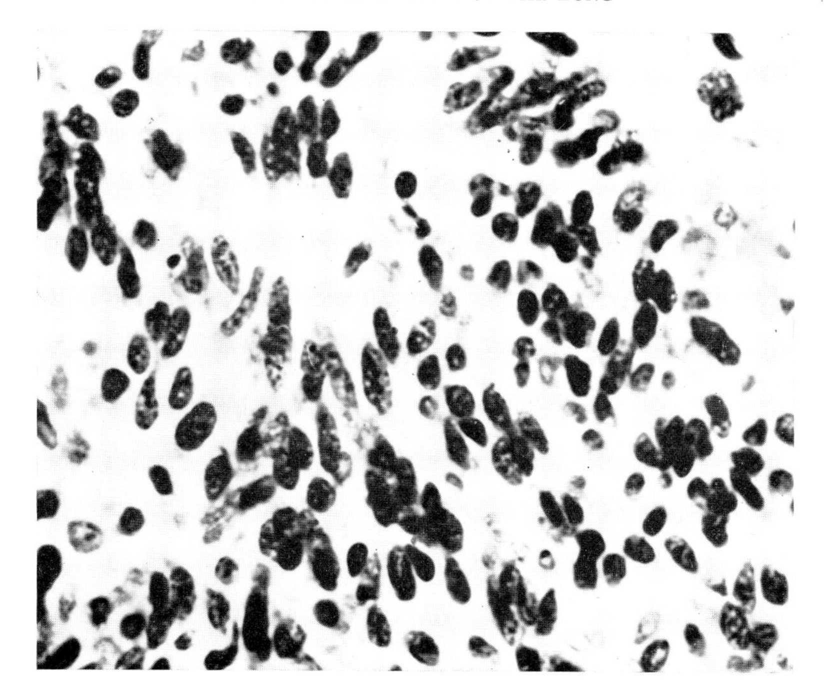

Fig. 112

HISTOLOGIC SECTION: OAT CELL CARCINOMA OF THE LUNG

The normal cell type from which oat cell carcinoma arises is not known. It is often called "small cell" carcinoma of the lung. The term "oat cell" is descriptive and refers to the shape of the nuclei, since they are long and oval. The photomicrograph of the section illustrates the variation in size of the cell, but the relatively small variation in the shape. The same picture is seen cytologically. These cells are unusual malignant cells both because of their small size and their lack of hyperchromasia. The first time they were seen in our laboratory they were referred to as "small, bland nuclei varying in size, of unknown origin."

The cells as seen either in sputum or in bronchial secretions often occur in groups. No cytoplasm is present. The nuclei do not appear as elongated after they have desquamated, as they do in the fixed section of tissue. They have an oval or round shape. The size variation is considerable but not nearly so marked as in the common undifferentiated cells. The identification of these cells as malignant depends on nuclear structure. They do not show the increase in chromatin that other malignant cells do, but in other respects they fulfill the criteria for cancer. Chromatin which is present is often condensed at the periphery, and chromatin structure is irregular.

DESCRIPTION OF OAT CELL CARCINOMA IN SPUTUM

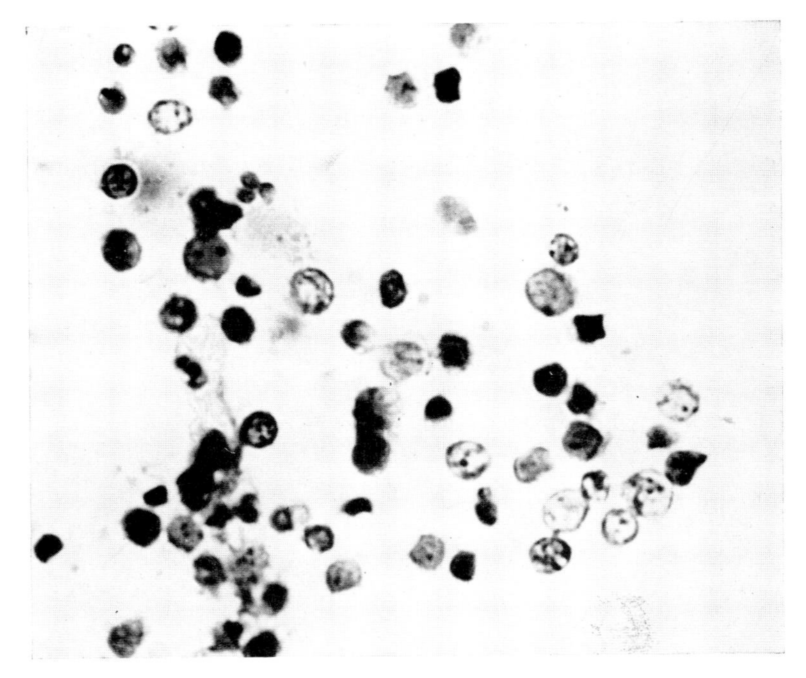

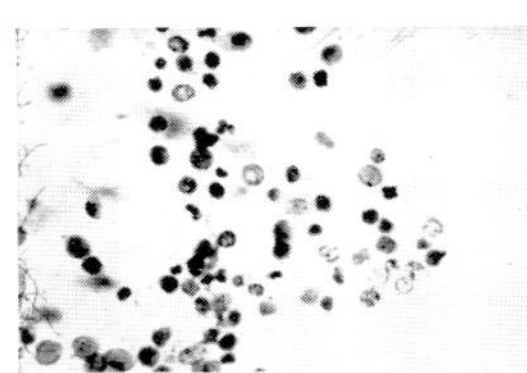

Low Power: Field of extruded nuclei varying only slightly in size and shape.

High Power:

A. Characteristics of Nucleus: 1. Sharp nuclear borders. 2. In general, the nuclei are not very hyperchromatic but instead are rather smooth or glazed in appearance. (See groups 2 and 3.) However, several of the nuclei do show more chromatin structure, such as the cells in group 5 which have small clumps and strands. Cell 1 looks quite active because the chromatin particles stand out in small clumps against the clear nuclear background, whereas cell 3 appears inactive as the nuclear background is glazed, and only two prominent clumps of chromatin are visible. 3. The nuclei, in oat cell carcinoma, tend to vary more in size than in shape, the usual shape being round or oval. (See group 4.) The size of the nuclei vary from that comparable to a leukocyte to three or four times larger. Compare the size relationship in group 5.

B. Characteristics of Cytoplasm: 1. There is little or no evidence of any cytoplasm. Occasionally small rims of cytoplasm may be seen but the cellular borders are always indistinct. (See cell 3.) 2. Staining reaction: purple nuclei.

C. General Characteristics of Group: Nuclei showing differences in chromatin pattern with clear nuclear borders, varying more in size than in shape, and with little or no cytoplasm.

PLATE 22

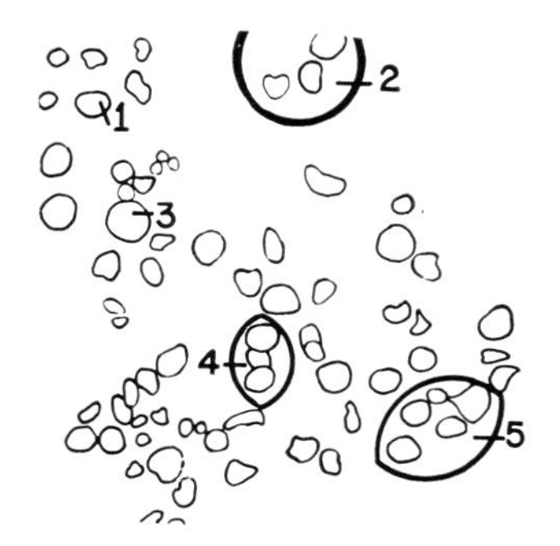

KEY TO OAT CELL CARCINOMA IN SPUTUM

1. Single undifferentiated cell with a sharp nuclear border and many tiny chromatin particles scattered throughout.
2. Three degenerate undifferentiated nuclei, varying slightly in size nd shape, with irregular nuclear borders and no obvious clumps of chromaatn.
4. Three overlapping undifferentiated cells, fairly uniform in size and shapie. Two of them appear dense because of the chromatin intensity.
5. Cluster of undifferentiated cells showing clumps and strands of chromatin and slight variation in size and shape. The rest of the field is filled with undifferentiated cells and leukocytes.

Fig. 113

OAT CELL CARCINOMA: IRREGULAR CHROMATIN PATTERN

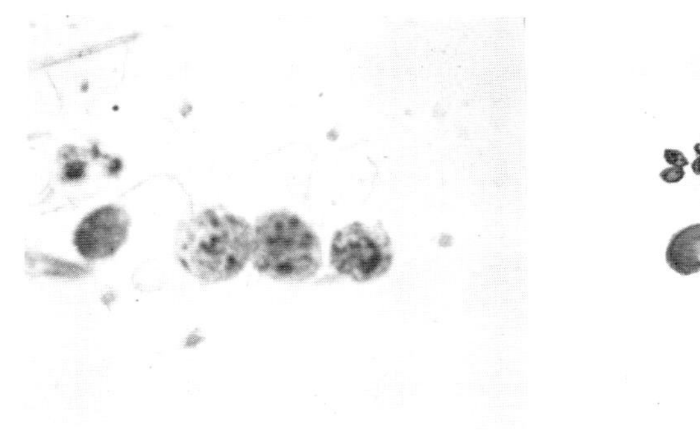

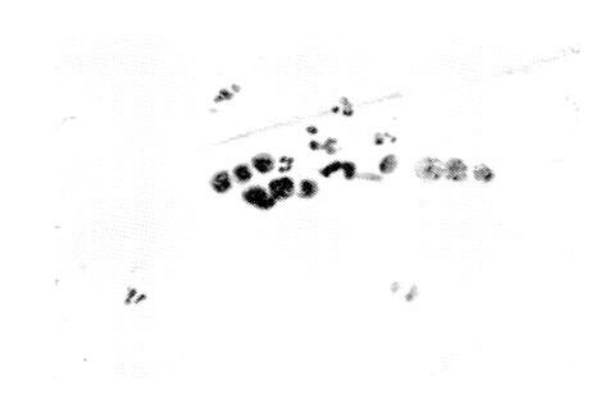

Low Power: Cluster of small regular nuclei.
High Power: These cells are extremely even for malignant cells, showing very little variation either in size or in shape. No cytoplasm is present. Careful examination of the nuclei will show the chromatin structure to be abnormal. The chromatin is in heavy clumps leaving intervening empty spaces. This nuclear pattern is typical of carcinoma cells.

Fig. 114

OAT CELL CARCINOMA: HYPERCHROMATIC NUCLEI

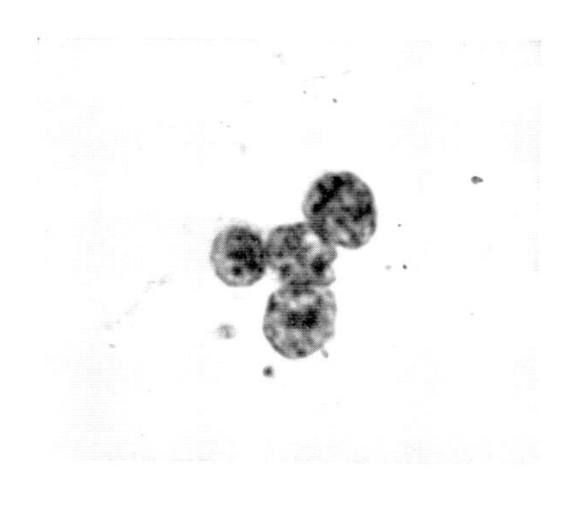

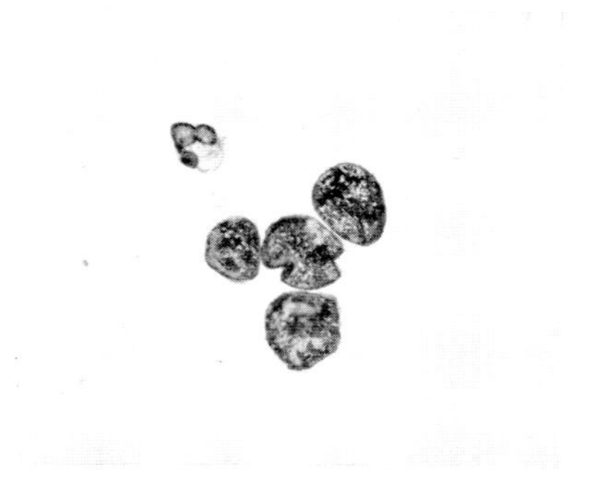

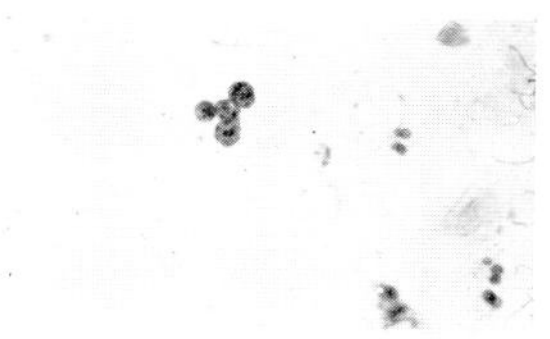

Low Power: Few leukocytes and one cluster of active nuclei.
High Power: These four nuclei are typically characteristic of the malignant cells found in oat cell carcinoma: (1) They are seen in little groups, (2) the size is relatively small, (3) there is more variation in the size than the shape, (4) the nuclei are hyperchromatic, but clumps of chromatin are not prominent and (5) there is no cytoplasm.

UNDIFFERENTIATED CARCINOMA OF THE LUNG

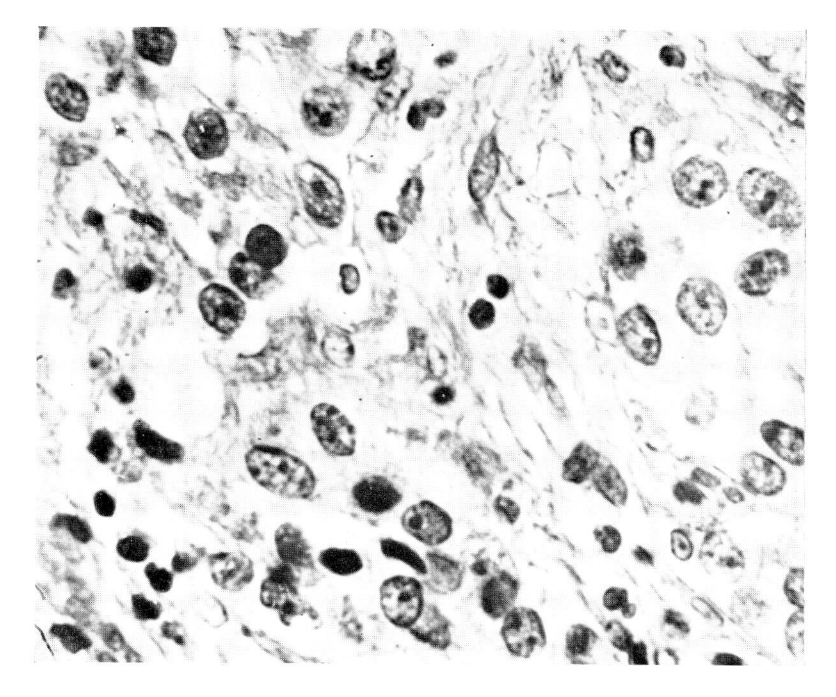

FIG. 115

HISTOLOGIC SECTION: UNDIFFERENTIATED CARCINOMA OF THE LUNG

Both histologically and cytologically an occasional carcinoma shows no definite characteristics to identify the epithelium from which it originated. As seen in the photomicrograph above there are no definite criteria to identify the type of carcinoma. It is obvious that the tissue is neoplastic, but since all the cells are undifferentiated it is not known whether the source was columnar or squamous epithelium. Such cases obviously exfoliate only cells which are undifferentiated.

The malignant cells seen cytologically mirror the picture seen above. A background of cytoplasm is present but no distinct cellular borders, so that cells from such a tumor fulfill the criteria of undifferentiation cytologically as well as histologically. The nuclei vary tremendously in size and shape; the nuclear arrangement is that of typical carcinoma cells. The chromatin pattern is distinctly abnormal, being in thick strands and heavy clumps which often leave vacant spaces. Chromatin may appear increased but nuclear pattern disarrangement is more common. Chromatin is concentrated at the periphery of the nucleus and wrinkling of the surface of the nucleus and irregularity of border are common. Large red nucleoli may be present.

DESCRIPTION OF UNDIFFERENTIATED MALIGNANT CELLS IN SPUTUM

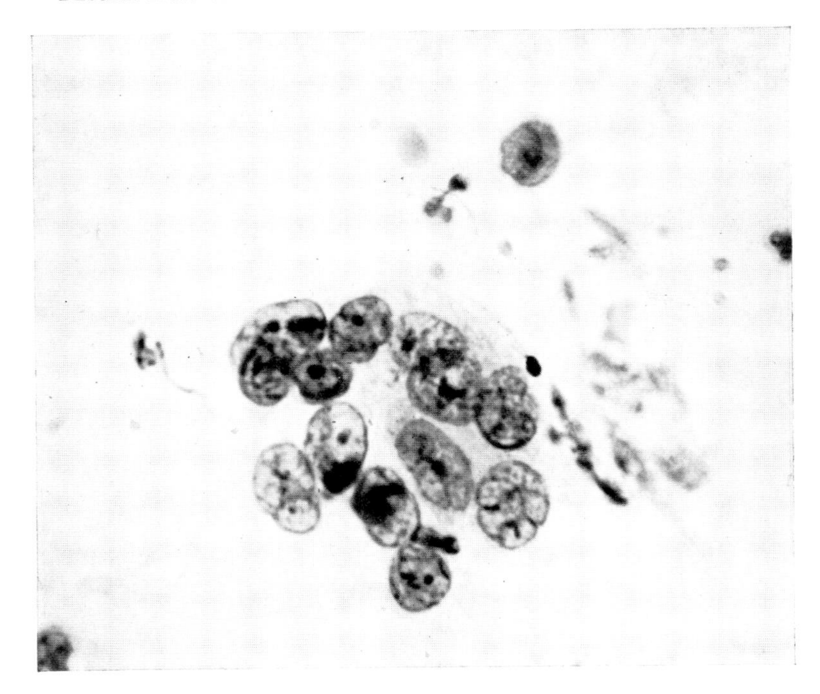

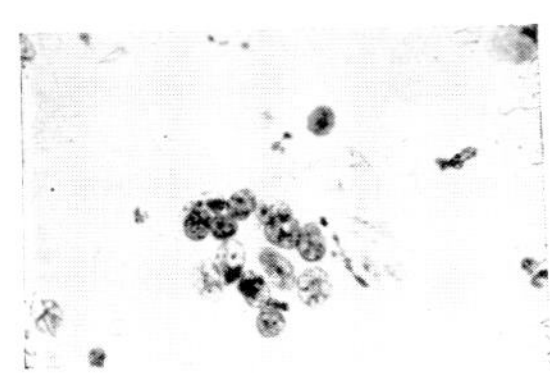

Low Power: Group of active nuclei varying in size and shape with no evidence of cytoplasm.

High Power:

A. Characteristics of Nucleus: 1. Sharp nuclear border. 2. There is variability in chromatin content. Compare cell 1, which has a coarsely granular hyperchromatic appearance, with cell 2, in which the chromatin is concentrated at the nuclear border. 3. There is great variation in the size of the nuclei. In determining size variation, the student should look for the smallest and largest nucleus in the field. (Compare cell 2 with cell 3.) 4. The variation in shape is not as obvious as the variation in size, but notice the unusual forms of cells 3 and 5, caused by the wrinkling of the nuclear borders.

B. Characteristics of Cytoplasm: 1. Absence of cell borders. Note that it is impossible to tell where one cell ends and the next begins. 2. The cytoplasm forms an indistinct background. 3. Staining Reaction: bluish purple nuclei.

C. General Characteristics of Group: A group of undifferentiated malignant cells with clear but irregular nuclear borders and prominent clumps of chromatin, showing more variability in size than in shape. Some of the nuclei are overlapping each other and are arranged in an unorderly manner. There is no cytoplasmic structure around each individual nucleus, only a faint background.

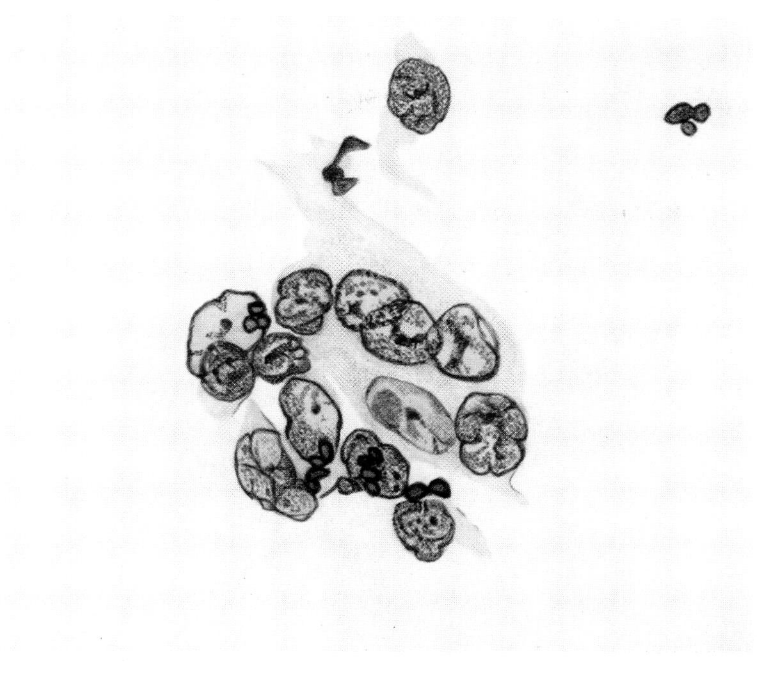

PLATE 23

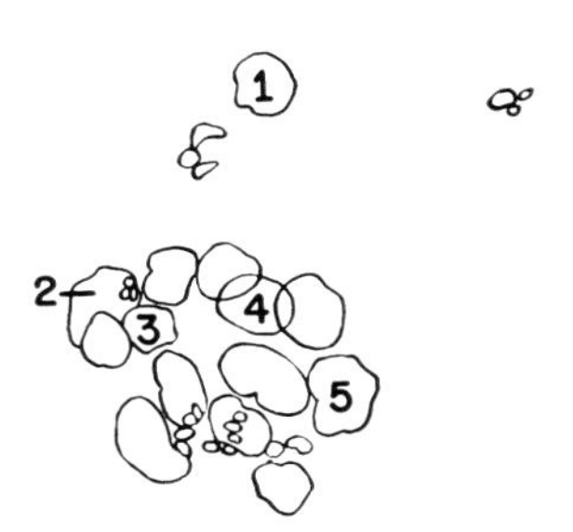

KEY TO UNDIFFERENTIATED MALIGNANT CELLS IN SPUTUM

1. Undifferentiated malignant cell showing a slightly indented nuclear border and many granules of chromatin.
2. Large, transparent appearing undifferentiated cell with one prominent clump of chromatin and a sharp nuclear border.
3. Small, wrinkled undifferentiated malignant cell.
4. Undifferentiated cell, with coarse clumps of chromatin superimposed on two other malignant cells.
5. Undifferentiated cell with a wrinkled or indented nuclear border and strands of chromatin.

FIG. 116

UNDIFFERENTIATED CARCINOMA OF LUNG: HYPERCHROMATIC NUCLEI

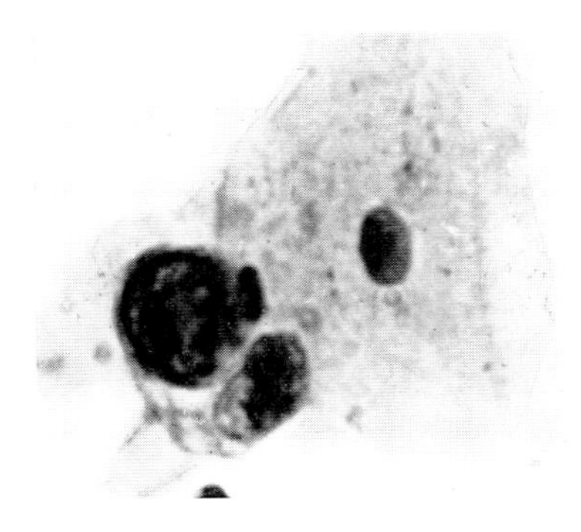

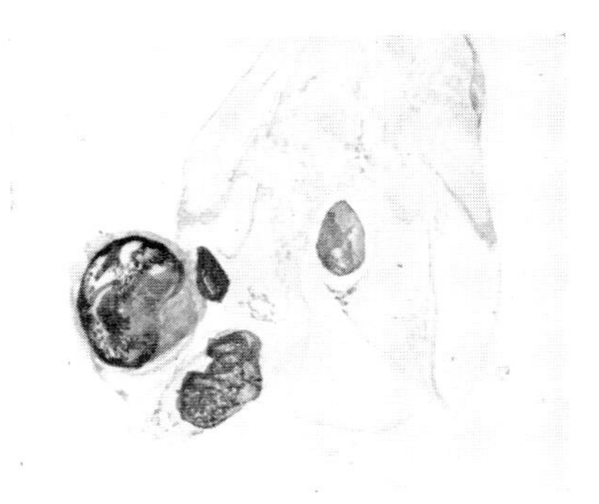

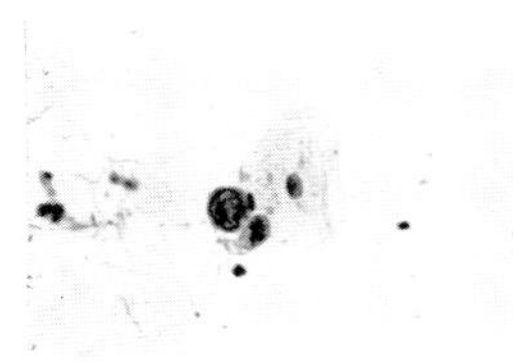

Low Power: Single superficial cell with two large nuclei superimposed.
High Power: The two large hyperchromatic nuclei are recognized as undifferentiated malignant cells because of the dense, irregular nuclear borders, thick clumps of chromatin and the extremely abnormal nuclear-cytoplasmic ratio. Contrast with features of the normal superficial cell shown in Fig. 99.

FIG. 117

UNDIFFERENTIATED CARCINOMA OF LUNG: DEGENERATE NUCLEI

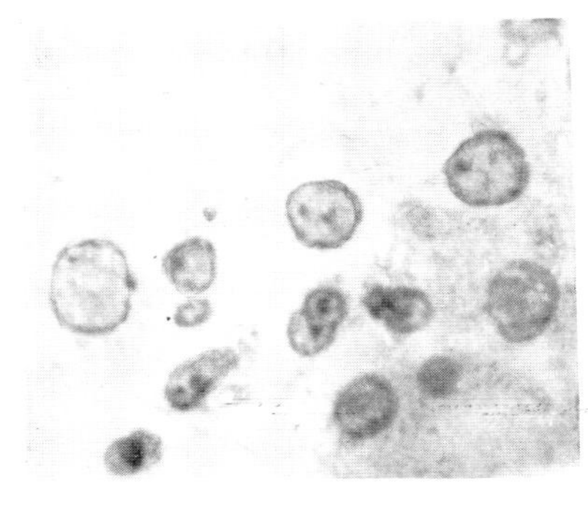

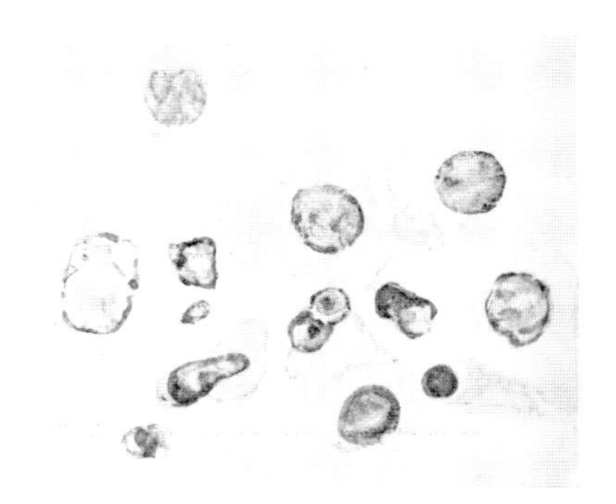

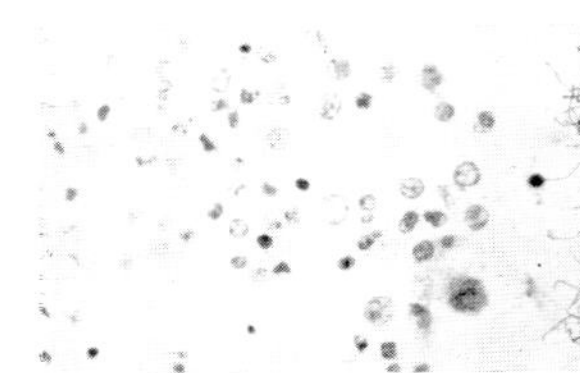

Low Power: Many bland nuclei showing variation in size and shape.
High Power: A group of degenerate undifferentiated malignant cells which appear transparent, as individual chromatin particles are not visible. Condensed particles of chromatin around the edge of the nuclei, producing a sharp nuclear border, differences in size and shape and chromatin content identify these as undifferentiated. These cells cannot be distinguished from oat cell carcinoma. (See Plate 22.)

Of these three types of pulmonary carcinoma the desquamated cells from adenocarcinoma are the only ones with evidence of differentiation. They commonly occur in groups in which the cytoplasm shows distinct vacuolization. Oat cells can sometimes be distinguished from other undifferentiated carcinoma cells. However, the distinction is often difficult; the differences are only general and are as follows: Oat cell nuclei are less hyperchromatic, vary more in size than in shape, and are found in sheets rather than tight clusters. Comparison of the previous plates will illustrate that these diagnostic points overlap considerably. It is often impossible to distinguish oat cells from other undifferentiated carcinoma cells, and if the malignant cells do not classically fit the criteria of oat cell carcinoma, they should be classified as undifferentiated.

Difficulties in Interpretation: Both adeno- and undifferentiated carcinoma cells should be readily identified as malignant. They are hyperchromatic, have an irregular nuclear pattern, sharp nuclear borders and vary considerably in size and shape. Those of adenocarcinoma often show the characteristic vacuolization. On the other hand, oat cells may be mistaken for the free nuclei of columnar epithelium if care is not taken. The distinction is almost entirely on nuclear pattern. The chromatin of the nucleus of oat cell carcinoma is irregular in arrangement, though there may be no increase in amount. Size variation may be a helpful point. In a group of degenerate columnar cells usually an occasional cell at the periphery of the group will retain the characteristic columnar pattern. Oat cell carcinoma nuclei will appear free of cytoplasm. With experience differentiation between the benign columnar cell and the oat cell can be made accurately.

General Criteria: Adenocarcinoma of the lung: Cells appear in tight clusters. Cytoplasm is usually present as an indistinct background and is vacuolated. If cellular borders are present, nuclei are eccentric in unevenly stained cytoplasm. Nuclei present an increase in chromatin, irregular chromatin pattern, sharp nuclear borders and variation in size and shape.

Oat cell carcinoma: Discrete, small, oval or round nuclei showing some variation in size but little in shape, occurring in large sheets. Cytoplasm is absent. Hyperchromasia of the nuclei varies from seemingly normal content to a definite increase. Nuclear pattern is irregular and chromatin is deposited at periphery of nucleus.

Undifferentiated carcinoma: Cells may occur either in tight groups or singly. Cellular borders are absent and cytoplasm either appears as a faint background or is entirely absent. Nuclei are hyperchromatic, irregular in size and shape. Chromatin arrangement is abnormal and nuclear borders are very sharp.

SUMMARY

In the examination of secretions from the respiratory tract, the student should remember that not only may there be malignant cells from a primary carcinoma of the lung, but there may be cells which are desquamated from pulmonary metastases. The metastatic cells retain the definite characteristics of the original tumor. In the cases seen in our laboratory the metastatic tumor has usually been either squamous or adenocarcinoma. (See Fig. 111.) One case of pulmonary metastases from a transitional cell carcinoma of the nasopharynx contained undifferentiated malignant cells in the sputum.

These cells had no distinguishing characteristics cytologically. However, the possible occurrence of malignant cells with the definite morphologic peculiarities of the parent tumor should always be considered in the interpretation of any unusual cell groups.

It has been our experience that smears of sputum and bronchial secretions present a more decisive picture than is true in other secretions. Cells which cause difficulty in interpretation are infrequent and these can be classified with fair assurance after experience. Benign cells are only rarely misjudged as malignant, as evidenced by the fact that there has not been a false positive report in our laboratory in sputum or bronchial secretion in over two years.

As in vaginal preparations, smears may occasionally be called negative erroneously. The error may be that no malignant cells were present, that they were present but not seen by the observer, or that they were seen but misinterpreted as benign. It is of course essential that all fields of the slide be covered methodically to avoid the second error. Misinterpretation of malignant cells as benign is extremely rare, since the malignant cells present such distinctive morphologic differences that they are usually readily identified. Absence of malignant cells represents the error inherent in the method.

It should be emphasized that attempts at a diagnosis should not be made if the preparation is not entirely adequate. We have already mentioned our criteria for satisfactory specimens of sputum. Histiocytes must be present to be certain that the secretion is one of sputum and not just saliva. In bronchial secretions it is somewhat more difficult to judge the adequacy of the specimen. We have preferred to call specimens of bronchial secretions unsatisfactory if there were very few or no columnar epithelial cells present. Specimens containing only blood and leukocytes have not been considered satisfactory for interpretation. It has been our experience that specimens of sputum have been satisfactory more consistently for diagnostic purposes than those from the bronchial aspirations. This is simply a technical factor, since the personnel of an operating room fluctuates and the smears have not been prepared in any consistent manner. We believe that examination of the sputum has the following advantages: The laboratory can exercise more rigid control over the preparation of the smear. Repeat specimens of sputum can be obtained with greater ease. Finally, patients presenting definite enough symptoms to justify bronchoscopy have too often far advanced disease. In any attempt to find early, operable tumors of the lung, we feel that the cytologic examination of the sputum offers greater possibilities than that of bronchial secretion. Ideally the combination of smears of both sputum and bronchial secretion will give the best results.

The cytologic examination of sputum and bronchial secretion yields excellent results and is a real addition to the diagnostic armamentarium for carcinoma of the lung. With extensive use more cases of early pulmonary carcinoma will be discovered and ultimately will increase the cure rate of this prevalent disease.

CHAPTER XIV

NORMAL COLUMNAR CELLS FROM GASTRIC MUCOSA

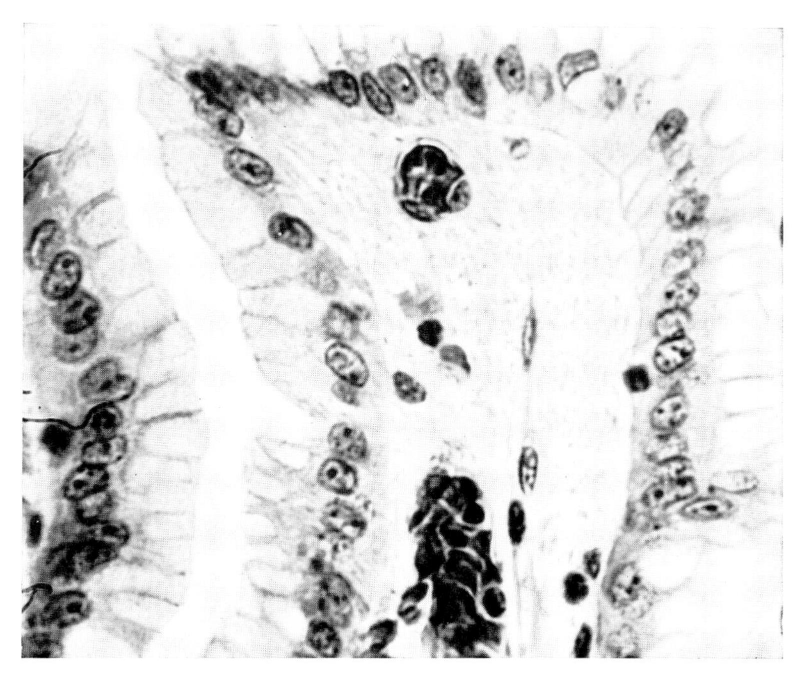

Fig. 118

HISTOLOGIC SECTION: NORMAL GASTRIC MUCOSA

Though the upper portions of the alimentary tract are lined by squamous epithelium at the cardia of the stomach, the epithelium changes abruptly to the columnar type. The histologic preparation above illustrates the typical columnar pattern of the gastric mucosa. The nuclei are fairly even in size, have a round or oval shape and are eccentric in position. The cytoplasm exhibits secreting vacuoles.

In no other secretion does degeneration take place as rapidly as in the gastric juice. Cytolysis of the cells is common because of the proteolytic enzymes present. The cells must be fixed very quickly so that cell detail will remain. (See chapter on technic.) As noted previously in the discussion of the columnar epithelium of the endocervix and bronchi, this type of epithelial cells is apt to exhibit rapid cytoplasmic degeneration under the best of circumstances. The student should become thoroughly familiar with the various stages of preservation of the normal desquamated columnar cells from gastric mucosa. The shape of the preserved cells is rectangular and in general appears somewhat wider and shorter than those desquamated from respiratory epithelium. The nuclear material is vesicular, usually with one or more small condensations of chromatin. The nuclei show little variation in size or shape.

DESCRIPTION OF COLUMNAR CELLS IN GASTRIC SECRETION

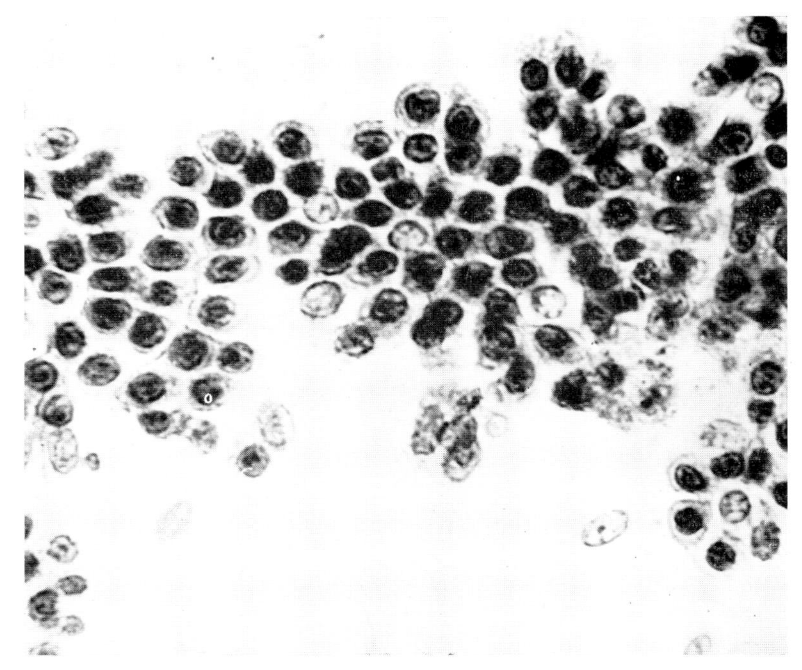

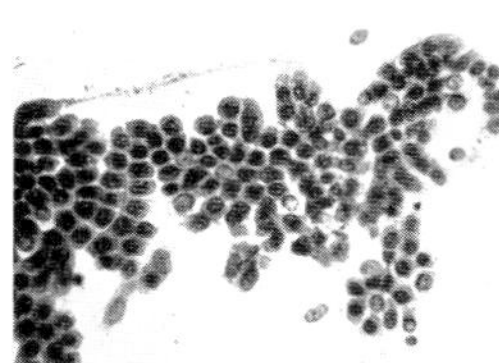

Low Power: A large sheet of evenly spaced cells.
High Power:

A. Characteristics of Nucleus: 1. Clear, well defined nuclear borders. 2. All nuclei are smoothly granular, showing little or no variation in chromatin structure. 3. There is more variation in shape than in size, the various shapes being oval, round or triangular. (See group 3.) In group 1 the small variation in size may be seen. 4. Group 4 shows three overlapping nuclei, but the majority of the nuclei are separated from each other by a rim of cytoplasm. 5. The nuclei are usually eccentric in relation to the cytoplasm. (See group 2.)

B. Characteristics of Cytoplasm: 1. The cellular borders depend on the preservation of the cell: clear, when fresh, indistinct when degenerate. 2. In group 2 the cytoplasm is square in shape, since the cells are seen from the side. In groups 3 and 5 the cells are seen from on top and the cellular borders give a honeycomb appearance to the cytoplasm. 3. The amount of cytoplasm depends on the state of preservation of the cell. In group 4 the cytoplasm has entirely disappeared. 4. Staining reaction: bluish purple nuclei with acidophilic or basophilic cytoplasm.

C. General Characteristics of Group: Many nuclei with little variation in shape, size or chromatin structure, separated from each other by rims of cytoplasm.

PLATE 24

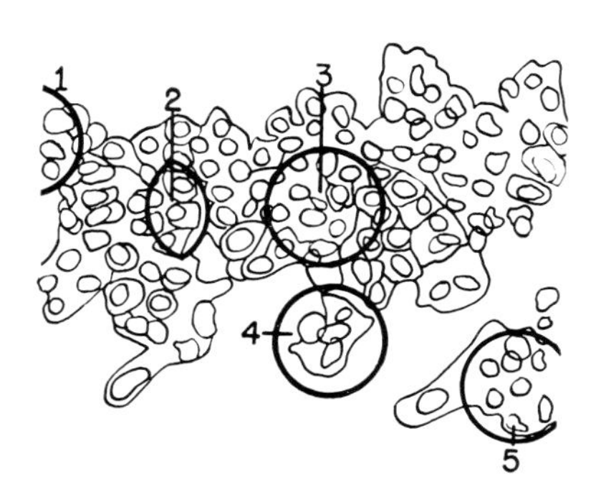

KEY TO COLUMNAR CELLS IN GASTRIC SECRETION

1. Three extruded columnar nuclei, varying slightly in size with no cytoplasm and similar chromatin structure.
2. Group of three columnar cells lined up side by side with small round nuclei eccentrically placed in square-shaped cytoplasm.
3. Collection of columnar cells seen from on top. The nuclei are evenly spaced and are surrounded by small rims of cytoplasm.
4. Three overlapping columnar nuclei with identical nuclear content.
5. A rosette formation of columnar cells as viewed from on top.

Fig. 119

GASTRIC COLUMNAR CELLS: WELL PRESERVED

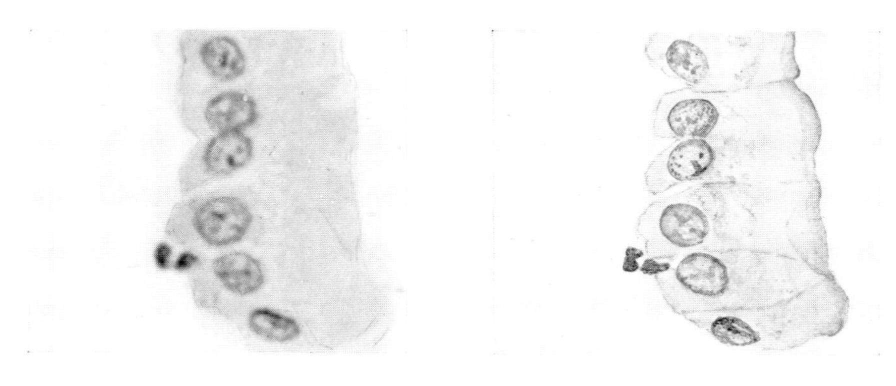

Low Power: Six well preserved cells evenly lined up in a clear field.
High Power: Careful examination reveals individual cellular borders and thickening of the cell wall at the broad end of the cytoplasm. All of the nuclei are arranged evenly in a unipolar position and are uniform in size, shape and chromatin content. Compare these columnar cells to those in the histologic section in Fig. 118.

Fig. 120

GASTRIC COLUMNAR CELLS: DEGENERATION OF THE NUCLEI

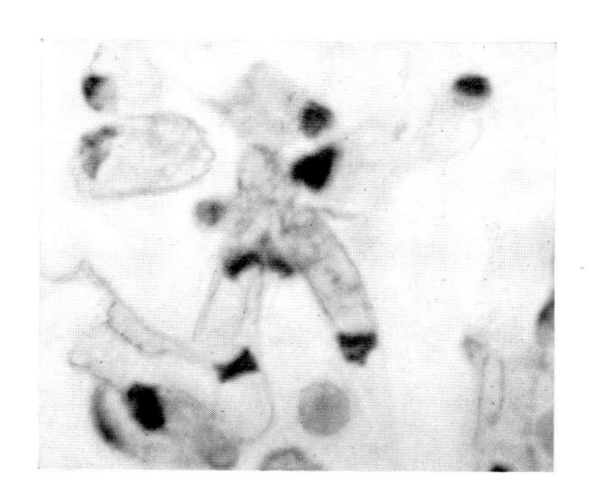

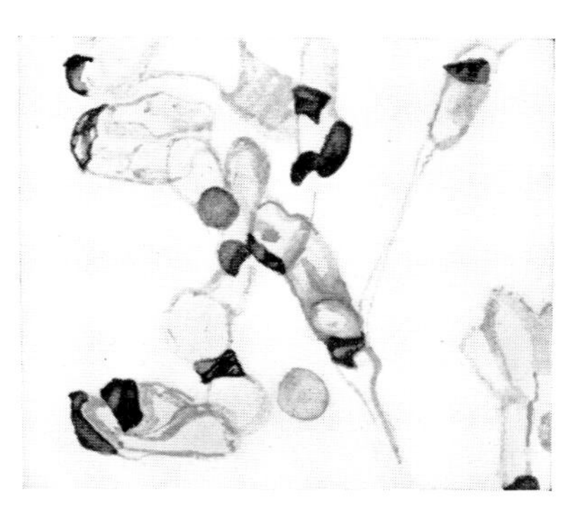

Low Power: Field of elongated cells, each with an eccentric dark nucleus.
High Power: The typical long shape of columnar cells is evident even though the cytoplasm is extremely irregular. The cytoplasm of each cell shows vacuoles, a sign of degeneration. The nuclei are located at one end of the cell and are dense and pyknotic in appearance, with irregular nuclear borders.

FIG. 121

GASTRIC COLUMNAR CELLS: LOSS OF CELLULAR BORDERS

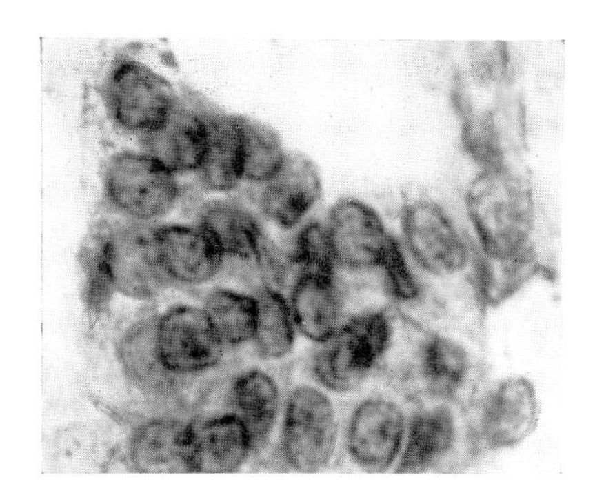

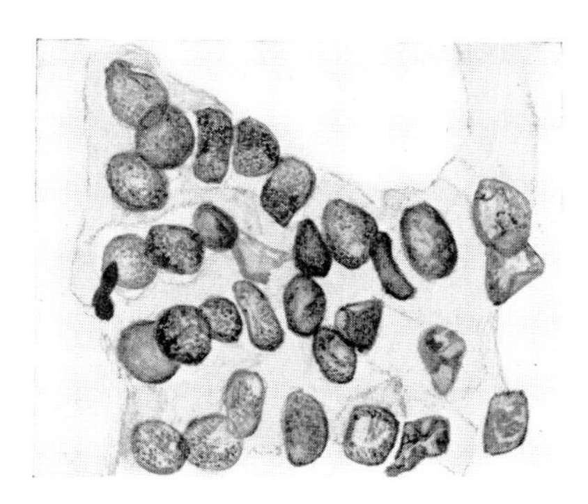

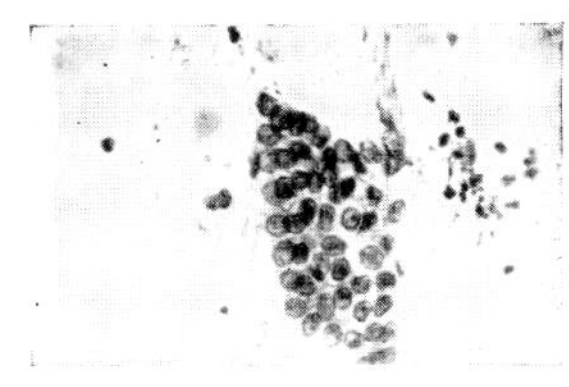

Low Power: Rectangular collection of uniform nuclei.
High Power: Individual cellular borders are lost, as the cytoplasm of columnar cells degenerates more rapidly than the nuclei. However, the upper portion of the group still retains slight palisade formation. The remainder of the cells, seen from on top, show only slight variation in size and shape. All of the nuclei are similar in chromatin structure.

FIG. 122

GASTRIC COLUMNAR CELLS: COMPLETE LOSS OF CYTOPLASM

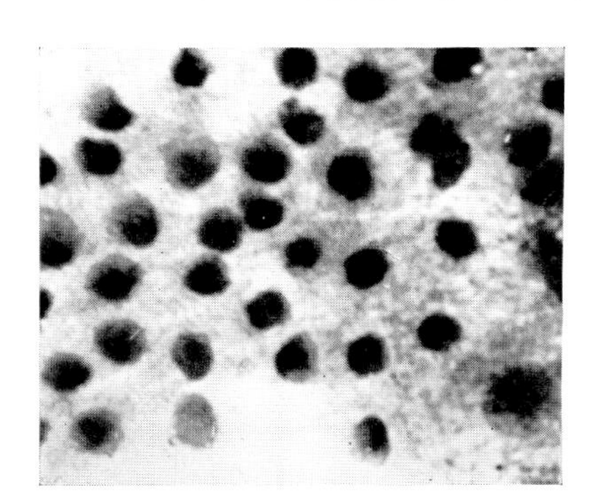

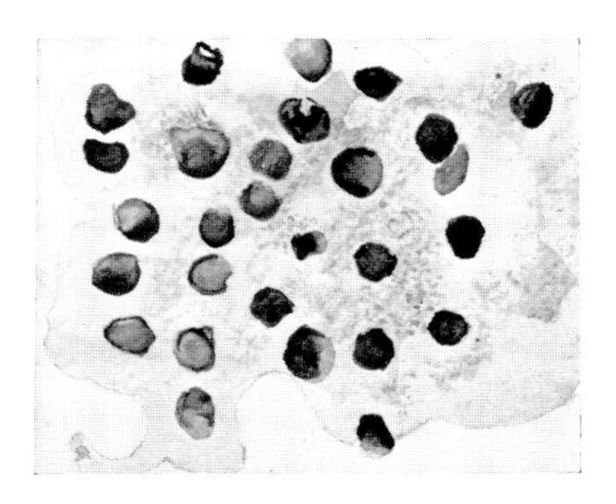

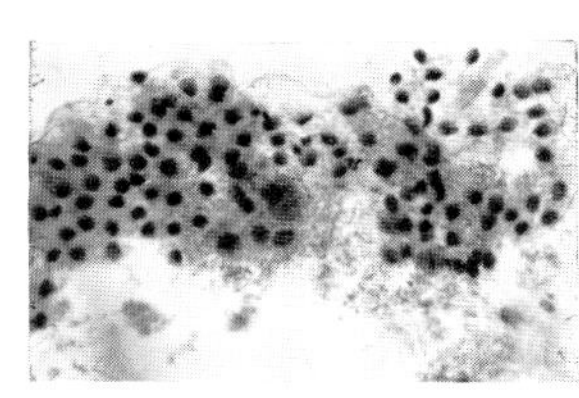

Low Power: Many hyperchromatic nuclei in a field of blood pigment.
High Power: The columnar cells in this group are more degenerate than those in the above figure. The cytoplasm has completely disappeared, leaving only the extruded nuclei, which are smooth and pyknotic in appearance. All of the nuclei are spaced evenly throughout the field and there is almost no variation in their size or shape.

Fig. 123

VARIATION: COLUMNAR CELLS FROM RESPIRATORY TRACT

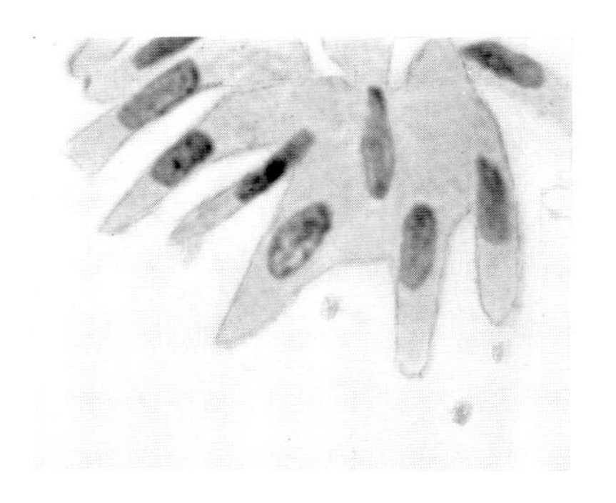

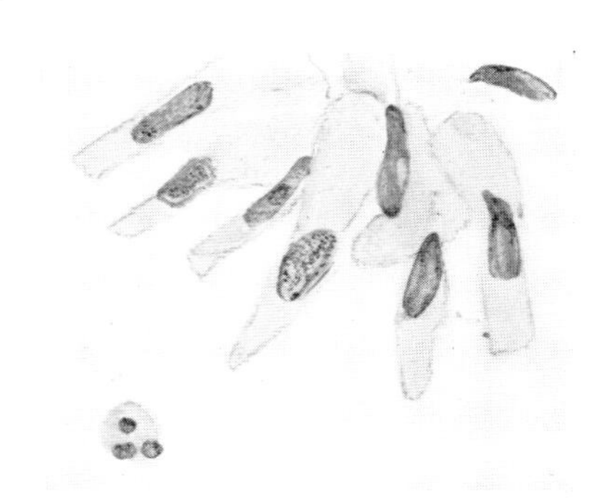

Low Power: Umbrella-like group of slender cells with oval nuclei.

High Power: We feel that these cells do not originate in the gastric mucosa but instead represent columnar cells from the respiratory tract. The cytoplasm and the nuclei are both elongated as compared to the rectangular ones from the gastric mucosa. (See Fig. 119.) Notice that the nuclei are all smoothly granular and more central in location.

* * * * *

The great majority of the normal complement of cells seen in gastric secretion do not desquamate from the gastric mucosa. The stomach acts as a reservoir for cells which are shed from the upper alimentary and respiratory tracts, and subsequently swallowed. The most numerous cells are the large, flat, squamous epithelial cells. (See Plate 16.) Columnar cells (Fig. 123), which occasionally may be well enough preserved to show cilia, indicating their origin is not from gastric mucosa, are also seen. Actually the great difficulty in the technic of gastric aspiration is to obtain cells originating from the gastric mucosa. They are always few in numbers and it is not uncommon for them to be absent entirely.

Preparations of gastric secretion contain cells of non-epithelial origin. The typical smear contains numbers of leukocytes, both lymphocytes and polymorphonuclears. Plasma cells are encountered fairly frequently. We have not been able to correlate the presence of large numbers of leukocytes with any specific pathologic entity. Blood may be present and is regarded as extremely suspicious. Fresh red blood cells are sometimes seen, but generally the blood is present either in the form of pigment or fibrin.

Histiocytes may be of two kinds. The typical small histiocytes with phagocytosed carbon particles are the most common. Small histiocytes are encountered which are similar in all respects to those seen in the vaginal secretion. They have a round or oval vesicular nucleus which is eccentrically placed in finely vacuolated cytoplasm. The foreign body giant cell is seen rarely.

It is, of course, essential that a fasting specimen be obtained, in order that the secretion will be clear and uncontaminated by food particles. In cases of obstruction it is sometimes impossible to obtain a specimen in which there is not some residue of food. Plant cells are easily recognized by their refractile cell walls. They appear much thicker than those of animal cells. The cells appear in groups and have a distinctive square shape. Muscle fibers from ingested meat may also be seen, and these retain the typical striated pattern. They usually occur singly.

If too many food particles appear in the smear it is not satisfactory for interpretation and the aspiration should be repeated. Smears are also unsatisfactory if degeneration has progressed until very few whole cells remain. We regard a smear from gastric secretion as not a true picture if the large flat squamous cells show no preservation. If the smear contains only the extruded nuclei of the squamous cells and leukocytes, it is much safer to call it unsatisfactory and to obtain another specimen.

Difficulties in Interpretation: The normal columnar cells desquamated from the gastric mucosa are not apt to be misinterpreted. They usually occur in groups, vary little in size and shape and have finely vesicular chromatin. The nuclear borders are well defined, but there is no condensation of chromatin at the periphery. The nuclear material has a smooth granular appearance. One or more condensations of chromatin are often present. Occasionally these normal columnar cells will show size variation. Only then will they be at all confusing. Absence of cellular borders and variation in size leads one to consider malignancy. However, as stated before, variation in size is not the most reliable criterion. The nuclear configuration of the enlarged gastric columnar cell is normal and without real evidence of nuclear abnormality the cells must be regarded as benign, despite differences in nuclear size.

Histiocytes in groups should not be questioned, since in smear of gastric secretion they have in our experience exhibited the definite characteristics of their cell type, *i.e.*, finely granular nuclei eccentrically placed in foamy cytoplasm.

General Characteristics of Columnar Cells from the Gastric Mucosa: Rectangular cells with faintly basophilic cytoplasm which may be vacuolated. Cells appear broader than the columnar cells of the respiratory epithelium. Cellular borders are usually not sharp. Nuclei are eccentric in position, round or oval in shape and vary only slightly in size. Nuclear material is finely granular with one or more small condensations of chromatin.

CHAPTER XV

CARCINOMA OF THE STOMACH

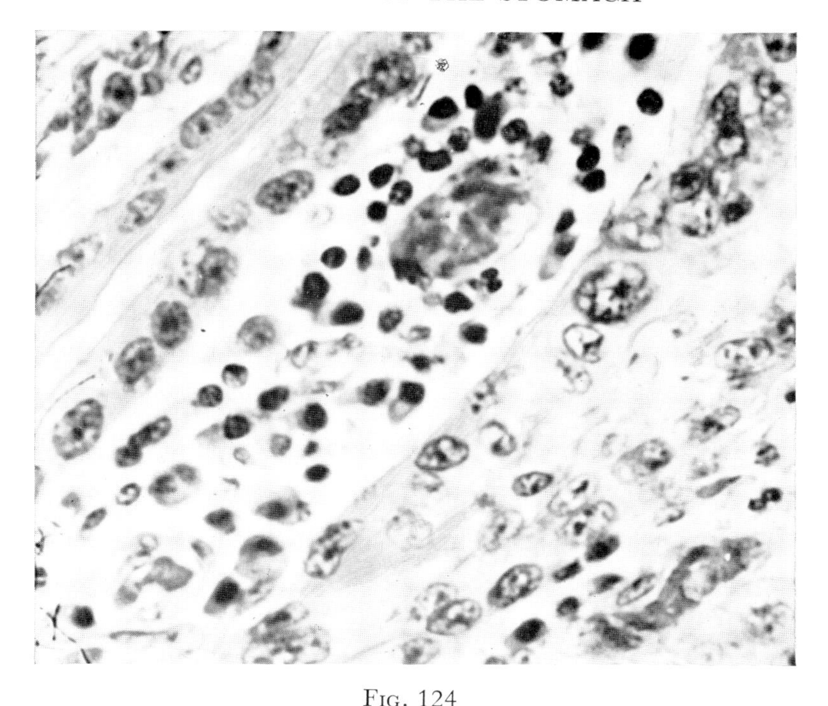

FIG. 124

HISTOLOGIC SECTION: GASTRIC CARCINOMA

In the photomicrograph above, the normal columnar epithelium of the gastric mucosa is shown on the left in comparison to adenocarcinoma of the stomach on the right. If single nuclei are examined in the histologic preparation, the same characteristics which are present in the smears are observed. Regularity in chromatin pattern and little variation in size or shape are seen in the normal cells on the left, while the malignant nuclei on the right exhibit marked variation in size and extreme irregularity in chromatin arrangement.

The carcinoma cells encountered in gastric secretion are predominantly undifferentiated. Only occasionally are cells found with well preserved cellular borders. The undifferentiated cells show a distinct irregularity in their grouping, not being evenly spaced as those from the normal gastric mucosa. There is considerable overlapping of nuclei. The chromatin has an extremely abnormal arrangement and is in heavy strands and thick clumps instead of finely granular. Nucleoli are often present. Variation in size is prominent, but variation in shape is not marked. Cytoplasm is present as an indistinct background and may have the pouched-out vacuoles so characteristic of adenocarcinoma. Differentiated cells are similar to those encountered elsewhere in adenocarcinoma. Cellular borders are present but are ill defined. Cytoplasm appears irregular rather than smooth and is often vacuolated. The characteristic malignant nucleus is eccentric in position.

DESCRIPTION OF MALIGNANT CELLS IN GASTRIC SECRETION

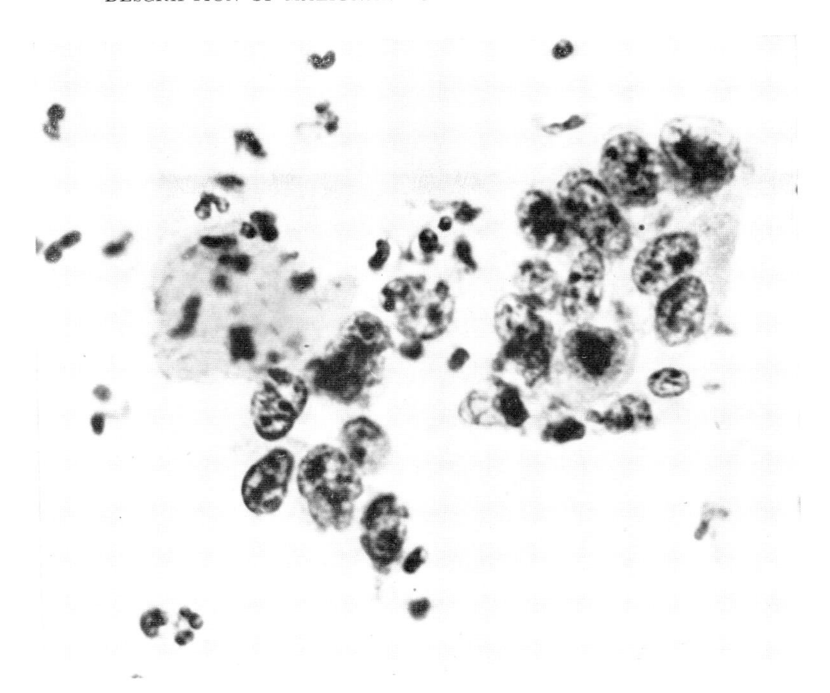

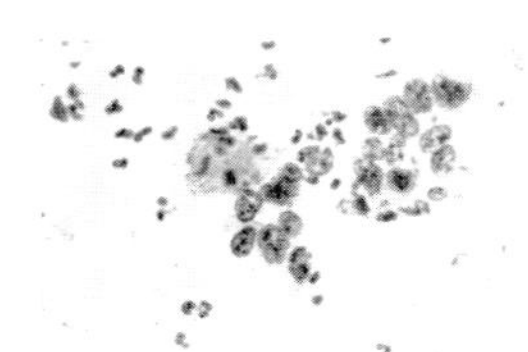

Low Power: Group of granular nuclei in field of leukocytes.

High Power:

A. Characteristics of Nucleus: 1. Sharp nuclear border. Chromatin is condensed at the periphery of the cell (as in group 1). 2. Extreme irregularity of chromatin arrangement. The large condensation and thick strands of chromatin leave spaces in the nuclei containing no nuclear material. (See group 1.) 3. Variation in size is prominent (see group 3), but variation in shape in much less obvious. 4. Degenerative changes are present. Notice the splitting of the nuclear border in upper cell of group 3. 5. There is considerable overlapping of nuclei. Compare groups 2 and 3 with the regular arrangement of normal columnar cells in Plate 24.

B. Characteristics of Cytoplasm: 1. Faint background of cytoplasm. 2. Cellular borders are entirely absent, classifying these cells as undifferentiated malignant. 3. Staining reaction: bluish purple nuclei with faintly staining basophilic cytoplasm.

C. General Characteristics of Group: Cluster of nuclei with sharp nuclear borders and irregular chromatin. There is some overlapping of nuclei and the general configuration of the group is irregular. Nuclei vary markedly in size, little in shape. Cytoplasm is an indistinct background and cellular borders are absent.

PLATE 25

KEY TO MALIGNANT CELLS IN GASTRIC SECRETION

1. Group of malignant nuclei, showing sharp nuclear borders.
2. Overlapping malignant nuclei with irregular chromatin pattern, indistinct cytoplasm.
3. Group of malignant nuclei, showing variation in shape, little in size. Nuclear borders are splitting, a sign of degeneration. Cytoplasm is an indistinct background.
4. Three malignant nuclei with extreme variation in size.

FIG. 125

GASTRIC CARCINOMA: GOOD PRESERVATION OF MALIGNANT CELLS

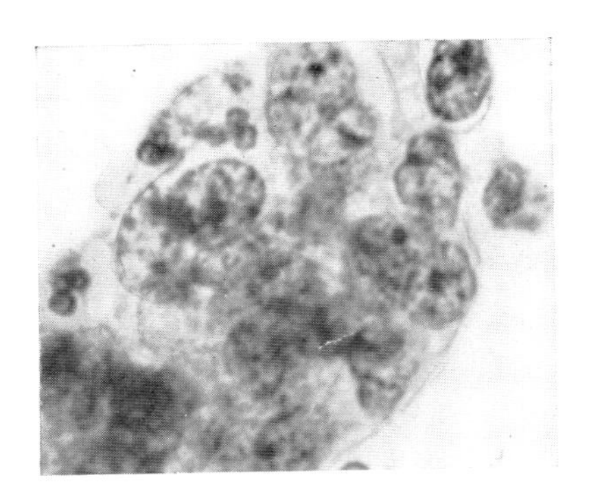

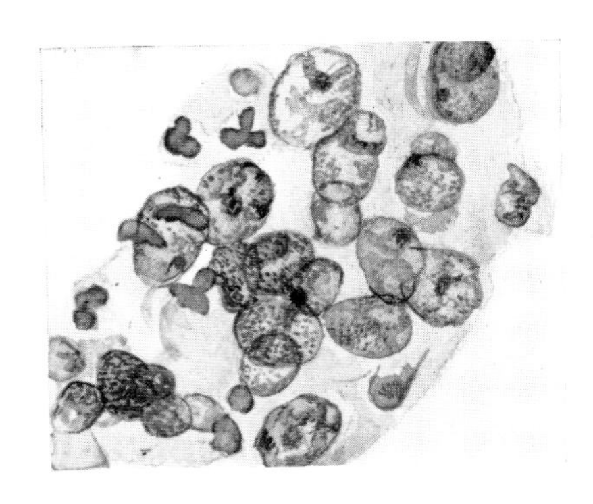

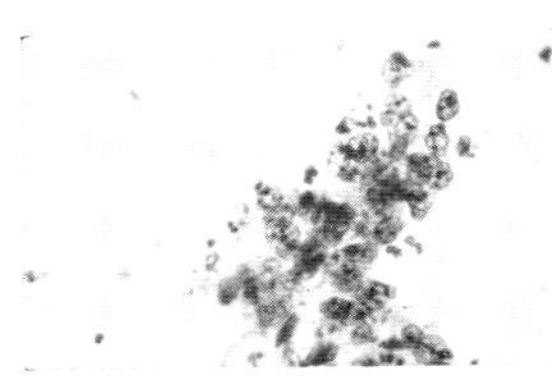

Low Power: Thick, messy group of nuclei on a background of blood.

High Power: These nuclei are well preserved, as individual particles of chromatin are clearly visible. The variation in size more than shape, the overlapping or piling qualities, the lack of cellular borders and presence of good nuclear borders identify this collection of nuclei as being typical of those seen in adenocarcinoma.

FIG. 126

GASTRIC CARCINOMA: LACK OF CHROMATIN STRUCTURE

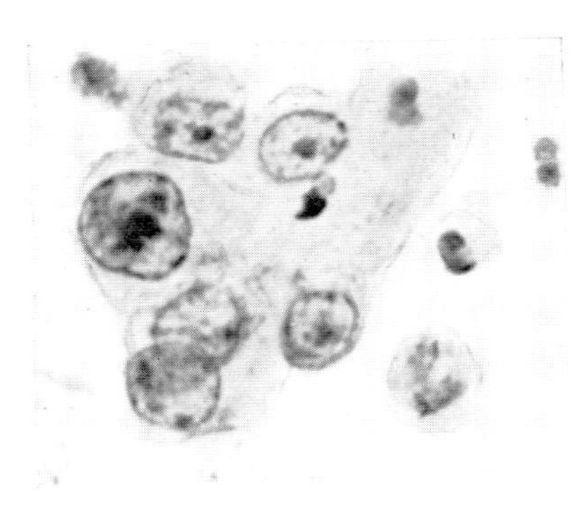

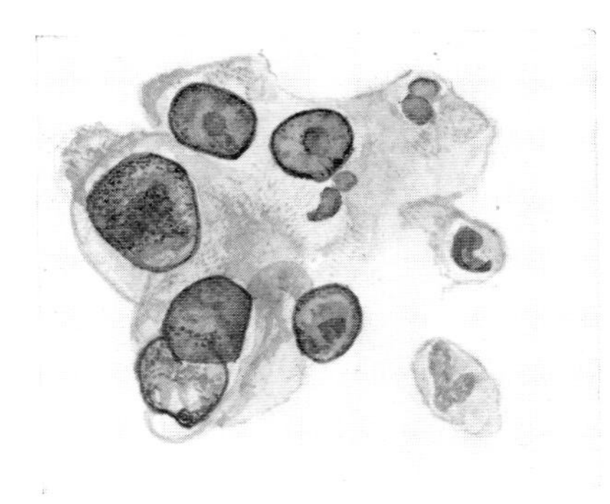

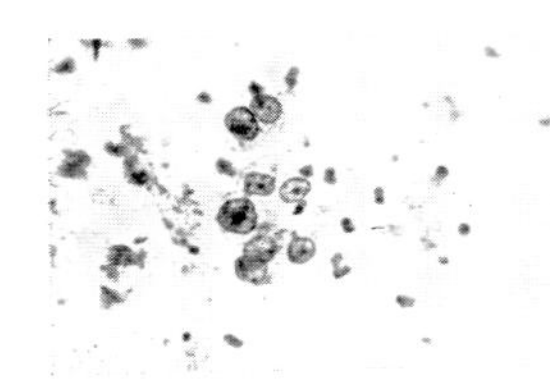

Low Power: Eight large nuclei in a field of leukocytes and mucus.

High Power: Again the cells are well preserved but individual chromatin particles are not as prominent as in the figure above. Rather they have condensed at the periphery of the nucleus, giving sharp nuclear borders. Nucleoli are present in two upper nuclei. Cytoplasm is present as a background, but definite cellular borders are absent.

Fig. 127

GASTRIC CARCINOMA: GLAZED MALIGNANT NUCLEI

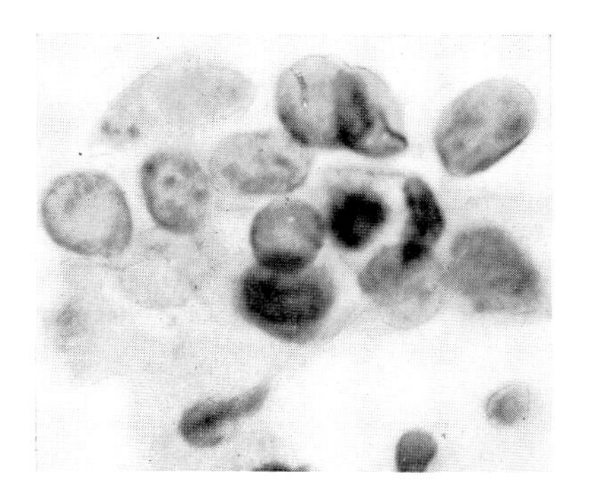

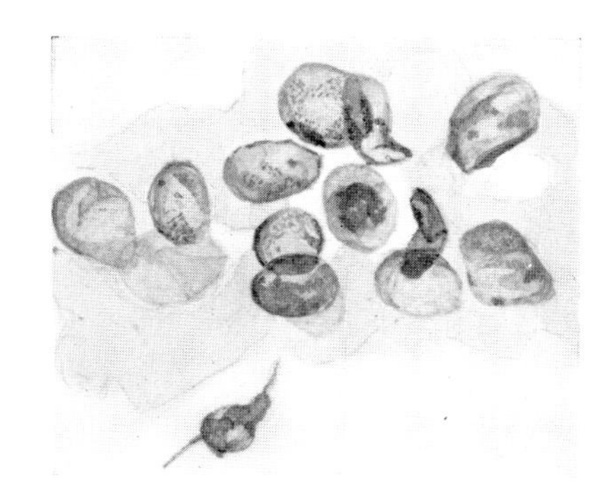

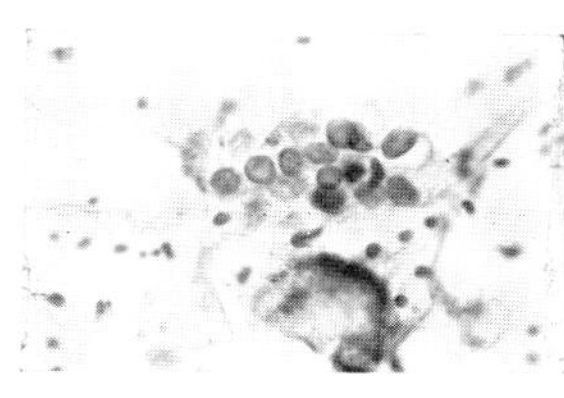

Low Power: Group of large nuclei which are smooth or glazed in appearance.
High Power: The glazed appearance represents degeneration but it is not as obvious under high power. In comparing these malignant nuclei to those in the two previous figures note that the chromatin particles are still irregular but do not appear increased, and that the nuclear borders are not as definite.

Fig. 128

GASTRIC CARCINOMA: MARKED DEGENERATION OF NUCLEI

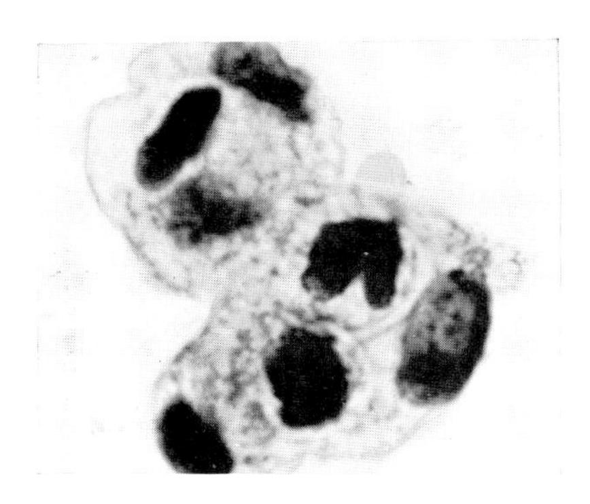

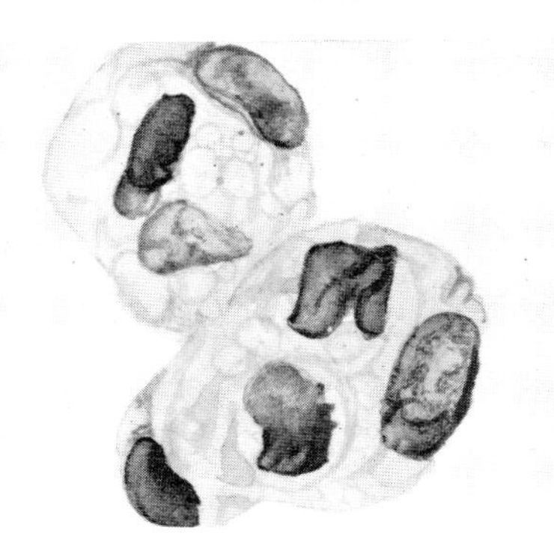

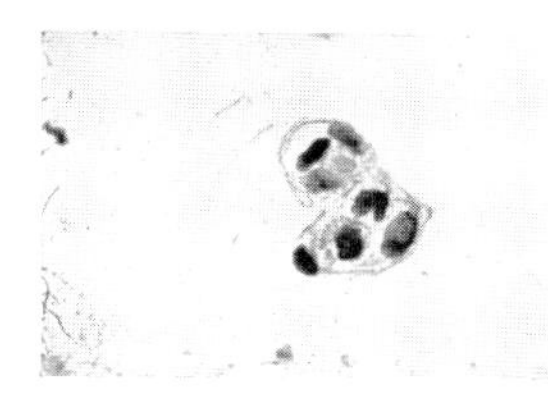

Low Power: Several large hyperchromatic nuclei with cytoplasm, in a clear field.
High Power: In this group the nuclei are even more degenerate than in the figure above. The chromatin particles have conglomerated in one mass, giving the nuclei a smooth but dark-staining surface. The cytoplasm is filled with many small vacuoles which are indicative of adenocarcinoma. Individual cellular borders are not distinguishable.

Fig. 129

GASTRIC SECRETION: CELLS FROM SQUAMOUS CARCINOMA OF THE ESOPHAGUS

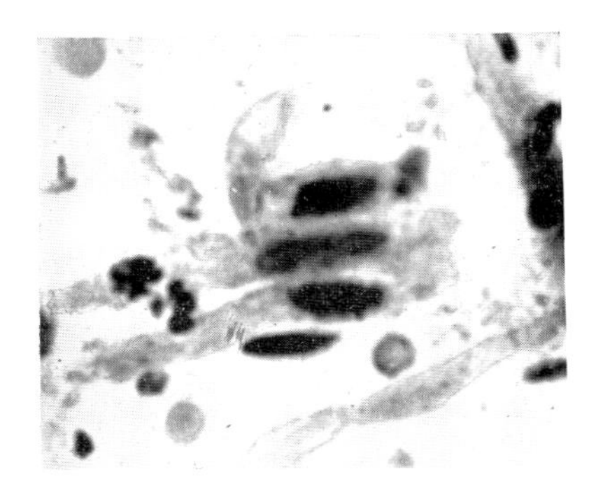

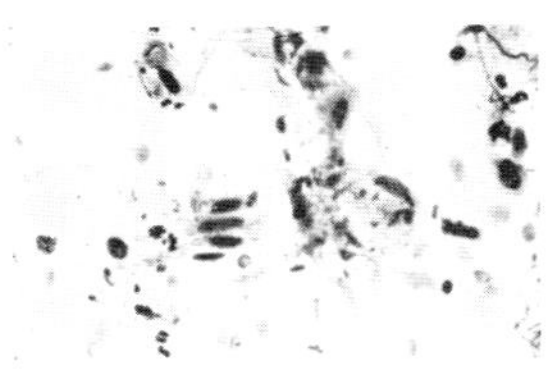

Low Power: Messy field with several large hyperchromatic cells.
High Power: The four elongated cells are typical differentiated fiber cells from squamous carcinoma. They show variations in nuclear structure from the increase of chromatin in the two lower nuclei to the irregularity of pattern in the longer nucleus. The cytoplasm is somewhat degenerate and cellular borders are indistinct. Red blood cells are scattered throughout the field.

* * * * *

We have been the most successful in identifying cells from tumors classified either as undifferentiated carcinoma of the stomach histologically or in those showing acinar arrangement of the cells and classified as adenocarcinoma. There has been an occasional case in which cells desquamated from a mucoid or signet ring carcinoma, but in general we have not found malignant cells in these cases as consistently as in either undifferentiated or adenocarcinoma. When cells have desquamated from a mucoid carcinoma we have not been able to classify them as cytologically distinct from other carcinoma cells encountered in gastric smears. We have been unsuccessful in identifying malignant cells in either the scirrhous type of carcinoma or in sarcoma of the stomach. This would be expected, since in both lesions the mucosa is frequently not extensively invaded by tumor. Cases of lymphoma of the stomach have not been diagnosed correctly by smear in our experience. We have not been able to distinguish any definite morphologic characteristics of the lymphocytes present in smears from patients with that diagnosis. In one case of lymphoma there were large numbers of lymphocytes present in the smear. Whether this fact alone should make one suspect lymphoma is not certain. Many more cases must be studied in order to determine whether large aggregates of lymphocytes should be regarded with suspicion.

As mentioned in the chapter on the normal cells of the gastric secretion, many cells are encountered whose source is not the gastric mucosa. This is true of malignant as well as normal cells. Particularly in carcinoma of the esophagus, malignant cells will desquamate, be carried down and collect in the gastric juice. In the few cases we have had of this type, the carcinoma has been of the squamous variety, which permits easy identification of its source as other than the gastric mucosa. The cells from squamous carcinoma are similar to those described in squamous carcinoma of the bronchi.

The typical positive gastric smear presents a large number of leukocytes, some evidence of blood, well preserved epithelial cells and occasional clusters of malignant cells with a few individual malignant cells. As in sputum specimens, there is often a great deal of mucus, and malignant cells are often found at the border of these strands. Frequently cocci in tetrad formation will be present in positive smears. There is not a specific correlation between the presence of these tetrads and the presence of malignant cells, but it occurs frequently enough to be regarded as a suspicious finding. Smears containing tetrads deserve a thorough search before being called negative.

Gastric smears are reported similar to other secretions: *i.e.*, negative, positive and doubtful. The doubtful smears are repeated and a real attempt made to put them either in the positive or negative category. If the malignant cells encountered have the characteristics of adenocarcinoma, the report reads: "positive, consistent with adenocarcinoma."

In general, the malignant cells seen in gastric secretion are similar to the undifferentiated or adenocarcinoma groups encountered in other secretions. Nucleoli are often prominent, and are large, smooth homogeneous bodies which stain acidophilic. The glazed appearance of malignant cells is rarely seen in any other secretion. (See Fig. 127.) It probably represents a definite form of degeneration. The nuclei have a very smooth dark appearance and chromatin particles are not as prominent as in well preserved malignant cells. However, there is still distinct abnormality in chromatin arrangement. In these glazed cells the nuclear material is usually concentrated at the periphery of the nuclei, producing a very sharp border.

Difficulties in Interpretation: The only normal cell which might be misinterpreted as positive is the columnar from the gastric mucosa. Since degeneration takes place rapidly, these cells seldom present a typical columnar pattern, but appear as nuclei with a faint background of cytoplasm, or cytoplasm may be entirely absent. In most instances these extruded nuclei are regular in size and shape, but if definite variation in size does occur they appear somewhat suspicious. Nuclear structure remains normal, so that even in enlarged nuclei identification of cells as benign presents minimal difficulty. In one instance we observed very small nuclei with distinctly granular chromatin. The nuclear abnormality was thought to be marked enough to justify a positive diagnosis. This was a "false positive" report. In retrospect the nuclei were too small in size with little variation in size and were probably from usually active benign columnar cells. The chromatin particles of the nuclei were extremely coarse instead of finely granulated, as usually seen, but on review they were seen to be arranged in a regular pattern. This example is given to emphasize that malignant nuclei present an irregularity of nuclear material. It occurs in heavy strands, coarse clumps of uneven size, and increased amounts at the periphery of the nucleus.

General Criteria for Identification of Malignant Cells in Gastric Secretion: Cells are of the undifferentiated type primarily. Cytoplam is present only as an indistinct background. Nuclei show irregularity of chromatin pattern, variation in size and shape, and often definite increase in chromatin material. Frequently the cytoplasm may contain the "pouched-out" vacuoles characteristic of adenocarcinoma. When single differentiated cells are encountered, the malignant nucleus is eccentric in ill defined irregular cytoplasm which may contain definite vacuoles.

CHAPTER XVI

NORMAL CELLS OF THE URINARY TRACT

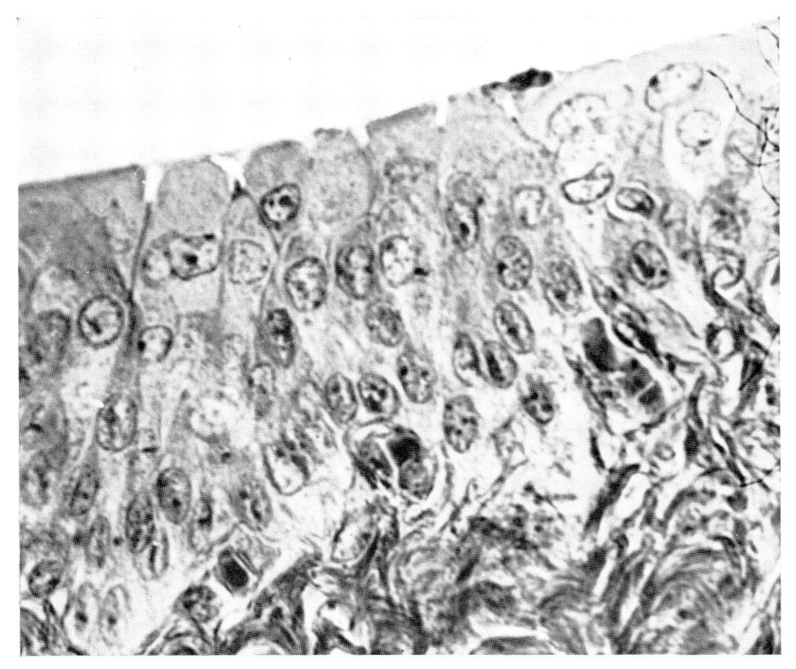

FIG. 130

HISTOLOGIC SECTION: TRANSITIONAL EPITHELIUM OF BLADDER

The transitional epithelium which lines the kidney pelves, ureters, bladder and urethra exhibits more variability in cellular size and shape than is observed in stratified squamous epithelium. As seen in the photomicrograph above of normal bladder epithelium, there are also moderate differences in nuclear size and shape. The chromatin appears in larger aggregates than is usual in normal epithelial cells.

The same peculiarities are seen in the desquamated cells from the epithelium of the urinary tract. The shape of the individual cells differs markedly from small round ones similar to the inner layer basal cell, to elongated forms almost approaching the thin rectangular appearance of columnar cells. Cellular borders are often quite irregular in appearance, as is the density of the cytoplasm. The nuclei of the transitional cells vary considerably in shape from the regular, round, smoothly granular one, to large irregular nuclei with condensations of chromatin. Variation in size is also common. However, the nuclear-cytoplasmic ratio is within normal limits, *i.e.*, the smaller cells have the smaller nuclei. Frequently cells are multinucleated. This is especially true in ureteral specimens where cells with as many as twenty nuclei are sometimes seen. The nuclei of these cells vary somewhat in size and chromatin pattern. Examples of these various cell types are presented on the following pages.

DESCRIPTION OF TRANSITIONAL EPITHELIAL CELLS IN URINE

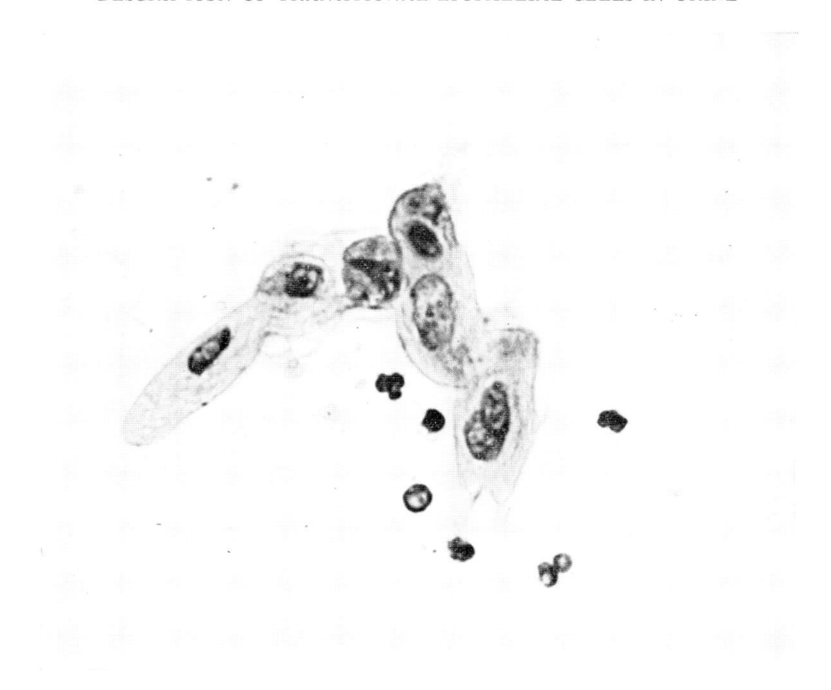

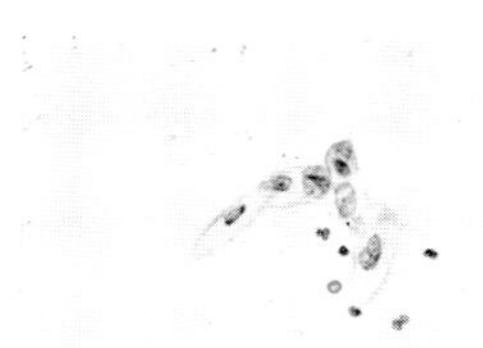

Low Power: Group of six cells with granular nuclei showing differences in size and shape.

High Power:

A. Characteristics of Nucleus: 1. Though nuclear border is well defined, chromatin is not concentrated at the periphery to any considerable extent. Borders show irregularity. (See cells 5 and 6.) 2. Variation in size and shape is a prominent feature. Compare nuclei 2 and 5. Despite the large nucleus of cell 5, the cytoplasmic-nuclear ratio is adequate, since the size of the entire cell is greater than that of nucleus 2. The shape differs from the triangular form (nucleus 3) to the smooth oval shape (nucleus 4). 3. All of the nuclei have essentially identical chromatin structure, *i.e.*, fine granules evenly distributed. Cell 4 is slightly denser than the rest, as the particles of chromatin are more compact. 4. Double nucleated cells are common. (See cell 6.)

B. Characteristics of Cytoplasm: 1. Hazy, uneven cellular borders which are often indistinct. (See cells 1 and 2.) 2. Cytoplasm is not smooth, exhibiting changes in density. 3. The size and shape of the cells correspond fairly closely to size and shape of nuclei. 4. Staining reaction: bluish purple nuclei with acidophilic or basophilic cytoplasm.

C. General Characteristics of Group: Cluster of cells with differences in size and shape of both nuclei and cytoplasm. Nuclei are vesicular in appearance. Cytoplasm varies in density and some cellular borders are indefinite.

PLATE 26

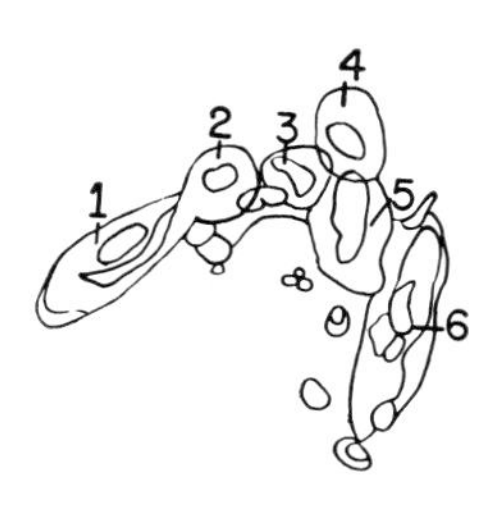

KEY TO TRANSITIONAL EPITHELIAL CELLS IN URINE

1. Elongated transitional cell with an oval nucleus and uneven cytoplasm.
2. Small round epithelial cell with a round vesicular nucleus. The cellular border of this cell is undistinct.
3. Round transitional cell with a triangular-shaped nucleus.
4. Transitional cell with a dense nucleus and a sharp nuclear border.
5. Epithelial cell with a large irregular nucleus showing evenly distributed particles of chromatin.
6. Wrinkled, elongated transitional cell with three vesicular nuclei.

The background is clear except for scattered leukocytes.

Fig. 131

TRANSITIONAL CELLS: INCREASE IN CHROMATIN

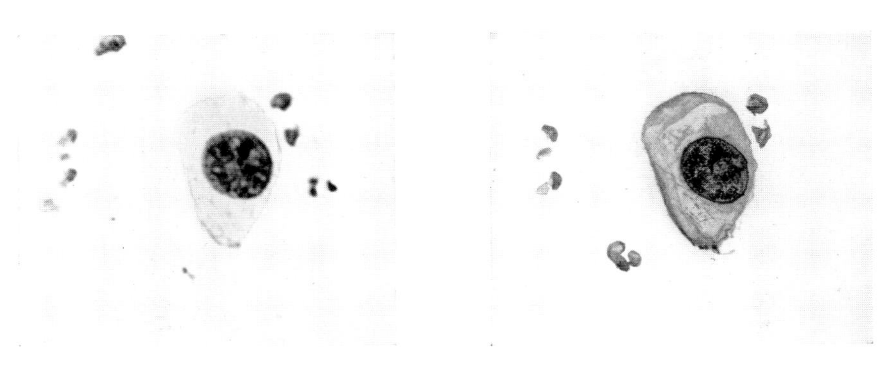

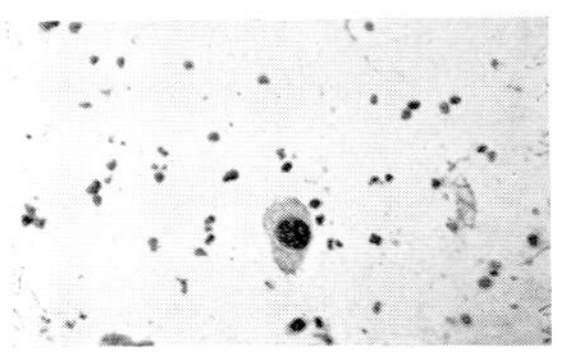

Low Power: Single cell with dark nucleus in field of leukocytes.

High Power: The nucleus of this epithelial cell is extremely hyperchromatic. It illustrates the great variability which transitional cells show. Careful examination of the chromatin content, however, shows that despite its increase the chromatin is evenly granular and that the hyperchromatic appearance is due to an increase of chromatin material.

Fig. 132

TRANSITIONAL CELLS: LARGE NUCLEI

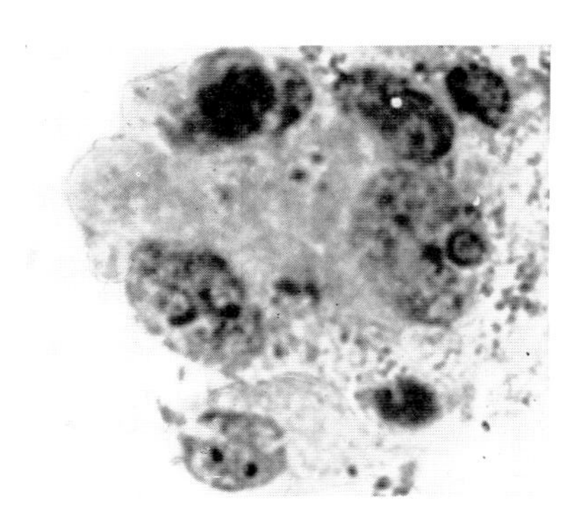

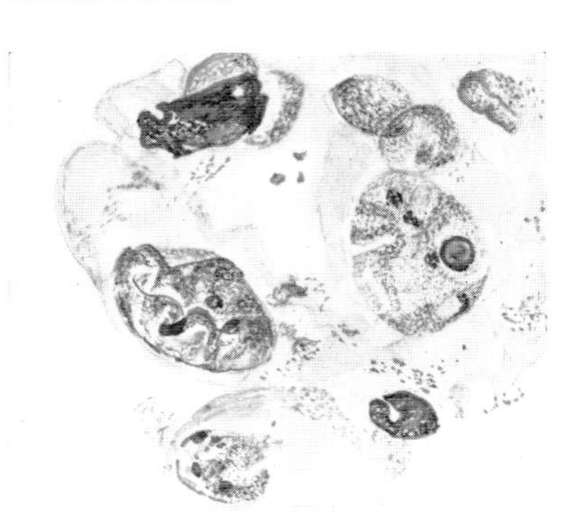

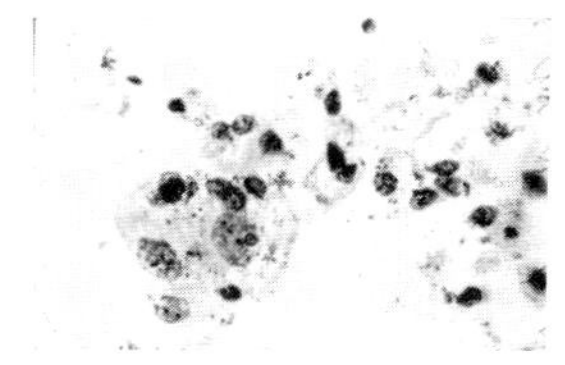

Low Power: Two large granular nuclei in field of dark nuclei.

High Power: The two large granular nuclei illustrate the large size which the nuclei of the cells of the urinary tract may assume. Their chromatin is smoothly granular and they are not suspicious except for size. They are somewhat degenerate, since definite cellular borders are not apparent. Messy background is typical of many urine smears.

FIG. 133

TRANSITIONAL CELLS: MULTINUCLEATED

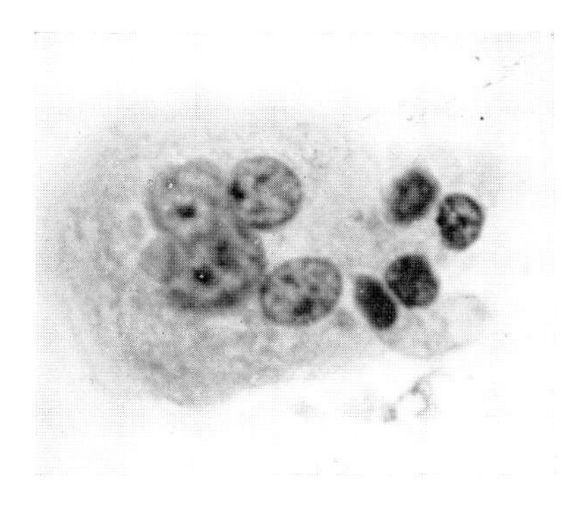

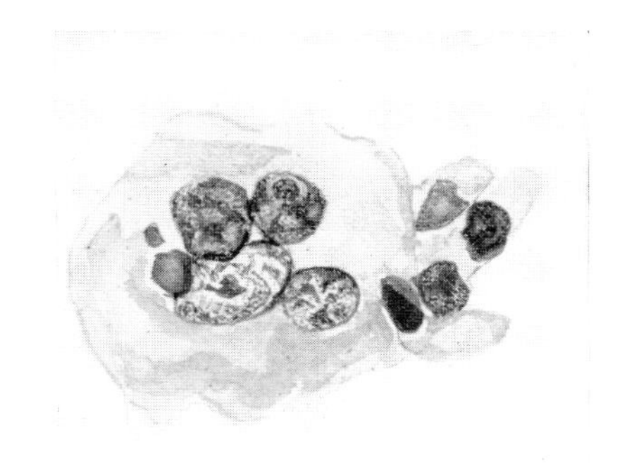

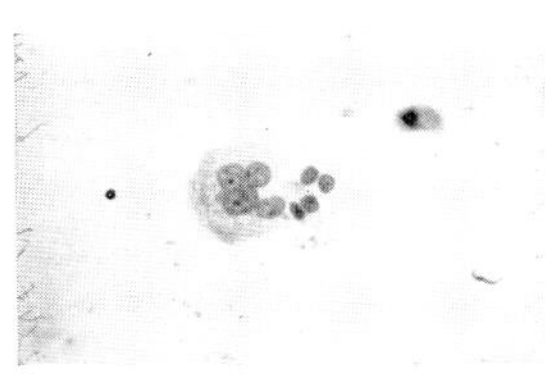

Low Power: Epithelial cell with four nuclei in clean field.

High Power: This cell illustrates a type of multinucleated cell occurring in urine. The nuclei vary somewhat in size, little in shape, and the chromatin structure in all is similar. It should be emphasized that the occurrence of multinucleated benign cells in urine specimens is extremely common, and is not indicative of malignancy.

FIG. 134

TRANSITIONAL CELLS: MULTINUCLEATED

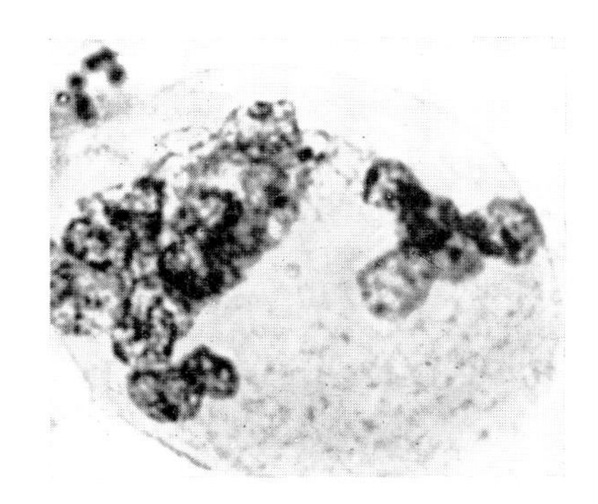

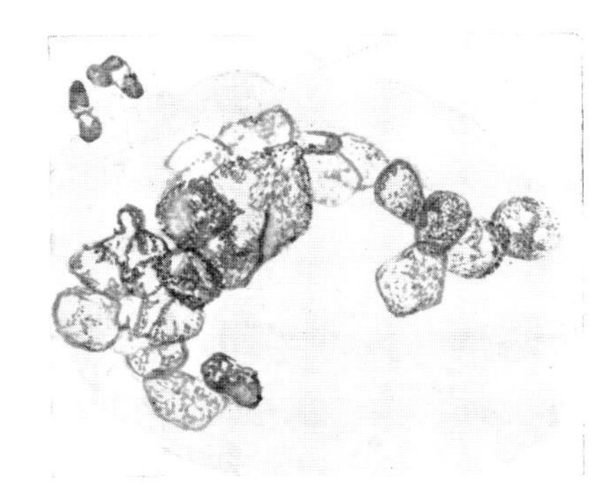

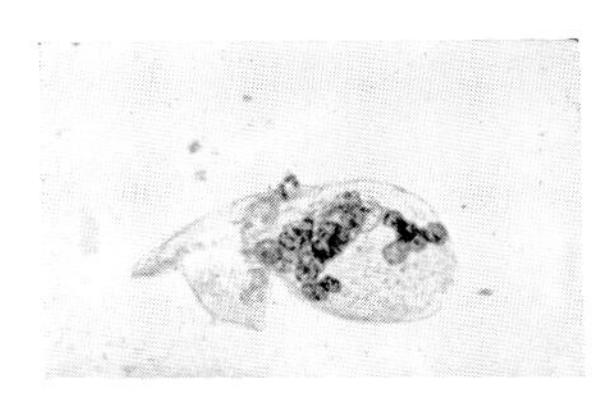

Low Power: Enlarged multinucleated cell.

High Power: Large multinucleated cells of this type occur fairly often in ureteral specimens. The cells are very large and have many even granular nuclei. Whether they are actually epithelial or histiocytic in origin is questionable. Their cellular border is quite distinct, which indicates an epithelial origin, but the arrangement of the nuclei suggests a foreign body giant cell.

The sediment of urine often contains no epithelial cells, and even when they are present they are seldom numerous. The typical specimen examined for the presence of malignant cells contains many leukocytes, primarily polymorphonuclears, occasional small histiocytes and scattered epithelial cells. In females there are often typical flat superficial squamous cells present. Blood may be observed either in the form of preserved red cells, fibrin or pigment. The foreign body giant cell is seen infrequently.

Since urine does not contain the mucus which is present in the vaginal and gastric secretion or sputum, a definite technical difficulty arises in the preparation of smears. There is no protein to act as a cohesive agent to fix the sediment on the glass slide. Therefore, often the entire sediment washes off during the process of staining. The use of egg albumin is helpful but still occasionally it is hard to retain all the sediment on the slide. Great care should be taken in the preparation and staining of smears from urine sediment in order to gain maximum effectiveness. (See chapter on technic.)

Difficulties in Interpretation: The transitional cells which line the urinary tract present more variation than any of the other types of epithelial cells described. To illustrate that lack of uniformity of these cells leads to real difficulty, the false positive diagnoses in urine smears in our laboratory have consistently been higher than in any other kind of fluid. In the past we have erroneously interpreted the unusual differences seen in comparing cell to cell as indicative of malignancy. In our opinion these differences are as follows: 1. Less uniformity in shape and size of both nucleus and cell. In Fig. 130, illustrating the normal transitional epithelium of the bladder, this variation may be seen in the normal histologic section. Differences in shape of cells may be explained on the basis that the epithelium of the bladder may be contracted or expanded depending on whether the bladder is full or empty. Both cell and nucleus may occasionally reach large proportions (see Fig. 132), and it should be emphasized that total size of cells is of no consequence in studying sediment of urine. 2. Less uniformity of chromatin structure. It must be realized that the nuclear pattern of the transitional epithelial cell is not as consistently regular in appearance as in other epithelial cells. There may be distinct strands of chromatin and occasional small clumps. These irregularities are not marked enough to be indicative of malignancy. Condensation of the chromatin at the periphery of the nucleus is most unusual. It is important to mention that occasionally in the enlarged nuclei, wrinkling of the surface of the nucleus suggests more chromatin aberration than is actually present. (See Fig. 132.) There may be an increase in the amount of chromatin (see Fig. 131), but the nucleus in this instance retains a regular pattern. 3. Presence of multinucleated cells: Epithelial cells are seen frequently with many nuclei. These occur more often in ureteral specimens of urine than in the voided specimens. The nuclei do not vary in chromatin content or arrangement. Numbers of nuclei in one cell are not indicative of malignancy.

General Criteria: Shape of cell varies from triangular to small round forms. Size of cells and nuclei is not constant, but nuclear-cytoplasmic ratio is within normal limits. Chromatin varies from a finely granular appearance to one showing distinct strands of chromatin.

CHAPTER XVII

CARCINOMA OF THE GENITO-URINARY TRACT

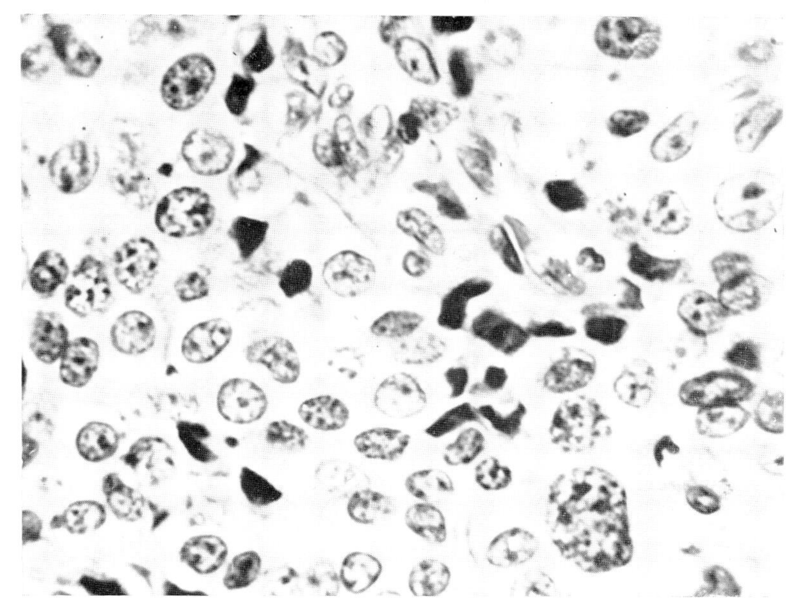

FIG. 135

HISTOLOGIC SECTION: CARCINOMA OF THE BLADDER

The section of bladder carcinoma photographed above shows marked abnormalities of nuclear structure, size and shape. However, such distinctive cytologic changes are not always present. Often the nuclei will show only a mild degree of abnormality and the histologic diagnosis of malignancy must take into consideration many factors other than the cellular abnormalities, *e.g.*, loss of normal configuration, invasion of the basement membrane and the presence of tumor cells in blood vessels.

Because not all bladder carcinomas present distinctive cytologic changes, the diagnosis of carcinoma of the bladder on the basis of desquamated cells is often difficult. We have preferred not to loosen our criteria in an attempt to include those cells which show only slight variance from the normal, but to require very distinct abnormalities of nuclear structure before diagnosing any cell in urine as malignant. It is obvious that in doing this we will miss some carcinomas but the false positive diagnoses should be kept to a minimum.

The malignant cells observed in urine are primarily undifferentiated. Cytoplasm is an indistinct background and cellular borders are absent. Nuclei present the characteristic appearance of malignancy. Occasionally differentiated cells will be encountered and these are like the third type differentiated cells or less often like the differentiated cells of adenocarcinoma.

DESCRIPTION OF UNDIFFERENTIATED MALIGNANT CELLS FROM BLADDER CARCINOMA

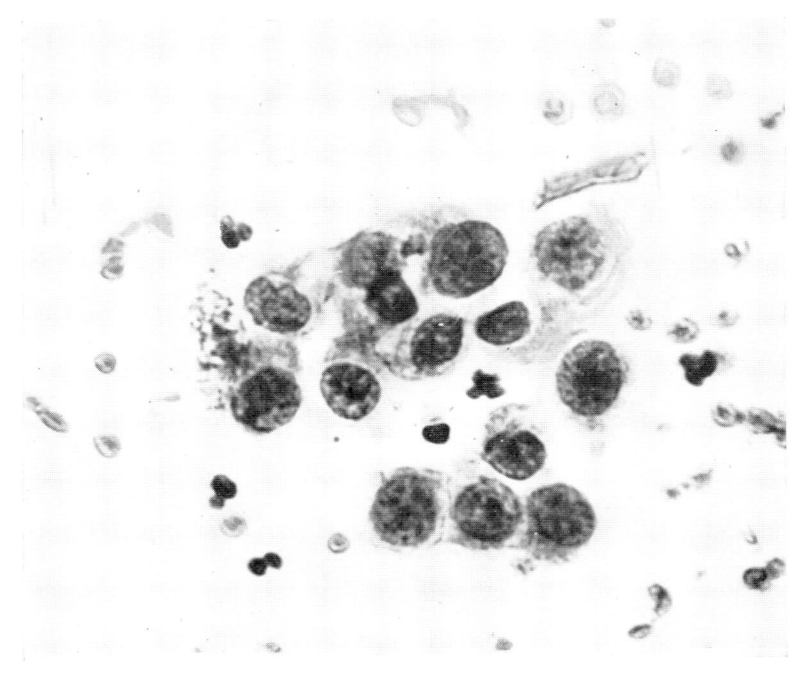

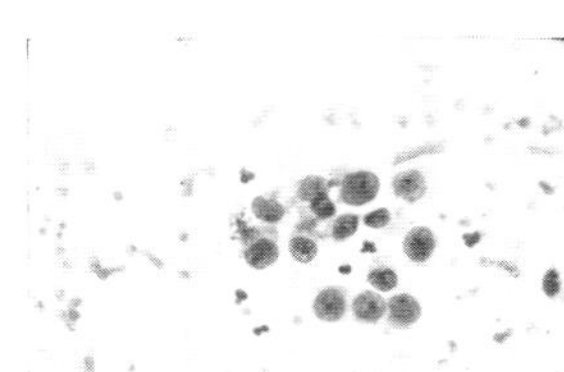

Low Power: Group of hyperchromatic nuclei without cytoplasm, showing variation in size and shape.

High Power:

A. Characteristics of Nucleus: 1. Chromatin is condensed at the periphery of the nucleus. (See cells 1 and 5.) 2. All nuclei show irregularity of nuclear material. The chromatin is in heavy clumps or thick strands. (See cells 4 and 5.) 3. The amount of chromatin present in the nuclei varies. Compared to cell 2, cell 1 is much more hyperchromatic. 4. Nuclei show marked variation in size. Compare the size of the largest nucleus in the field (cell 1) with that of the smallest (cell 6). 5. Shape of the nuclei is not constant, though there is less variation in shape than in size.

B. Characteristics of Cytoplasm: There is no definite cytoplasm present. It is merely an indistinct background. No cellular borders are apparent, classifying these cells as typical undifferentiated malignant cells. Staining reaction: bluish purple nuclei of varying intensity of color with indistinct lightly staining cytoplasm.

C. General Characteristics of Group: Nuclei showing marked variation in chromatin content and nuclear pattern. There is more variation in size than in shape. Cytoplasm is an indistinct background.

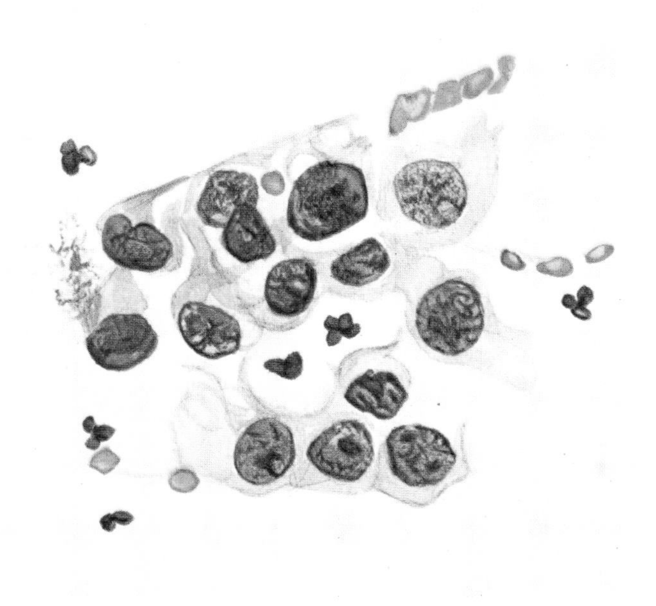

PLATE 27

KEY TO UNDIFFERENTIATED MALIGNANT CELLS FROM BLADDER CARCINOMA

1. Large undifferentiated malignant nucleus with prominent clump of chromatin and sharp nuclear border.
2. Malignant nucleus showing less chromatin than others in the field.

3 and 4. Undifferentiated nuclei with irregular chromatin pattern.

5. Smaller nucleus with sharp border and irregular chromatin arrangement.
6. Small nucleus with increase and irregularity of chromatin.

   All cells in group are undifferentiated malignant. Background of field is composed of red blood cells and leukocytes.

FIG. 136

BLADDER CARCINOMA: LARGE MALIGNANT NUCLEI

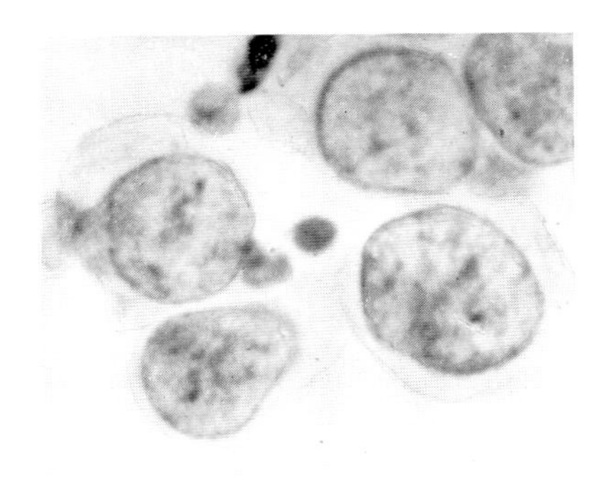

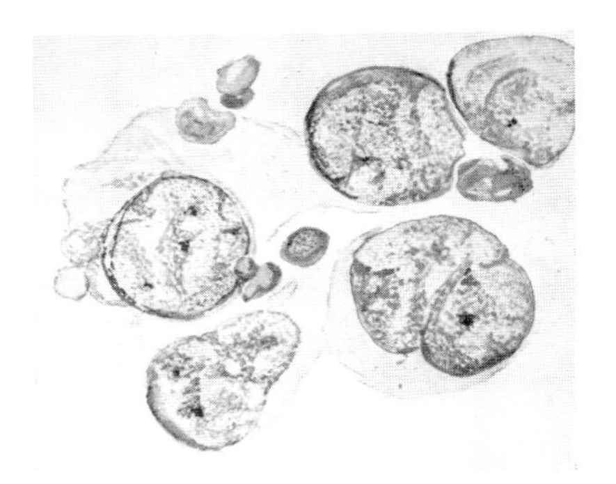

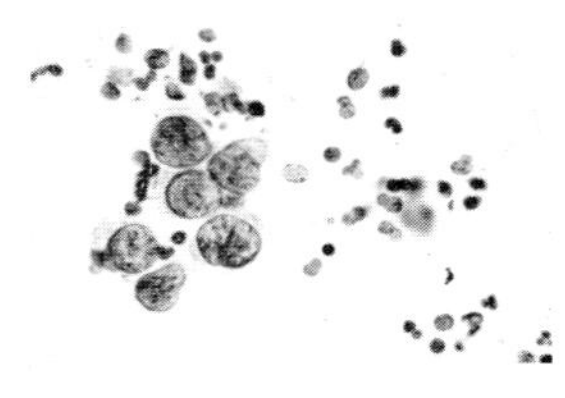

Low Power: Collection of six extremely large nuclei in a field of leukocytes.
High Power: The nuclei are not hyperchromatic but the chromatin is irregularly distributed within the sharp nuclear borders and occasional clumps of chromatin are visible. The nuclei also show abnormal increase in size, compared to the size of the leukocytes, and slight variation in shape. Note that the cellular borders are indefinite.

FIG. 137

BLADDER CARCINOMA: UNDIFFERENTIATED MALIGNANT CELLS

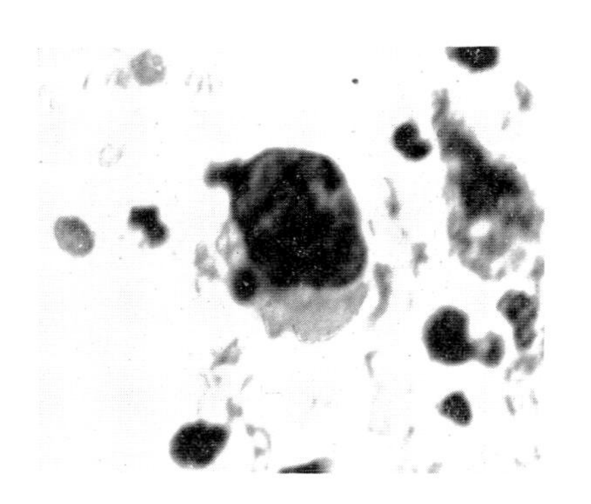

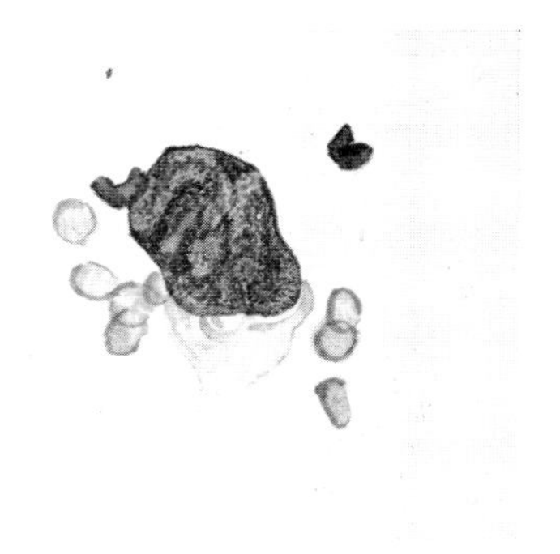

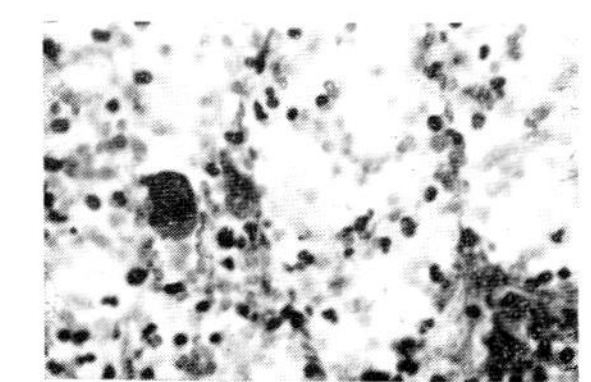

Low Power: Large hyperchromatic nucleus in a background of blood and leukocytes.
High Power: The nucleus is approximately seventeen times larger than the blood cells surrounding it. The cytoplasm has degenerated, leaving only a faint amount with an indistinct cellular border. The nuclear border is irregular and dense and the chromatin particles are unevenly distributed throughout the nucleus, giving the nuclear surface a wrinkled appearance.

Fig. 138

BLADDER CARCINOMA: ECCENTRIC NUCLEI

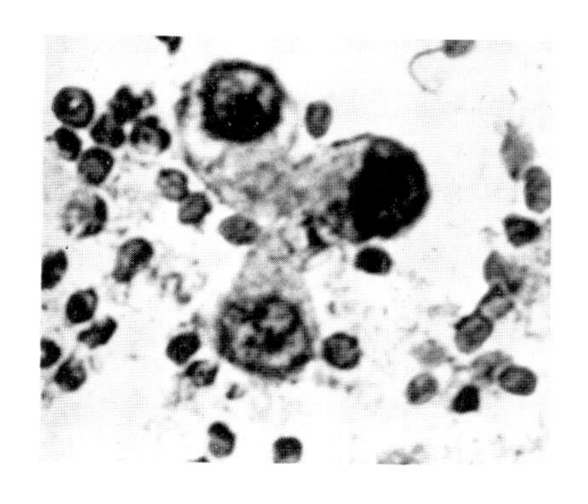

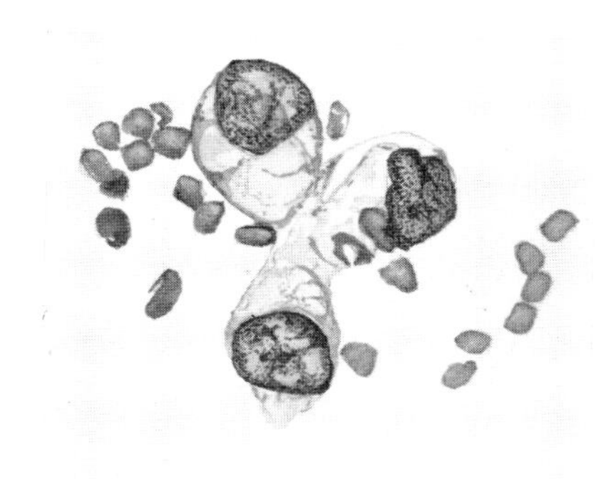

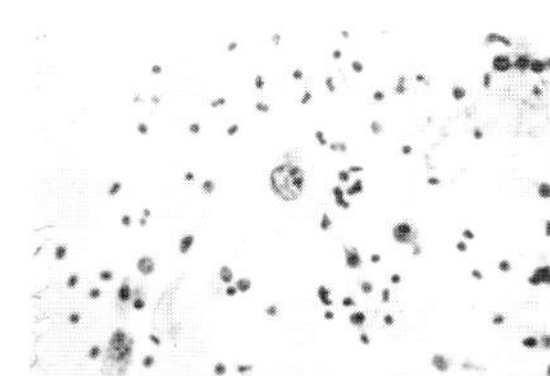

Low Power: Three enlarged hyperchromatic nuclei in a messy field of blood and leukocytes.
High Power: These malignant cells are similar to those found in adenocarcinoma (see Fig. 110), with eccentrically located nuclei, irregular chromatin structure and sharp uneven nuclear borders. The cytoplasm appears vacuolated, and individual cellular borders are lost where the cytoplasm of the three cells merges.

Fig. 139

BLADDER CARCINOMA: THIRD TYPE DIFFERENTIATED CELL

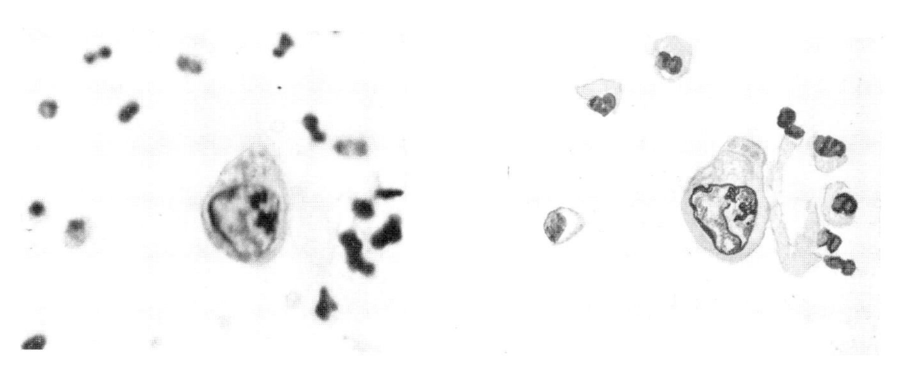

Low Power: Large discrete cell with active clumps of chromatin.
High Power: A typical third type differentiated cell exhibiting all of the characteristics of malignancy, *i.e.*, thick wrinkled nuclear border, clumps and strands of chromatin, adequate cytoplasm with a clear cellular border and in general an abnormal nuclear-cytoplasmic ratio. Compare this cell to the third type malignant cells found in the vagina, Fig. 56.

The cells which desquamate from bladder carcinoma are as stated previously either undifferentiated or similar to the third type differentiated cell. Primary squamous carcinoma of the bladder does not occur commonly. In one case of this type seen in our laboratory the cells were similar to those seen in squamous carcinoma of the cervix. On the other hand, it should be remembered that extensions of a carcinoma of the cervix to the bladder is not uncommon. These tumors show both the differentiated and undifferentiated cells of squamous carcinoma.

The rate of desquamation of tumor cells from bladder carcinoma varies a good deal. The majority of tumors shed only occasional cells. The usual positive smear contains many leukocytes, scattered transitional cells and rather infrequent groups of malignant undifferentiated nuclei plus occasional single malignant cells. However, an occasional case will show marked desquamation of malignant cells. In two cases of early bladder carcinoma so many tumor cells were cast off that the sediment contained only rare cells of any other type. Both these cases were unsuspected clinically and serve as excellent examples of the real advantage of the cytologic method. Early lesions have an intact well preserved surface and appear to shed many more recognizable cells than the long-standing lesions whose surfaces may be necrotic.

Difficulties in Interpretation: The real difficulty in interpretation of urine sediment is in distinguishing transitional cells whose nuclear structure is unusually active from differentiated carcinoma cells. As has been mentioned, the transitional cells show greater variation in chromatin content and arrangement than is usual in normal epithelial cells. Therefore, third type differentiated cells must show very marked aberrations in nuclear structure to be regarded as positive. There must be extreme irregularity of the chromatin pattern, *i.e.*, the nuclear content must be unevenly distributed in large clumps and heavy strands. Chromatin should be condensed at the periphery of the nucleus. Finally, the cytoplasmic-nuclear ratio should be abnormal. It should be emphasized that this type of differentiated cell must fulfill these criteria, since in urine sediment so many border line cells appear which are probably variants of the normal transitional cell rather than malignant.

We have only encountered multinucleated tumor giant cells rarely in bladder carcinoma. On the other hand, benign multinucleated cells are common. They often have as many as twenty or thirty nuclei. Therefore, in urine sediment, multinucleated cells are usually benign rather than malignant.

The most reliable basis for the diagnosis of bladder carcinoma cytologically is the presence of groups of undifferentiated malignant nuclei. These groups of nuclei are not tightly clumped but are usually spread discretely. (See Plate 27.) When malignant nuclei are present in groups the additional criteria of variation in size and shape may be taken into consideration.

General Criteria for Identification of Malignant Cells from Bladder Carcinoma: Undifferentiated cells predominate. They occur usually in groups, showing marked variation in size and some variation in shape. Cytoplasm occurs as an indistinct background and cellular borders are absent. Nuclear aberrations are distinct. The nuclear material is in heavy clumps of unequal size or in heavy strands. The chromatin is condensed at the periphery of the nucleus. Third type differentiated malignant cells may be present. They have the same type of nuclear abnormalities as described above and may be recognized by their distinct cellular border and abnormal nuclear-cytoplasmic ratio.

CARCINOMA OF THE KIDNEY

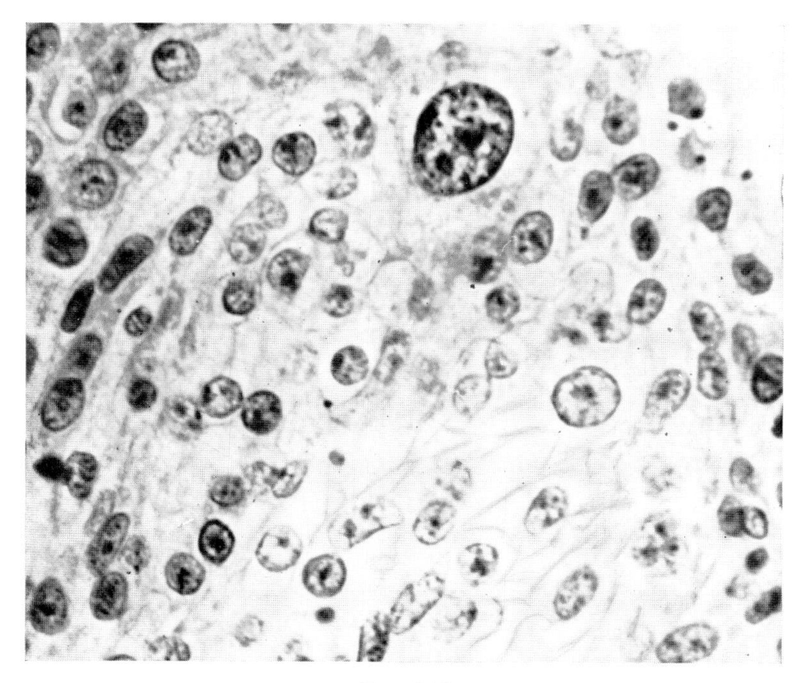

FIG. 140

HISTOLOGIC SECTION: CARCINOMA OF RENAL PELVIS

The photomicrograph above is of a histologic section from a carcinoma of the renal pelvis. There are moderate differences in nuclear pattern, size and shape in the majority of the cells seen. Obviously, the huge malignant nucleus near the top of the figure accentuates such differences. It has marked irregularity of chromatin pattern, the condensations being large and irregular, leaving empty spaces in the background. The nuclear border is very sharp, owing to the condensed chromatin. It is just these same characteristics so typical of undifferentiated malignant cells that are found in the desquamated cells from kidney carcinoma. The cells are undifferentiated. Cytoplasm is seldom present except as a background with no definite cellular borders.

We have not been able to distinguish clear cell carcinoma (hypernephroma or renal cell carcinoma) from other kidney carcinoma. One case, illustrated on the following pages, showed large cells with clear cytoplasm histologically but the desquamated cells from the tumor were indistinguishable from the other undifferentiated cells of kidney carcinoma.

Though we have not seen enough cases of kidney carcinoma to state any definite percentage, it is our experience that well preserved cells appear in the urine sediment less consistently than from tumors of the bladder. Thus the false negative error is higher.

DESCRIPTION OF MALIGNANT CELLS FROM KIDNEY CARCINOMA

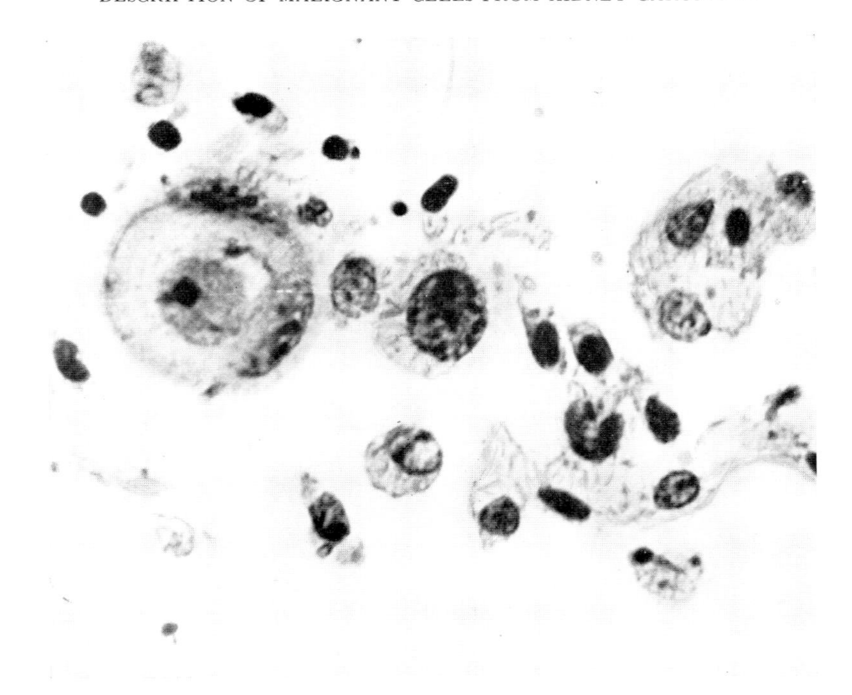

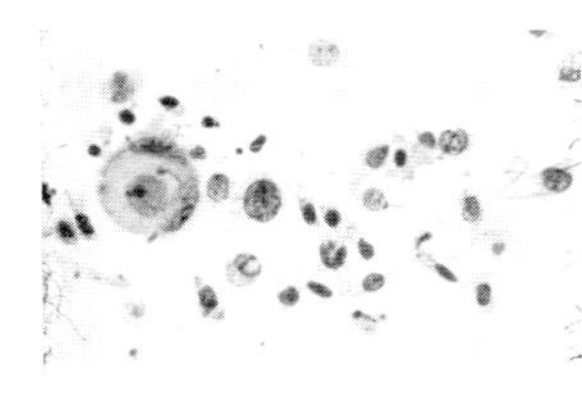

Low Power: A collection of cells, large and small, varying in staining intensity.

High Power:

A. Characteristics of Nucleus: 1. Clear, regular nuclear borders. 2. The normal transitional cells in the field have pyknotic or vesicular nuclei (see group 4), whereas the malignant nuclei are more hyperchromatic (cell 3), show prominent clumps of chromatin (giant cell 1,) and have sharper nuclear borders (see cell 2). 3. The malignant cells are also much larger than the normal cells: compare the nucleus of cell 3 with those in group 4. However, the variation in shape among the malignant nuclei is similar to that of the normal nuclei, *i.e.*, round, oval or triangular. 4. Cell 1 is a malignant giant cell, which is a collection of several malignant nuclei within the same cytoplasm.

B. Characteristics of Cytoplasm: 1. Obscure cellular borders of the malignant cells and relatively clear borders of the normal cells. 2. All of the normal cells have adequate, wrinkled or folded cytoplasm (see group 4), while there is little or no cytoplasm around the undifferentiated malignant cells (see cells 2 and 3). 3. The giant cell 1 shows more differentiation and has adequate cytoplasm with a definite cellular border. 4. Staining reaction: bluish purple nuclei with basophilic cytoplasm.

C. General Characteristics of Group: Normal and malignant cells; the malignant are larger and exhibit greater irregularity in chromatin structure.

PLATE 28

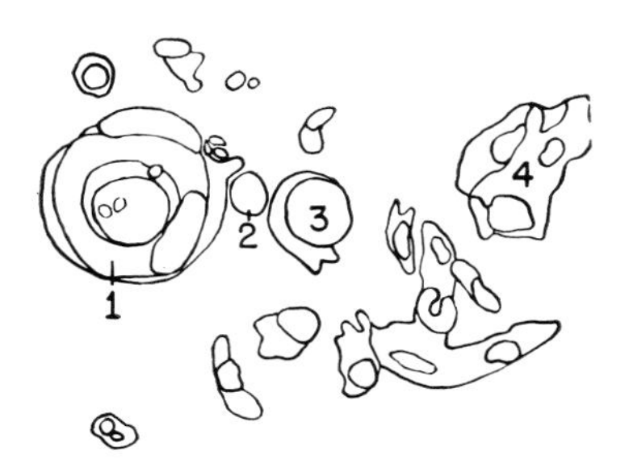

KEY TO MALIGNANT CELLS FROM KIDNEY CARCINOMA

1. Large malignant giant cell with three nuclei varying in shape, with irregular chromatin.
2. Single undifferentiated nucleus with the condensation of chromatin particles mainly at the nuclear border.
3. Large hyperchromatic undifferentiated nucleus with an irregular chromatin pattern and no definite cellular border.
4. Collection of three transitional epithelial cells with wrinkled cytoplasm and indistinct cellular borders.

FIG. 141

CARCINOMA OF THE KIDNEY: UNDIFFERENTIATED MALIGNANT CELL

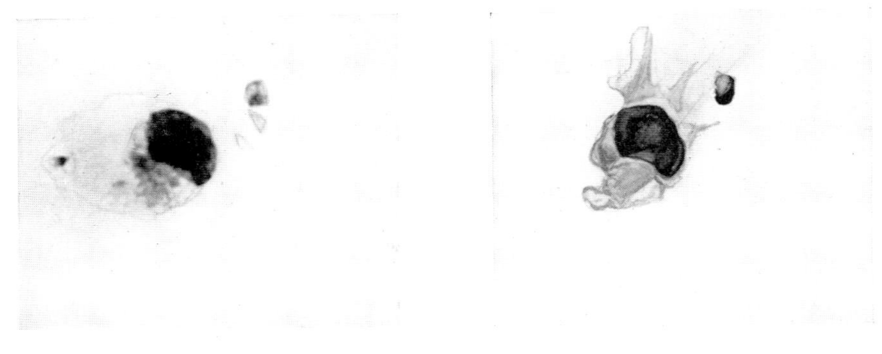

Lower Power: Single hyperchromatic nucleus.

High Power: This is an example of a single undifferentiated malignant cell from a kidney tumor. Its lack of a cellular border and inadequate cytoplasm classify it as undifferentiated. The intensely hyperchromatic nucleus showing irregularity in chromatin pattern and sharp nuclear border identifies the cell as malignant. Single cells may be found fairly often in positive smears of urine.

FIG. 142

HYPERNEPHROMA OF THE KIDNEY: UNDIFFERENTIATED MALIGNANT CELLS

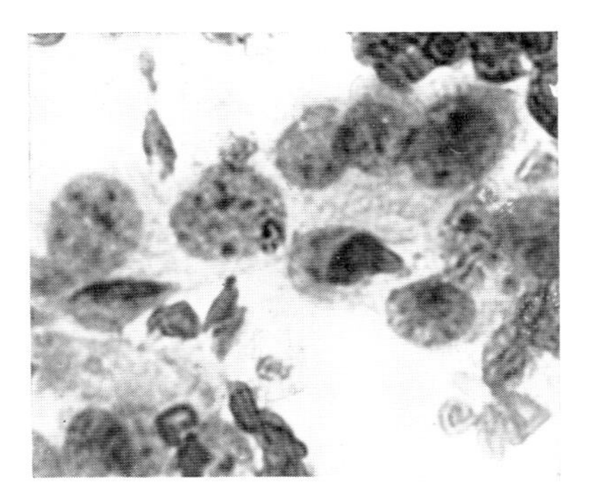

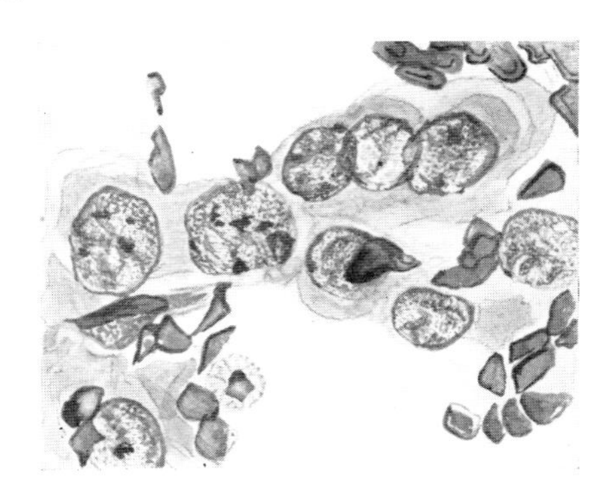

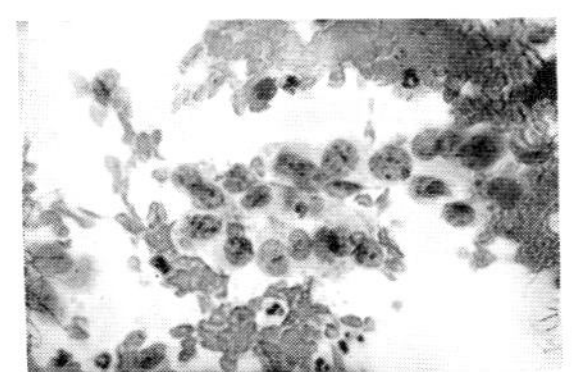

Low Power: Group of granular nuclei in a field of blood.

High Power: This group of malignant nuclei shows marked irregularity in chromatin pattern and sharp nuclear borders. Variation is more in size than in shape. Cytoplasm is an indistinct background. We have not been able, from desquamated cells, to distinguish between hypernephroma and other carcinomas of the kidney.

Often the examination of urine specimens obtained from each ureter at cystoscopy is required. This may be extremely helpful if a kidney tumor is suspected but has not been definitely localized. Theoretically, the finding of malignant cells in one ureter specimen and the lack of cells in the specimen from the opposite ureter should be very specific information. However, it should be pointed out that this may not be true in all instances. If a bladder tumor is located near either ureteral orifice, it is not unlikely that contamination of the ureteral specimens takes place as the tube is inserted. We have found definite malignant cells in one ureteral specimen, none in the other when the patient had a carcinoma of the bladder and no lesion in either kidney. We do not mean to indicate that the examination of sediment from ureteral specimens is not of practical value, but we do feel that it should be realized that there is the possibility of contamination from a bladder carcinoma near the ureteral orifices.

Difficulties in Interpretation: For the most part the same difficulties are encountered here as in bladder carcinoma. It is essential that the cells show distinct nuclear aberrations of chromatin content and distribution before a positive diagnosis is made. In the examination of ureteral specimens the large multinucleated, normal epithelial cells are encountered. (See Fig. 134.) As has been stated previously, these cells are benign despite their size and numbers of nuclei. The chromatin content of such cells varies little and there is not marked difference in size and shape. Compare these cells (Fig. 134) to the multinucleated tumor giant cell in Plate 28, which exhibits differences in size, shape, arrangement and amount of chromatin material.

General Criteria for Identification of Malignant Cells from Carcinoma of the Kidney: The desquamated cancer cells are undifferentiated, occurring both in groups and singly. They have the typical appearance of such cells, having an indistinct background of cytoplasm, no cellular border, increase in chromatin content, abnormal nuclear pattern and sharp nuclear border. Malignant giant cells may be found. These cells have several abnormal nuclei surrounded by cytoplasm with a definite cellular border.

CARCINOMA OF THE PROSTATE

In general, the examination of the urine sediment for the presence of cancer cells from carcinoma of the prostate has been the least satisfactory of that of any of the secretions. The false negative error is the highest for any type of tumor. This lack of accuracy may be on an anatomic basis. If the tumor of the prostate is not so situated that the secretions from it flow directly to the urethra to be carried out by the urine, few if any malignant cells will be found. We have not found the examination of prostatic smears to yield any greater accuracy. It is our impression that the best method is the collection of a voided urine specimen directly after prostatic massage.

The malignant cells which we have recognized as desquamating from carcinoma of the prostate are undifferentiated. They occur in groups rather than singly. They usually do not show the vacuolization which is characteristic of adenocarcinoma. However, the nuclei do appear in tight clusters with a great deal of overlapping, in contrast to those seen in bladder carcinoma where the nuclei are spread discretely. The cells vary more in size than in shape. A positive diagnosis depends on finding groups of cells with definite abnormal alteration in the nuclear structure.

FIG. 143

PROSTATIC CARCINOMA: UNDIFFERENTIATED MALIGNANT CELLS

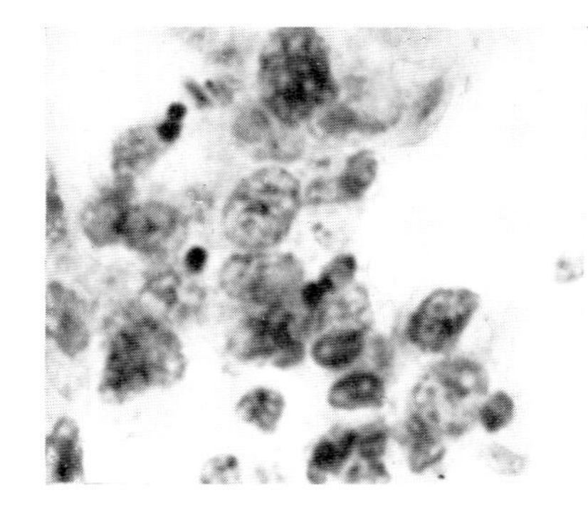

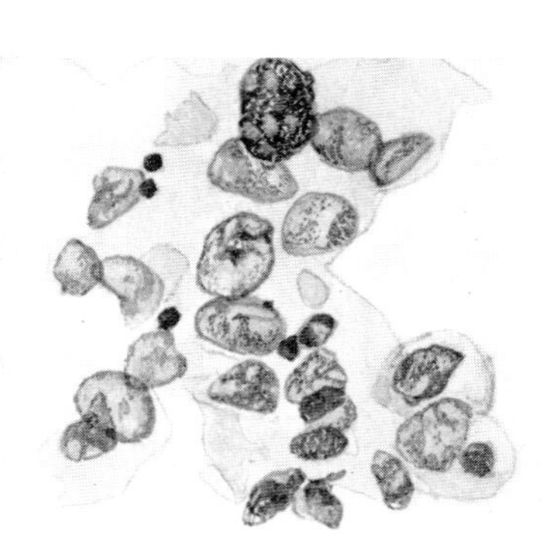

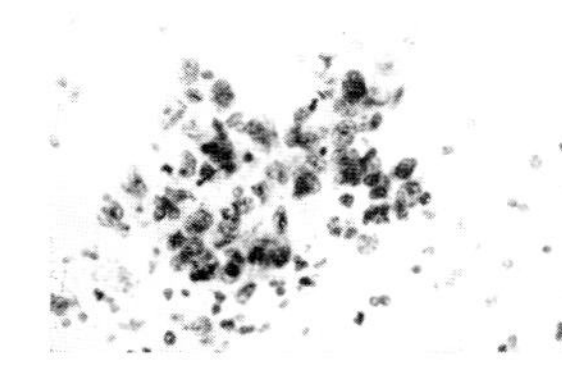

Low Power: Group of granular nuclei in field of leukocytes.

High Power: This group fulfills all the criteria for undifferentiated malignant cells and is typical of the type of cluster seen in cells from prostatic carcinoma. There is marked variation in size, less in shape. Chromatin varies both in the amount present and in its distribution. Nuclear borders are distinct but cellular borders are absent.

Fig. 144

PROSTATIC CARCINOMA: UNDIFFERENTIATED MALIGNANT CELLS

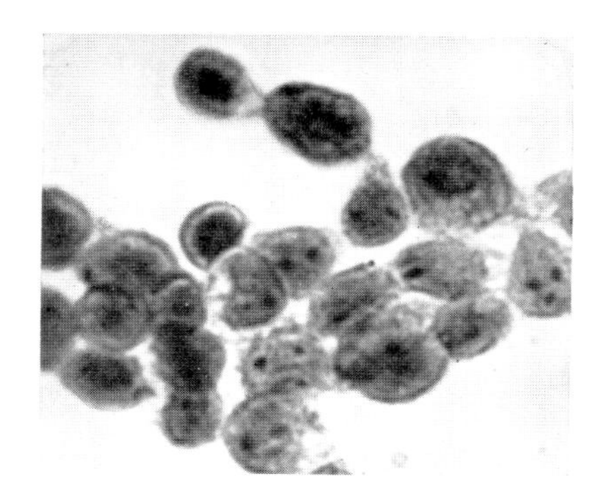

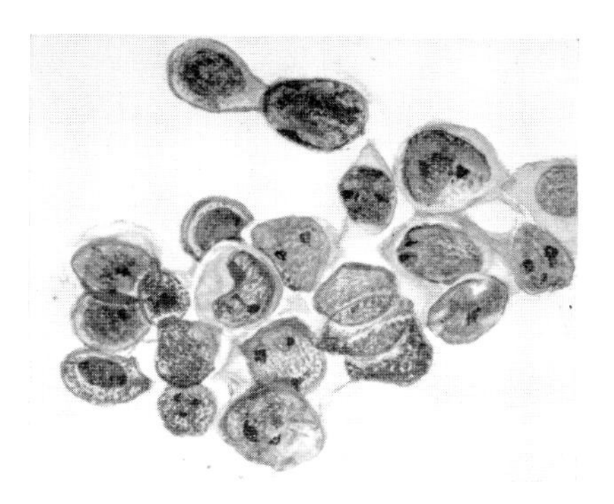

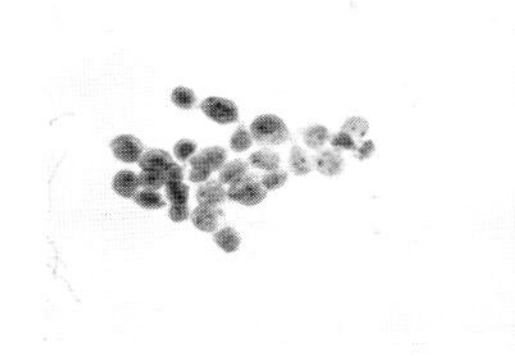

Low Power: Groups of large nuclei, some with rims of cytoplasm.

High Power: These malignant nuclei are not hyperchomatic and the chromatin particles are fairly evenly distributed. However, they do show prominent clumps of chromatin, overlapping qualities and variation in size and shape. The nuclei have small rims of cytoplasm surrounding them, but distinct cellular borders are absent. These cells appear fairly well preserved.

Fig. 145

PROSTATIC CARCINOMA: UNDIFFERENTIATED MALIGNANT CELLS

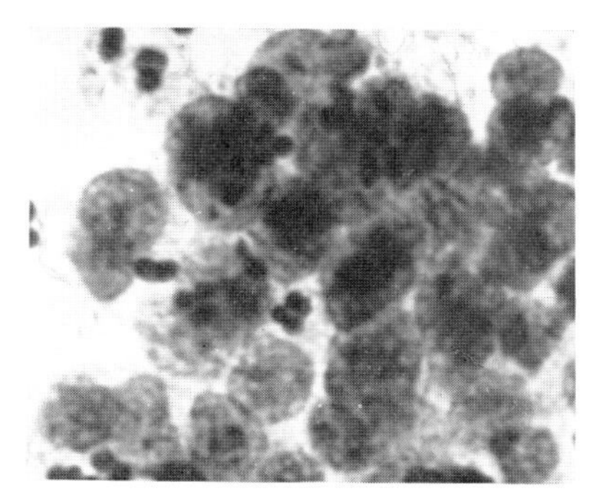

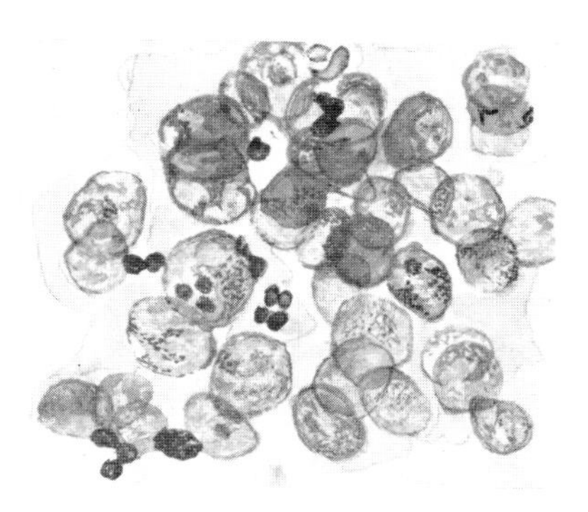

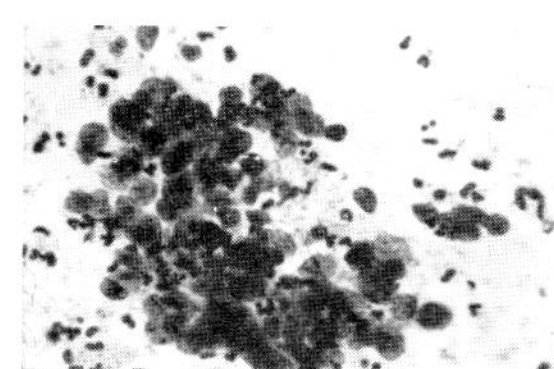

Low Power: Thick mass of overlapping nuclei irregularly arranged.

High Power: In comparison to the figure above, these malignant nuclei are more degenerate, since individual chromatin particles are not distinct. However, the nuclear borders are well outlined. The nuclei are similar to those above in their variation in size, shape and tendency to pile on one another. Notice the absence of cytoplasm.

Carcinoma of the Genito-Urinary Tract

Often the cytologic laboratory is asked to examine the urine sediment of patients with carcinoma of the prostate treated by stilbestrol. The sediment of the urine from these patients contains numerous large flat superficial cells. These are like those encountered in vaginal smears. With this exception the smears examined for possible prostatic carcinoma are similar in all respects to other urine sediment.

Difficulties in Interpretation: One obstacle in the diagnosis of prostatic carcinoma from desquamated cells is the problem of identification. The Grade I carcinomas represent a real problem and we have not been able to distinguish these shed cells in most instances from those of benign prostatic hypertrophy. We are faced with a problem similar to that encountered in carcinoma of the endometrium. Both organs are lined by columnar epithelium which in many instances becomes hyperplastic. Such hyperplasias shed cells which may appear unusually active. On the other hand, the single cells from a low-grade carcinoma may not appear abnormal enough to justify a diagnosis of positive. For these reasons we have found it practical to base our positive diagnoses only on groups of overlapping nuclei showing marked aberrations of nuclear structure.

General Criteria for Identification of Cells from Prostatic Carcinoma: Cells are undifferentiated. The positive diagnosis is based on the presence of tight clusters of nuclei showing overlapping. Cytoplasm may be present as a background or be entirely absent. Nuclei show abnormality of chromatin pattern, vary markedly in size, some in shape.

SUMMARY

We have discussed these three common tumors of the genito-urinary tract separately because this seemed a clearer way to present the various types of malignant cells found in urine sediment. However, this does not mean that it is possible to distinguish them cytologically. They have some points of difference, *i.e.*, malignant undifferentiated cells from bladder carcinoma are more apt to be spread discretely than to be in tight clusters as the cells from prostatic carcinoma. Still, the differences are not specific enough to justify an attempt to classify the type of tumor. We report smears from urine sediment as positive, doubtful or negative. The doubtful smears are repeated and are usually called either negative or positive after the second or third examination.

Though we have emphasized that the expected accuracy of urine smears is not as high as in other secretions, it should be pointed out that this does not necessarily exclude its usefulness. A negative smear bears little weight, but a positive smear necessitates careful search for a tumor of the urinary tract. If used in this way, early carcinomas will be discovered by the cytologic examination of urine sediment.

CHAPTER XVIII

NORMAL CELLS OF PLEURAL AND PERITONEAL FLUID

The examination of pleural and peritoneal fluids for the presence of malignant cells is a technic which has been used extensively for years. The ease of aspiration of fluids and the necessity for knowing whether the accumulation of fluid was due to metastatic implants has kept the cytologic examination of pleural and abdominal fluids a necessary procedure in routine pathologic laboratories. For the most part, specimens have been centrifuged, and the button obtained fixed and put through dehydration processes, infiltrated with paraffin and sectioned. It was natural, as more workers became interested in the exfoliative cytology of other secretions and fluids, that cytologists should take up the examination of pleural and abdominal fluids.

In our laboratory we have prepared smears of these fluids in the same way as we prepare gastric and urine smears. A fresh adequate sample is centrifuged, the supernatant fluid poured off, and the sediment placed on a glass slide. We feel that this method of preparation is far superior to the method of sectioning for the following reasons: 1. There is better preservation of cells, since the fluids are centrifuged immediately. 2. The process of fixation, dehydration and embedding causes marked shrinkage of the cells and cellular characteristics are less easily identified. If great care is taken to fix specimens immediately, smear preparations of pleural and ascitic fluid will yield exceptionally satisfactory smears for purposes of interpretation.

The serous membranes are covered by a thin epithelial layer, the mesothelium. Thus, mesothelial cells are the normal cellular component of serous fluids. These cells differ in some respects from other thin, flat, squamous epithelial cells observed in other body secretions. They have a less distinct cellular border. Their cytoplasm is less homogeneous, often almost as irregular in appearance as that of the histiocyte. The nucleus is usually central in position but may be eccentric. Nucleoli are prominent, but as a rule only one is present. The nuclear material is granular in appearance, with occasional small aggregates of chromatin. Nuclear-cytoplasmic ratio is within normal limits. These cells are spread discretely and only occasionally group in tight clusters.

Though mesothelial cells are the normal ones usually encountered in either pleural or ascitic fluid, some fluids may show other types of normal cells to be in the majority. These cells are not epithelial in origin. Some fluids may contain only polymorphonuclear leukocytes. These are easily recognized by their distinctive characteristics. Others will contain no cells except lymphocytes. (See Fig. 148.) A third type of smear is that which contains only histiocytes. These are in all respects similar to the small histiocytes encountered elsewhere. The cellular border is not sharp. The cytoplasm has a finely vacuolated appearance. Nuclei are eccentric in position and are round, oval or kidney-shaped. Blood is often present in serous fluids aspirated for the question of malignancy. It is usually in the form of preserved red blood cells. Preservation of the red cells depends on speed of fixation. If cytolysis of the red cell takes place, the smear will have a background of blood pigment. Examples of the various types of normal cells found in pleural and ascitic fluids are shown on the following pages.

DESCRIPTION OF NORMAL CELLS IN SEROUS FLUIDS

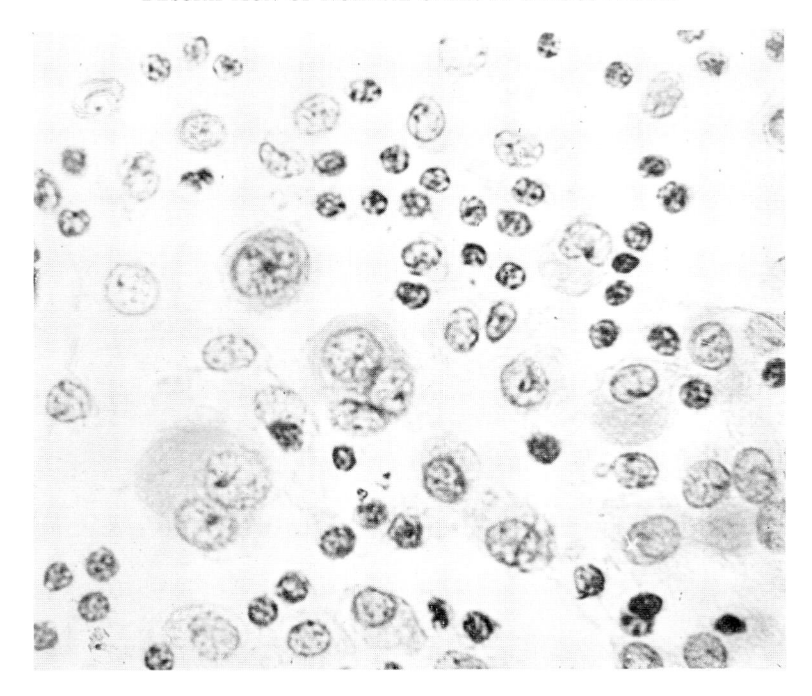

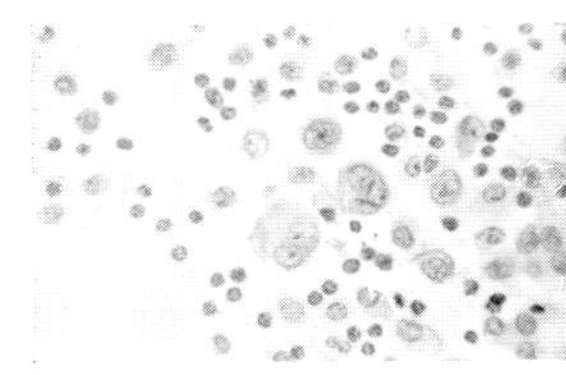

Low Power: Field of many bland nuclei, varying in size and shape.

High Power:

A. Characteristics of Nucleus: 1. Clear nuclear borders. 2. In viewing all of the cells in the field together it is evident that none of the nuclei are hyperchromatic, instead they are lightly stained with evenly distributed particles of chromatin. Occasionally the nuclei exhibit a prominent clump of chromatin (see cell 1) or a nucleolus (see cell 6). The mesothelial cell nucleus in cell 2 is more active than the rest, but the chromatin particles are still arranged regularly throughout the nucleus, forming one clump in the center. 3. Scattered throughout the field are small, round, typical lymphocytes (see group 5), but notice their even regular size. 4. Mesothelial cells in pleural fluids will vary greatly in size and shape. Compare the size of cell 2 with the double-nucleated cell in group 1, for example. 5. Cell 4 has the typical bean-shaped nucleus of a histiocyte.

B. Characteristics of Cytoplasm: 1. Indistinct, irregular cellular borders. 2. Often the cytoplasm of several cells will merge together (see group 1), or contain two or three nuclei (see cells 3 and 6). 3. Staining reaction: Light purple nuclei with faint acidophilic or basophilic cytoplasm.

C. General Characteristics of Group: Many bland nuclei varying in size and shape with light-staining irregular cytoplasm.

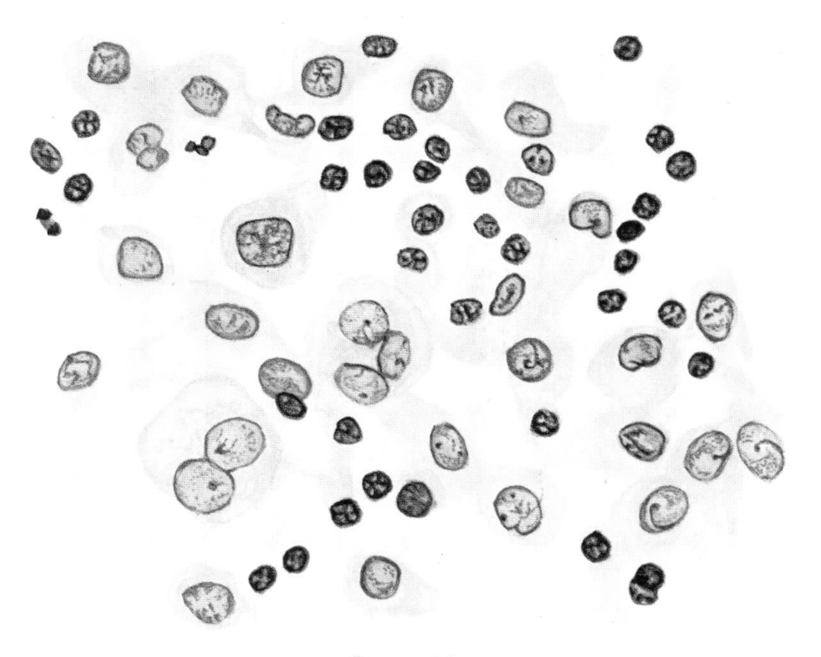

PLATE 29

KEY TO NORMAL CELLS IN SEROUS FLUIDS

1. Three mesothelial cells with merging cytoplasm. Two of the cells have round nuclei and the other is double nucleated.
2. Large active mesothelial cell with a thick nuclear border and darker-staining nucleus than other cells in the field.
3. Three mesothelial cell nuclei enclosed in the same cytoplasm, each with a nucleolus.
4. Typical bean-shaped histiocyte nucleus with elongated foamy cytoplasm.
5. Six small hyperchromatic lymphocytes, uniform in size and shape.
6. Large double-nucleated mesothelial cell with bland light-staining nuclei.

Fig. 146

ASCITIC AND PLEURAL FLUIDS: MESOTHELIAL CELLS

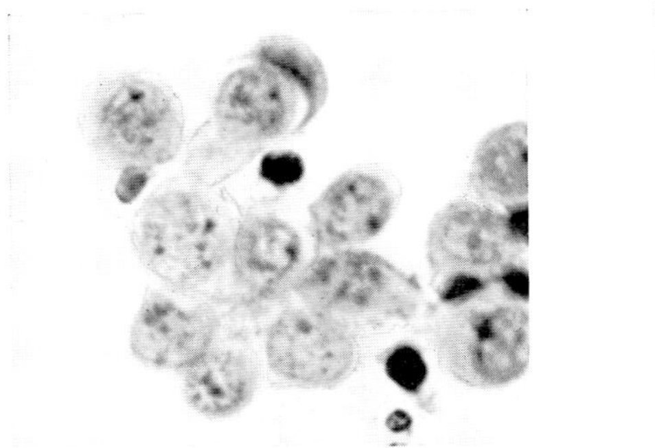

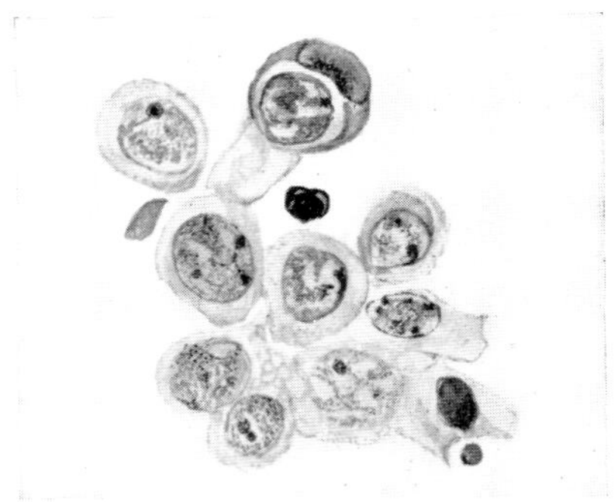

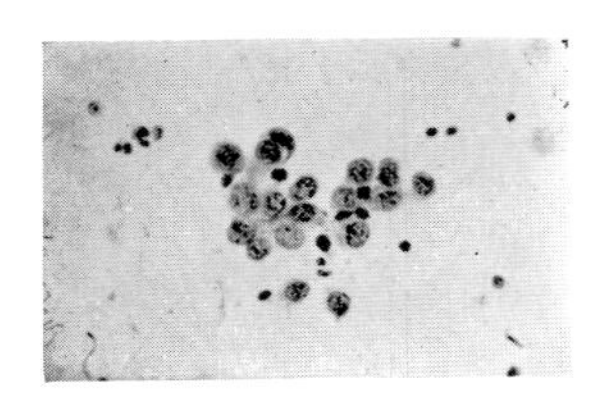

Low Power: Group of densely clustered nuclei. *High Power:* These mesothelial cells are unusual in that there is not as much cytoplasm present as in the typical cell shown in Plate 29. The nuclei, however, are even in size and do not vary in chromatin content, the nuclei being smoothly granular with one prominent clump of chromatin. The cell at the top has been phagocytosed by a histiocyte.

Fig. 147

ASCITIC AND PLEURAL FLUIDS: MESOTHELIAL CELL IN MITOSIS

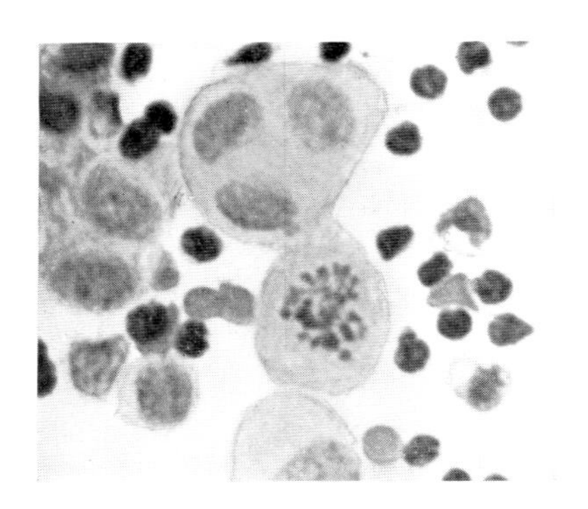

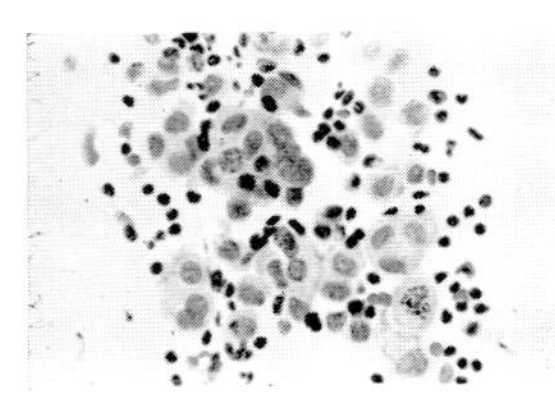

Low Power: Field of bland cells with cytoplasm. *High Power:* In the center of the field is a mesothelial cell in mitosis. The rest of the mesothelial cells have evenly distributed nuclear particles and several of the cells have merging cytoplasm. The two nuclei on the left both show nucleoli. The small dark-staining free nuclei are lymphocytes. Red blood cells are also present.

Fig. 148

ASCITIC AND PLEURAL FLUIDS: LYMPHOCYTES

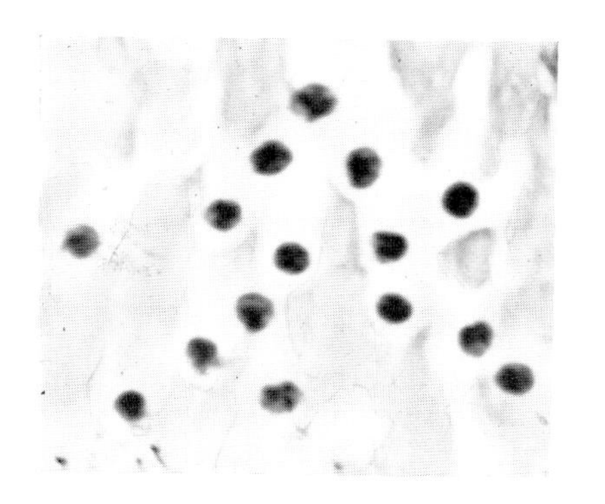

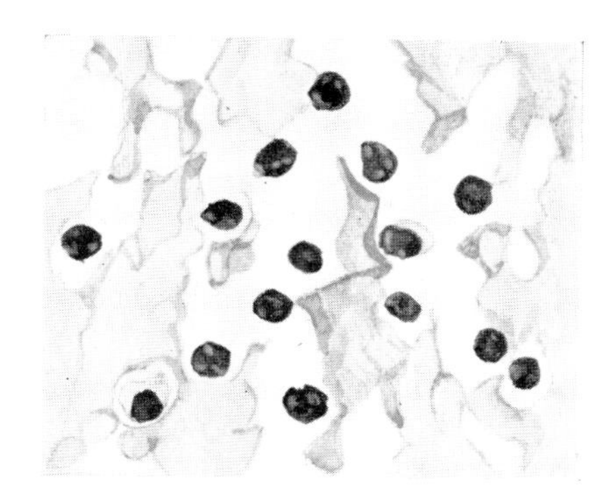

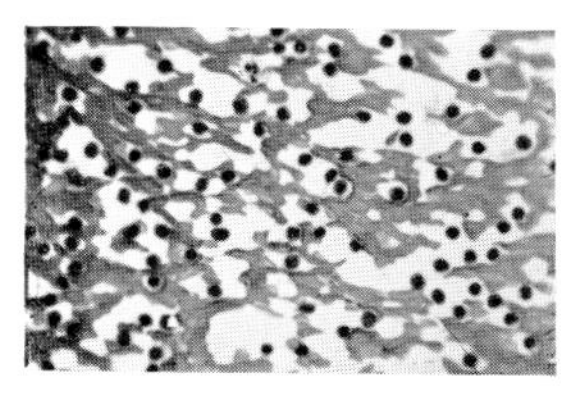

Low Power: Blood and leukocytes.

High Power: Fairly frequently fluids from serous surfaces contain no mesothelial cells but only leukocytes. It is unusual to see one where lymphocytes are the only cells present and this field is an example of such an instance. The cells are easily recognized, as they retain their characteristic deep purple stain, slight amount of cytoplasm and regular size.

Fig. 149

ASCITIC AND PLEURAL FLUIDS: HISTIOCYTES

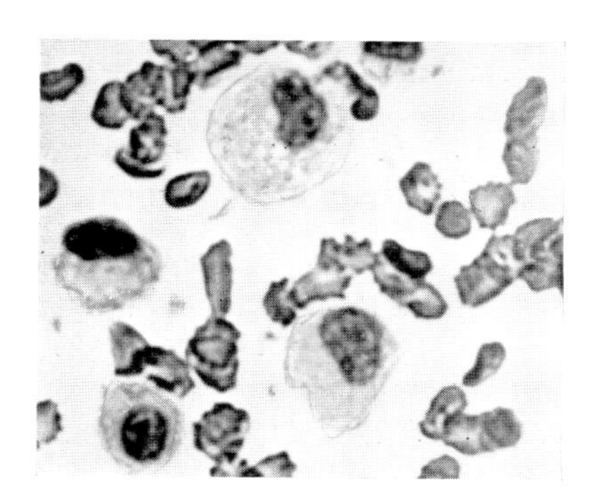

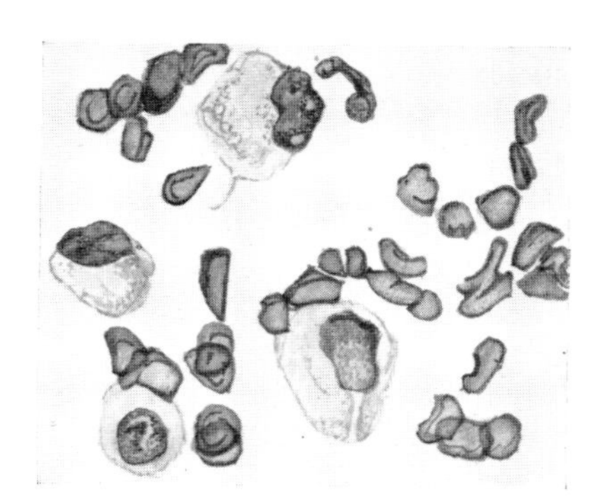

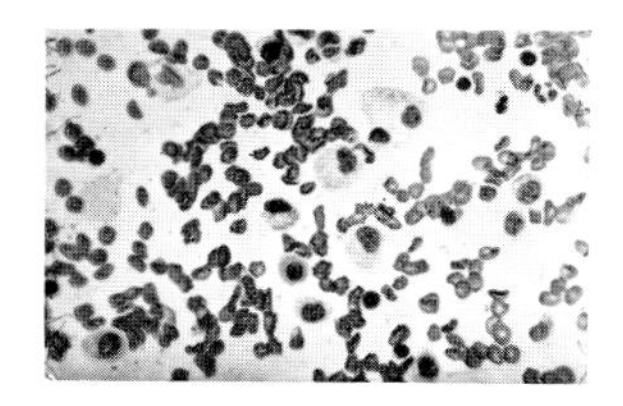

Lower Power: Histiocytes, leukocytes and blood.

High Power: Even under low magnification the cells present in this field may be recognized as typical histiocytes. They fulfill all the criteria for histiocytes mentioned previously: round, bean-shaped or oval nuclei, eccentric position of the nuclei in the cytoplasm, which appears foamy or finely vacuolated. Occasionally only these cells are found in serous fluids.

Normal Cells of Pleural and Peritoneal Fluid

Since the cells present in either pleural or peritoneal fluid show the same characteristics and since we have not been able to see any distinct differences between the two, they have been described here together.

The mesothelial cells of the serous cavities show a great deal more activity than is common in other epithelial cells. The presence of prominent true nucleoli and the granularity of the chromatin is unusual in epithelial cells. Double and triple nuclei are common in desquamated mesothelial cells. Mitotic figures are seen frequently and are not an indication of malignancy.

Difficulties in Interpretation: The unusual nuclear activity of the mesothelial cells seen in serous fluids appears to the uninitiated observer as definitely suspicious of malignancy. Since in histologic sections the presence of large numbers of mitotic figures is an extremely suspicious finding, it is natural that the inexperienced cytologist should attempt to carry over this criterion to interpretations of exfoliated cells. Though it appears to be a paradox, the presence or absence of mitoses in desquamated cells bears no weight in the decision of whether the cells encountered are benign or malignant. Mitotic figures are encountered as often if not more often in benign than in malignant cells. In no fluid is this of more importance than in those aspirated from serous cavities. In our experience, fluid from a patient with cirrhosis of the liver may contain many more mitotic figures than fluid from a patient with metastatic carcinoma. The presence of large numbers of cells in mitoses is a striking finding but it must be emphasized that this alone should not influence the diagnosis. It is impossible to tell whether or not a cell in mitoses is malignant. One must judge its potential characteristics by comparison with adjacent cells and by this indirect evidence classify the cell with the mitotic figure.

The distinction between mesothelial cells and histiocytes is often difficult. The cytoplasm of the mesothelial cell is irregular in appearance, while that of the small histiocyte is finely vacuolated. The nuclei of mesothelial cells are central in position, but those of histiocytes are eccentric. Mesothelial cells contain one prominent nucleolus. Histiocytes have small clumps of chromatin. Nuclear material of mesothelial cells is more granular than that of histiocytes. The preceding criteria are in general the characteristics which differentiate between the two types of cells. However, there is occasional overlapping of characteristics and it is not always possible to distinguish with certainty the cell type. Still, this overlapping of cellular characteristics seems to be unimportant, since we have not been able to correlate the presence of either type of cell with any definite pathologic entity. The important point is that these cells can be identified as benign.

General Criteria for Identification: Mesothelial cells are round in shape with lightly staining irregular cytoplasm. Nuclei are commonly central in position, round or oval in shape with definite nuclear border, granular chromatin and a nucleolus. Cells are often multinucleated.

CHAPTER XIX

MALIGNANT CELLS IN PLEURAL AND PERITONEAL FLUID

In considering the malignant cells found in serous fluids, we immediately see a marked difference in approach. We are not dealing here with primary malignant growths which arise from the mesothlieum lining the serous cavities. Primary tumors of the serous surfaces are rare. We have never seen such a tumor. Though primary lesions of the serosa are seldom seen, metastatic carcinomas are frequent. Therefore, in interpretation of pleural and ascitic fluid we are concerned predominantly with metastatic lesions, and occasionally with direct extensions of tumor such as may occur in carcinoma of the ovary or carcinoma of the lung. In other secretions the first consideration was the type of cell present in primary carcinomas. A classification could be based on a limited number of cell types common to a definite primary tumor. In serous fluids we have cells desquamated from many types of lesions.

Even though serous fluids may contain malignant cells from a variety of tumors, a classification of a sort may be used. This is possible only because it is rare to see distinct differentiated malignant cells other than those from adenocarcinoma. For instance, we have never seen the differentiated cells of a squamous carcinoma in either pleural or ascitic fluid. This may be suggestive that only malignant cells of the immature embryonal type produce this kind of metastases. The only types of malignant cell showing any degree of differentiation are those from a metastatic adenocarcinoma. These are groups of differentiated cells with apparent cell borders and distinct vacuolization of the cytoplasm. Thus the two types of cells encountered in both pleural and ascitic fluid are: 1. Typical undifferentiated malignant cells, exhibiting the following characteristics: increase in nuclear content, irregular arrangement of the chromatin, sharp nuclear border, indistinct background of cytoplasm but no cellular borders, and variation in size and shape. 2. Differentiated cells from adenocarcinoma. These exhibit the same nuclear characteristics as listed above, plus a distinct vacuolization of the cytoplasm. Cellular borders are usually present. Large nucleoli are often present in both types of malignant cells and the abnormal nucleus may contain two or more of uneven size.

The smear of pleural or ascitic fluid shows great numbers of malignant cells in most instances. They are far more numerous in proportion to the normal cells present than in preparations from other areas.

An important point to keep in mind is that malignant cells as they occur in fluids from serous surfaces are in groups. We have depended primarily on the presence of large groups of cells fulfilling the above criteria for the diagnosis of positive. The groups of malignant cells desquamated from adenocarcinoma are tightly clustered. Undifferentiated nuclei may be more discrete in arrangement, showing little overlapping but still retaining the impression of grouping. Occasionally numbers of malignant cells may be present which are so characteristic that there is no question as to their origin. In general, however, the finding of large groups is a much safer criterion. It is hazardous to make a diagnosis of positive on only occasional scattered nuclei which appear to vary from the normal. The diagnostic accuracy will be good and false positive diagnoses will be avoided if a positive diagnosis depends on the presence of groups of nuclei showing definite aberrations in nuclear morphology.

DESCRIPTION OF ADENOCARCINOMA IN SEROUS FLUIDS

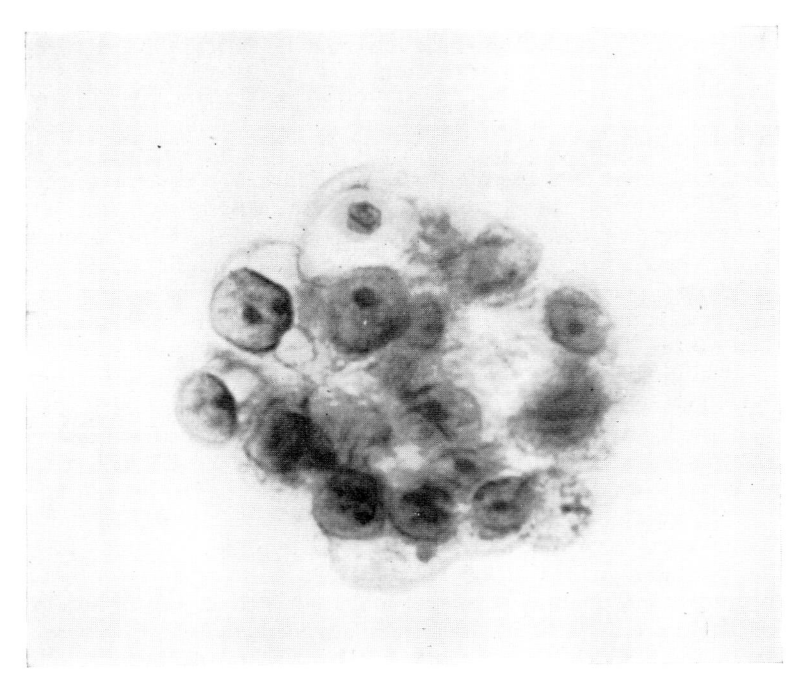

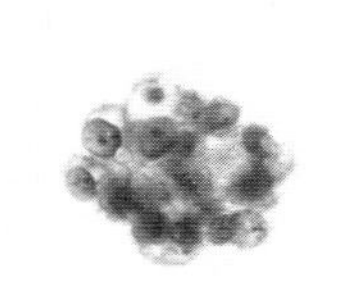

Low Power: Rosette formation of large hyperchromatic nuclei.

High Power:

A. Characteristics of Nucleus: 1. Sharp, well defined nuclear borders. 2. There is variation in the chromatin density of the nuclei, starting with the two nuclei in cell 4, which are fairly transparent with small particles of chromatin unevenly distributed, to cell 6, which has denser particles of chromatin, some collected in irregular clumps, to cells 1 and 2, which have darker clumps of chromatin and thicker nuclear borders, to group 3 showing dark nuclear backgrounds with speckled particles of chromatin, to group 5, which are degenerate with smooth nuclear backgrounds. 3. The nuclei vary more in size than in shape. Compare the sizes of the two nuclei in groups 3 and 4. However, in group 5 the slight variation in shape is illustrated. 4. One of the characteristics of adenocarcinoma is the piling or overlapping of the nuclei. (See groups 3 through 5.)

B. Characteristics of Cytoplasm: 1. Cellular borders usually present but indefinite. 2. Cell 6 has no cytoplasm, whereas cell 1 shows four vacuoles in the cytoplasm and a clear cell border. Cell 2 has a typical "pouched-out" vacuole seen in adenocarcinoma, which fills the entire cytoplasm. 3. Staining reaction: bluish purple nuclei with basophilic or acidophilic cytoplasm.

C. General Characteristics of Group: Piled vacuolated cells, varying more in size than in shape, with various chromatin densities.

Note: from ascitic fluid of same patient as Fig. 66 illustrating adenocarcinoma in vaginal secretion.

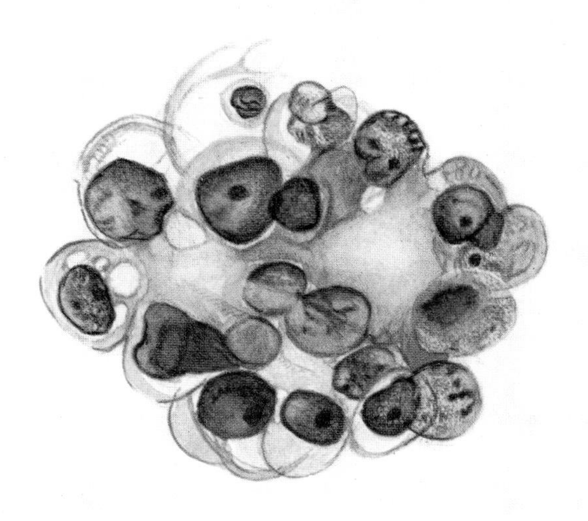

PLATE 30

KEY TO ADENOCARCINOMA IN SEROUS FLUIDS

1. Oval differentiated carcinoma cell with four vacuoles in the cytoplasm.
2. Large round differentiated nucleus with several clumps of chromatin and a big "pouched-out" vacuole filling the cytoplasm.
3. Two malignant nuclei with speckled particles of chromatin illustrating variation in size.
4. Two degenerate differentiated nuclei overlapping each other with indefinite cellular borders.
5. A group of three malignant nuclei showing variation in shape and smooth density in chromatin structure.
6. Single round differentiated nucleus with a clear nuclear border, prominent clumps of chromatin and no visible cytoplasm.

   The rest of the group consists of differentiated cells, also varying in size, shape and chromatin content.

FIG. 150

METASTATIC CARCINOMA: UNDIFFERENTIATED MALIGNANT CELLS

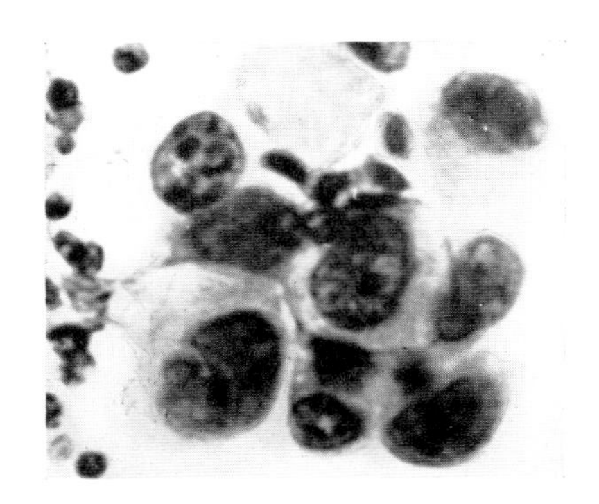

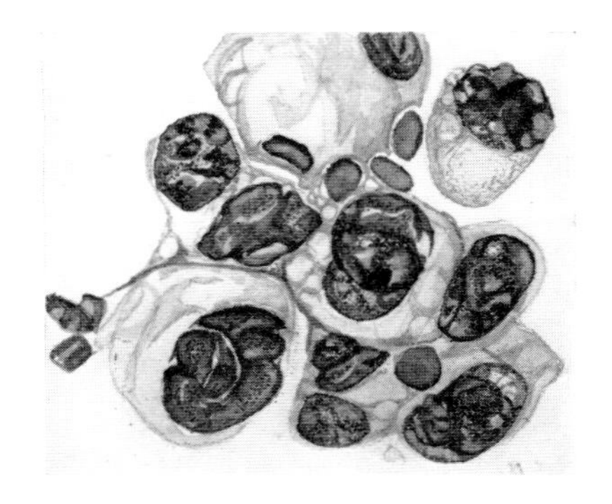

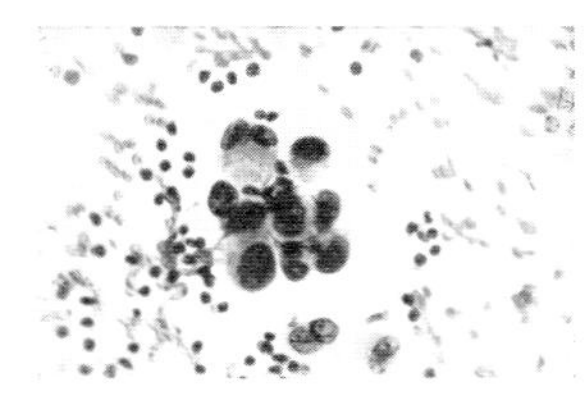

Low Power: Group of hyperchromatic nuclei in field of blood.

High Power: These cells are undifferentiated malignant, since none of the nuclei have a definite cell wall surrounding them. However, there is more of an impression of borders than is usual in such a group. The nuclei have all the common characteristics of malignancy. The variation in size and shape is striking.

FIG. 151

METASTATIC CARCINOMA: SINGLE MALIGNANT CELL

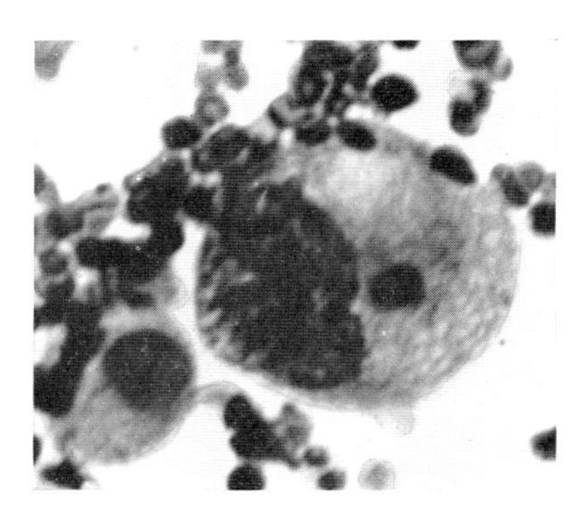

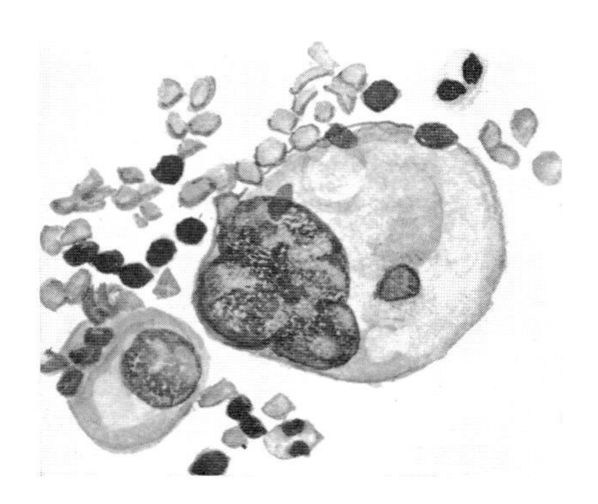

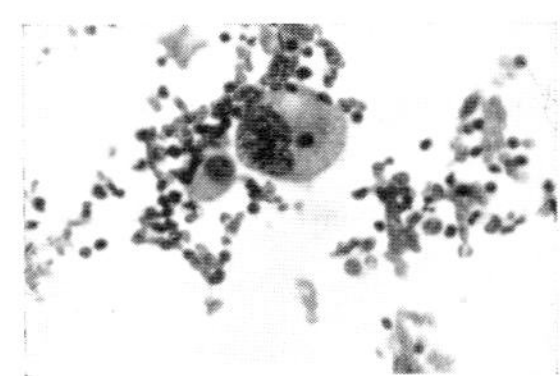

Low Power: Large distinct cell with an eccentric hyperchromatic nucleus.

High Power: This is an example of a single cell which may be identified as malignant. The nucleus appears lobulated rather than multinucleated. Chromatin is increased, irregular in arrangement and has condensed at the nuclear border. Compare chromatin arrangement to that of adjacent mesothelial cell where it is even in pattern, though the aggregates are larger than usual.

FIG. 152

METASTATIC CARCINOMA: CELLS FROM ADENOCARCINOMA

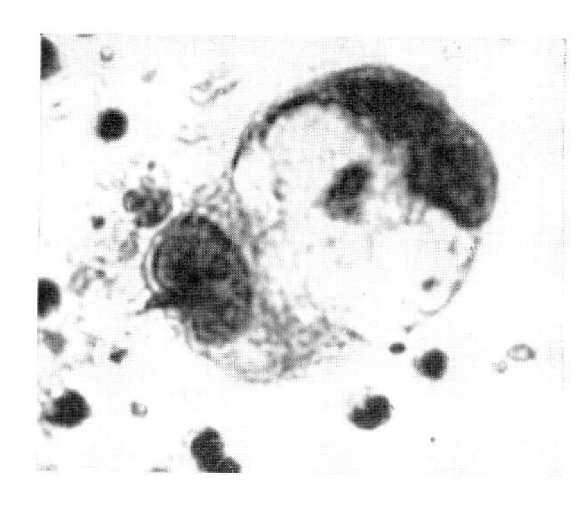

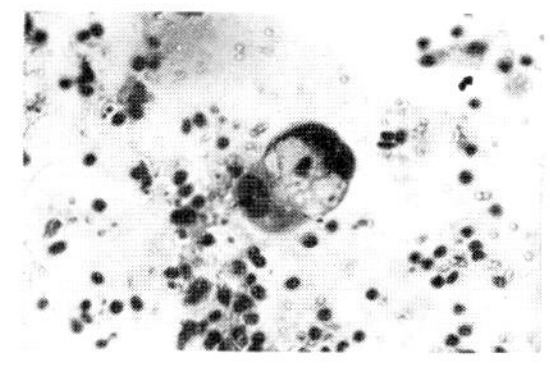

Low Power: Large vacuolated cells with three nuclei.

High Power: The three nuclei of this multinucleated cell fulfill the usual criteria of malignancy, *i.e.*, increase and irregularity of chromatin content, sharp nuclear borders and variation in size and shape. The irregular vacuolization is characteristic of adenocarcinoma as is the unevenness in stain of cytoplasm in left-hand border. Background contains fresh blood and leukocytes.

FIG. 153

METASTATIC CARCINOMA: CELLS FROM ADENOCARCINOMA

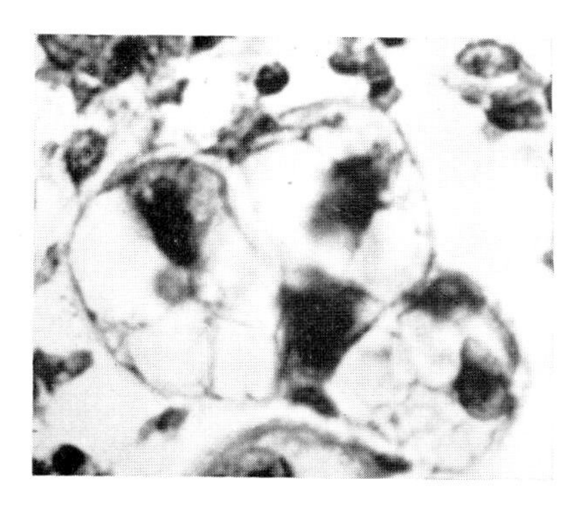

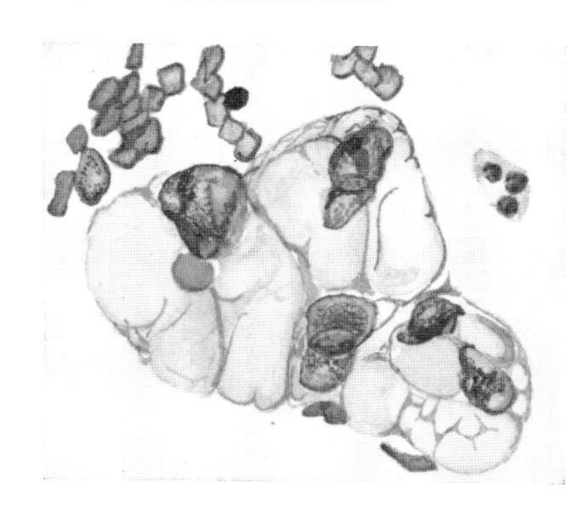

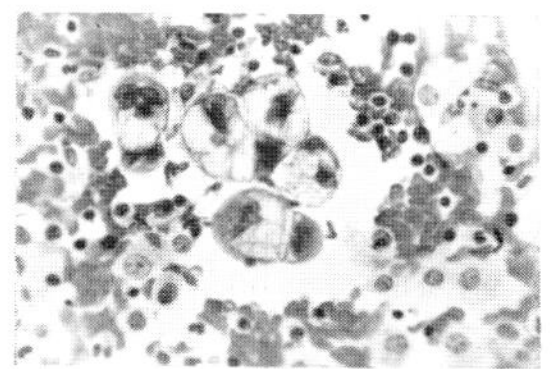

Low Power: Cells showing extreme vacuolization in field of mesothelial cells, leukocytes and blood.

High Power: Comparison of this group with those cells shown in Plate 11, illustrating adenocarcinoma of the endometrium, will show how identical are the characteristics of cells shed from a metastatic lesion with those from the primary tumor. The vacuolization, position and appearance of nuclei are typical of adenocarcinoma.

Since a positive diagnosis on pleural or ascitic fluid is difficult to confirm except by postmortem studies and because in addition a positive diagnosis labels the carcinoma as inoperable, we feel that the cytologist should be extremely conservative in his interpretation of serous fluids. If the cells present are not typically malignant, the specimen should be repeated, and it should be emphasized again that the positive diagnosis depends on finding large groups of malignant cells. A diagnosis of positive on rare cells which vary from the normal is not justified. Since immediate confirmation is not always possible, the cytologic report must often stand alone as an indication of metastatic carcinoma. It may of course be confirmed by a positive diagnosis by cell block, but such a diagnosis is subject to the same limitations as the diagnosis by a smear of the sediment from the fluid. Therefore the positive diagnosis will only be of value if kept extremely accurate.

Difficulties in Interpretation: The normal mesothelial cells of pleural and peritoneal fluid often appear unusually active. Their nuclei are larger than is common in normal epithelial cells. Too, the nuclei may have small clumps of chromatin plus a prominent nucleolus which to the inexperienced observer appears suspicious. They are often multinucleated. These are the only cells which present any difficulty. They may be distinguished from the malignant cells by the following points: 1. They do not have an increase of chromatin. 2. Though there may be an occasional clump of chromatin the remainder of the nucleus appears finely granular. 3. The chromatin is not concentrated at the periphery of the nucleus. 4. The nucleus is surrounded by an adequate amount of cytoplasm and a definite cellular border is present. 5. The cells are usually spread fairly evenly throughout the smear and do not occur in tight overlapping clusters.

General Criteria for Identification of Malignant Cells in Serous Fluids: The cells are undifferentiated with the exception of groups which show the characteristic vacuolization of adenocarcinoma. An occasional cell will show a distinct cell border but this is an exception. The cells appear as a dense cluster of nuclei with distinct aberrations of nuclear structure. The chromatin is increased, is irregularly distributed and concentrated at the periphery of the nucleus. There is variation is size, more so than in shape.

SUMMARY

We report the cytologic findings on fluids in the same manner as in other secretions, *i.e.*, positive, negative and doubtful. We attempt to catalogue the doubtful diagnoses by repeating the smears and in most instances calling it either negative or positive.

Though the cytologic examination of serous fluids is of no aid in discovering early tumors, since we are dealing here with metastatic lesions, we do feel that by its careful use the question of whether or not accumulations of pleural or ascitic fluid are due to the presence of metastatic lesions can be made with great accuracy.

CONCLUSION

The cytologic method depends on the satisfactory preparation of smears. A smear must be properly fixed and stained to be adequate for interpretation. Stains may vary from laboratory to laboratory, but the essential point is that the stain be satisfactory and that the microscopist be familiar with it. Too often an attempt at interpretation is made when the smear is not satisfactory. If, as sometimes happens, the smear is understained or overstained and cellular detail is not distinct, the smear should be repeated. Even with a great deal of experience smears cannot be interpreted properly if they are not well fixed and stained. Many errors will be avoided if one requires excellent technical preparations.

It has been shown that the undifferentiated malignant cells are similar in all types of carcinoma. Distinction between various tumors on the basis of *undifferentiated* cells is impossible. It would be extremely difficult to determine from the desquamated cells of adenocarcinoma whether the tumor originated in the endometrium, lung or stomach. They all have essentially the same characteristics. The cells from the same histologic type of tumor are remarkably similar despite a difference in anatomic location. It is the normal cellular components which vary markedly from secretion to secretion. The normal cells seen in urine sediment are much different from those of sputum. In teaching we have found that far too often the beginner in cytology spends too little time in learning to identify normal cells. The accuracy of interpretation will not be high unless the cytologist is familiar with the characteristics of the normal cells and the various forms they may assume. The primary question to be answered in interpretation is whether the cell is malignant or whether it merely shows unusual variation from the normal but is still benign. This question cannot be answered adequately unless the interpreter is completely familiar with the normal cells of each secretion.

A final word about interpretation is pertinent. Anyone who looks at any appreciable numbers of smears will be impressed with the great variation seen, both in the general characteristics of the smear and in individual cells. We have attempted to describe in the previous pages the most common variations of each cell type and to give the general criteria which have proven satisfactory in our interpretation. We have not included any borderline cells, since we believe that questionable cells would not facilitate learning. Only with considerable experience can such cells be catalogued. However, many smears contain cells which are borderline. Whenever the identification of a cell is in question, the cytologist should review the criteria for all the cells that may occur. Often this process of elimination is helpful. When suspicious cells are encountered, often further search reveals cells which can be definitely identified as positive. If the cells in question cannot be identified, and occasionally they cannot be, the smear should be repeated.

Finally, to classify cells as cancer cells there must be aberration in the nuclear pattern, either great increase in amount of chromatin material or a true alteration of the nuclear pattern. To label a cell malignant by cytologic examination definite abnormal nuclear changes must be present. Since definite alterations of nuclear division and growth are present in cancer, it is logical that these abnormalities be mirrored in nuclear morphology. Thus we must depend on nuclear structure for our diagnosis of malignancy.

The cytologic diagnosis of malignancy depends upon finding and identifying well preserved carcinoma cells. If a tumor is far advanced, there may be some interference in blood supply and the surface undergoes necrosis. Since cells desquamate only from the surface of tumor, the cells in this instance will often be too degenerate to identify as malignant, and thus as far as a cytologic diagnosis is concerned the case must be classified as a "false negative," though obvious tumor was present. It should be remembered that a negative smear *never* rules out the possibility that cancer may be present. On the other hand, if the tumor is small the surface cells will be preserved, so that identification of tumor cells may be made. This is the great advantage and value of the cytologic method. It offers a real opportunity to discover early cases of carcinoma. With accurate interpretation and careful clinical application the cytologic diagnosis should appreciably increase the cure rate of malignant disease.

CHAPTER XX

TECHNIC

To make the proper diagnosis on any smear, it is extremely important that all the technical details are performed with care. It is easier to prepare smears if the proper equipment is at hand. The secretion must be placed on the slide in a way to present the clearest possible picture and must be fixed immediately, since the method depends on wet fixation of the cells. Finally, the stain must be an excellent one for nuclear differentiation. The method of preparation of the various secretions is described below.

THE VAGINAL SMEAR

Material:

1. A slightly curved, 8-inch glass pipette with a capillary opening at one end.
2. A 1-oz. rubber bulb is attached to glass pipette.
3. A clean glass slide plainly marked by a diamond pencil with the type of secretion, patient's name and date.
4. A paper clip is placed on the end of the slide so that more than one smear may be placed in the bottle of fixative at the same time and still remain separate from adjacent slides.
5. A bottle of fixative containing equal parts of 95 per cent alcohol and ethyl ether.

Method:

The rubber bulb is compressed, and the pipette is then inserted into the vagina as far as possible in order to reach the posterior fornix. The bulb is released, and the pipette withdrawn slowly. The secretion is then blown on the slide, spread thinly with a rotary motion of the pipette, and dropped immediately into the bottle of fixative before drying occurs.

Precautions:

1. The pipette must be absolutely dry, and the patient must not have douched the same day. These precautions are necessary since the presence of water in the vaginal smear destroys the cellular detail.
2. The slide must be fixed immediately. It must not be allowed to dry. When the cells dry on the slide before fixation, they lose all differential staining characteristics and all stain a brilliant orange.
3. It is important that not too much secretion be placed on the slide. Often the inexperienced appear to think that the purpose of the smear is to collect as much secretion as possible. This is not so. The vaginal smear merely represents a sampling of the secretion which collects in the posterior fornix. A smear to be adequate for interpretation must be thinly spread. If too much secretion is placed on the slide, it is impossible to have a thin preparation. A small amount of secretion should be placed on the slide and spread evenly with a rotary motion of the tip of the pipette. We prefer this method of spreading the secretion to that of smearing it with another slide or with the gloved finger, since we feel it gives less distortion of the cells.

4. It is necessary to have a capillary opening in the tip of the glass pipette to obtain adequate preparations. The secretion of the vagina is in some instances highly tenacious. It is usually drawn up in the glass tube with moderate ease, but it may be difficult to blow out, as it adheres to the side of the tube. It is almost impossible to blow out if the opening of the tube is not limited in diameter. If the opening is large, the air passes by the secretion, and to deliver any on the slide is difficult. The glass pipettes are cleaned by soaking them in soap solution. After several hours in soap solution they are rinsed thoroughly in running water, boiled for twenty minutes, and rinsed in alcohol. Air is blown through them, either using a rubber bulb or compressed air, until they are perfectly dry.

Discussion:

It has been our policy to prepare one vaginal smear from each patient. Whether more taken at the same time will yield a higher percentage of correct diagnosis is debatable. We have found that one smear usually represents a fair sample of the cellular components in the vaginal secretion. In preparing teaching collections, we have frequently taken multipe smears from a carcinoma patient at one time. Only rarely in the preparation of such positive teaching slides are malignant cells absent from a slide.

We have relied primarily on vaginal smears for our cytologic diagnosis of uterine cancer. Three other types of smear may be taken from the female genital tract: cervical, endocervical and endometrial. The cervical smear is obtained by inserting a speculum and, with the cervix under direct vision, material is aspirated with a pipette and rubber bulb directly from both the surface and the os of the cervix. The secretion is then blown on a glass slide and immersed immediately in fixative. The endocervical and endometrial smears are made by direct aspiration of the endocervical canal or endometrial cavity with a small cannula attached to a syringe. The material obtained is placed on a slide and immediately fixed. The same precautions of fixing the cells before drying apply here as well as in vaginal smears.

In our laboratory we have attempted to keep the cytologic diagnosis of uterine cancer as practical as possible. Technicians have been trained to take vaginal smears. The cervical, endometrial and endocervical smears require additional equipment and, in our experience, are not as practical as the vaginal smear. We have reserved the other types of preparations for those cases which require further study. If the vaginal smear is equivocal, examination of other types of smear may be helpful. We are not convinced that the cervical smear will yield additional correct diagnoses. It has the advantage of aspirating cells directly from the surface of a suspected area which can be visualized, but has the disadvantage of collecting only those cells which are free but still localized to the given area at the moment of aspiration. The vaginal smear, on the other hand, gives a sample of all cellular components desquamated over some period of time; but has the disadvantage that occasionally malignant cells, such as those coming from an adenocarcinoma of the endometrium, will apparently degenerate. Therefore, both methods have certain advantages and disadvantages. We have found that well prepared vaginal smears are adequate in most cases. However, none of these types of smears should be used to the exclusion of the others.

In recent years Ayre has advocated scraping the cervix with a wooden spatula. We have had no experience with this method and are not competent to judge its advantages or disadvantages.

THE SPUTUM SMEAR

Material:

1. The sputum is collected in a paper sputum cup and sent to the laboratory within an hour.
2. Two wooden applicators.
3. Three glass slides, marked with patient's name and date by a diamond pencil. Each slide has a paper clip at end marked with name and date to avoid contact of slides after they are placed in fixative.
4. A bottle of fixative containing equal parts of 95 per cent ethyl alcohol and ethyl ether.

Method:

The sputum specimen is examined first for any blood-flecked areas. If these are present, portions of them are placed on a slide. If no such areas are seen, an attempt is made to secure representative portions of the sputum, *i.e.*, both the thick and thin portions. With the aid of two wooden applicators, these samples are spread evenly on three marked slides and immersed immediately in fixative of half 95 per cent ethyl alcohol and half ethyl ether.

Precautions:

The two factors to be considered in the preparation of sputum are (1) that the specimen is fresh and (2) that it has been coughed from "down deep" and represents a true specimen of sputum and not just saliva. The mucus present in sputum preserves the cells remarkably, but better preparations are obtained if the specimen is prepared on slides as soon as possible after it has reached the laboratory. In the chapter on normal cells in sputum, we have discussed how essential it is that the specimen be composed of sputum, not saliva. Any sputum which does not contain histiocytes microscopically is regarded as unsatisfactory.

SMEARS OF BRONCHIAL SECRETIONS

Material:

1. Two marked glass slides prepared with paper clips.
2. Fixative.

Method:

The secretions obtained at bronchoscopy are collected either by aspiration into a glass suction trap, by washing the bronchi with 1 to 2 cc. of saline, or on a cotton plug in the aspirator itself. If the secretions are obtained in a suction trap and there is enough secretion that it will not become dry in the cup, it is preferable to send it immediately to the laboratory where two smears are made in the same manner as sputum smears. However, if the secretions obtained either in the suction cup or by washing the bronchi are small in amount, the smears should be prepared immediately in the operating room. This is also true of the secretions obtained on the cotton swab. The material is placed on clean glass slides and spread thinly (wooden applicators are useful), and the slides are placed immediately in fixative.

Technic

Precautions:

It is essential that the material placed on the slide be actually secretion from the bronchus. Great care should be taken in this respect in order that the smear contain epithelial cells from the bronchus and not just red blood cells and leukocytes. As in the other secretions, it is important that the slide be placed immediately in fixative.

Discussion:

Whether sputum or bronchial secretions are used for the diagnosis of pulmonary carcinoma is a question to be decided by individual clinics. In our experience the diagnostic accuracy has been better in sputum examinations than in those of bronchial secretions. We have found that this was entirely a technical difference. The laboratory could exercise more control over the preparation of sputum smears and therefore these slides were far more satisfactory than those of bronchial secretions. This is an example of how important the technical preparation of the smear may be.

Other than technical factors, both types of secretion have certain advantages and disadvantages. Sputum has the advantage of being easy to repeat, but the disadvantage of being more dilute than the bronchial secretions. Often, by the time patients have pronounced enough symptoms to warrant bronchoscopy, the disease is far advanced, so that, theoretically, sputum offers an advantage in this respect. However, it should be pointed out that patients may have an unproductive cough so that it is not always possible to obtain adequate amount of sputum for examination. The best diagnostic results would probably be obtained if the examination of both sputum and bronchial secretion were done whenever possible.

SMEARS OF GASTRIC SECRETION

Material:

1. Size No. 14 Levin tube, which has several openings in addition to the ones already present.
2. A 50 cc. syringe.
3. Saline.
4. Two 50 cc. centrifuge tubes.
5. Wire loop.

Method:

The No. 14 Levin tube is introduced through the nostril of the patient for a distance of 75 cm. While the tube is being inserted, the patient is allowed to drink one glass of tap water. The stomach is then evacuated as completely as possible with a 50 cc. syringe attached to the Levin tube. This is the first specimen. For the second specimen 100 cc. of normal saline is injected through the tube, partially withdrawn, and reinjected several times. The tube is then clamped off and the patient is told to lie prone, supine and in both lateral positions, to sit up and finally to walk, if ambulatory. The stomach is again aspirated until empty.

The specimens are brought immediately to the laboratory, since, if there is more than half an hour's delay, the cells are digested and no definite cellular characteristics can be identified. Both samples of fluid are entirely centrifuged in 50 cc. centrifuge tubes for twenty minutes at 2000 r.p.m. The supernatant fluid is poured off and two marked glass slides for each specimen are spread thinly with egg albumin. With a wire loop the sediment is spread evenly on the slides and put immediately into fixative of half 95 per cent ethyl alcohol and half ethyl ether.

Precautions:

In no other fluid is it as important that preparations of the secretion be made immediately. Specimens must not sit around the laboratory for any length of time. The cells of the gastric fluid degenerate very rapidly, and to obtain satisfactory specimens, the secretion must be centrifuged, placed on slides and put in fixative, allowing as little time to elapse as possible. Specimens are considered unsatisfactory microscopically if the flat squamous cells are not well preserved.

Discussion:

We have not found any remarkable difference between the water or saline aspiration. In an occasional case, malignant cells will be present in one and not in the other, but in the great majority of cases which are positive cytologically, malignant cells are present in both. We have studied many different methods of collecting gastric secretion, such as use of 5 per cent alcohol, use of clove oil as an irritant perhaps to produce more desquamation of cells, and addition of sodium bicarbonate to neutralize the acidity. We found that the perfection of the specimen microscopically seemed to have no correlation with any of the above factors, but was closely correlated with the time which elapsed between the aspiration of the secretion and its fixation. Either tap water or saline lavage gives adequate specimens. The cells of the gastric secretion in situ do not seem to be affected by a solution which is not isotonic.

We have discussed previously the gastric specimens from patients with obstruction. If too many food particles are present, the specimens are of course unsatisfactory for interpretation. Adequate preparations may be obtained if the stomach of an obstructed patient can be completely emptied of all material, and then a second aspiration done twelve hours later.

It cannot be overemphasized that time is the most important factor in the preparation of gastric smears. The cells must be in fixative as soon as possible after aspiration. To insure immediate fixation, some workers have advocated adding fixative directly to the specimen of gastric secretion. We dislike this method of fixation, since the protein present in the specimen is precipitated and the precipitate makes centrifuging and smearing of the sediment difficult. We prefer, if at all possible, to have the fresh specimen sent to the laboratory immediately.

SMEARS OF URINE SEDIMENT

Material:

1. Two 50 cc. centrifuge tubes.
2. Wire loop.
3. Egg albumin.
4. Three glass slides marked with patient's name and date prepared with paper clips.

Method:

A fresh voided or catheterized specimen of urine is centrifuged at 2000 r.p.m. for fifteen minutes. If the volume of the specimen is greater than 100 cc., the supernatant fluid is poured off and more urine added to the same tubes so that all the sediment is collected in two tubes. Two or three drops of egg albumin is added to the sediment and mixed with a wire loop. Egg albumin is also spread thinly on three marked glass slides. The sediment is spread on the slides by means of the wire loop, and the slides are placed immediately in the fixative in the same manner as other smears.

Precautions:

The sediment of urine usually contains no protein to help in fixing the cells to the slide. It is difficult to retain the sediment on the slide during the process of staining. Egg albumin on the slide is of considerable help, and the addition of egg albumin to the sediment directly will improve results. However, the smears of urine sediment should be handled with care and watched particularly during the washes with running water.

Discussion:

As in other body fluids, the specimen of urine should be centrifuged and the sediment placed in fixative while the specimen is still fresh. If it is impossible to send the fresh specimen to the laboratory, we have found that the addition of 10 cc. of formalin to 90 cc. of urine will adequately preserve the cells. We have found no definite difference between catheterized and voided specimens, and for routine purposes have preferred the freshly voided morning specimen.

SMEARS OF SEDIMENT OF SEROUS FLUIDS

Material:

1. Two 50 cc. centrifuge tubes.
2. Wire loop.
3. Three slides marked with patient's name and date.

Method:

After aspiration of the fluid, it is sent directly to the laboratory where an adequate sample is centrifuged for 15 minutes at 2000 r.p.m. We do not attempt to centrifuge the entire specimen in most instances. The fluid is shaken, and then approximately 100 cc. centrifuged. The sediment is evenly spread on three slides by means of a wire loop and placed in fixative. Egg albumin preparation of the slides is usually not necessary.

Precautions:

We feel that it is essential that the specimen of pleural or ascitic fluid be fresh. It is centrifuged as soon as possible after reaching the laboratory. Specimens which have remained in the refrigerator overnight lose much of their cellular detail.

FIXATION OF SMEARS

All the types of smears are fixed in the same fixative, a solution of half ethyl ether and half 95 per cent ethyl alcohol. The slides should remain in the fixative for at least half an hour before staining. However, if immediate staining is not desired, they may stay in the fixative as long as two weeks without deterioration. If the slides are to be mailed without staining, they may be taken out after an hour's fixation, dried, wrapped in paper and mailed. This does not give quite as definite cellular detail as in slides which are fixed and immediately put through the staining procedure, but they are adequate for interpretation. If desired, a thin coating of glycerin may be put on the smear, covered by another slide and mailed, according to the directions of Ayre.[1]

After the smears are removed from the fixative, it should be filtered and put in fresh, clean bottles for re-use. Filtration is necessary because particles of secretion often are present free in the solution.

STAINING PROCEDURE

All the smears are stained in exactly the same way, according to the original description of Papanicolaou.[2] We have found that this is an excellent stain, giving especially good nuclear differentiation. The procedure is as follows:

Slides are transferred directly from the fixative to

70 per cent ethyl alcohol	10 dips
50 per cent ethyl alcohol	10 dips
Distilled H_2O	10 dips
Harris Alum Hematoxylin	3 minutes
Running tap H_2O	1 minute
0.5 per cent HCl	5 dips (no more)
Running tap H_2O	4 minutes
Dilute solution of Lithium Carbonate	1 minute
Running tap H_2O	1 minute
50 per cent ethyl alcohol	10 dips
70 per cent ethyl alcohol	10 dips
80 per cent ethyl alcohol	10 dips
95 per cent ethyl alcohol	10 dips
Orange G-6	1 minute
95 per cent ethyl alcohol	10 dips
95 per cent ethyl alcohol	10 dips
EA-50	2 minutes
95 per cent ethyl alcohol	10 dips
95 per cent ethyl alcohol	10 dips
95 per cent ethyl alcohol	10 dips
Absolute alcohol	4 minutes
Xylol	5 minutes

Slides are mounted in neutral Canada Balsam.

Harris Alum Hematoxylin is made as follows:[3]

Hematoxylin	1 gm.
Absolute alcohol	10 cc.
Ammonium or potassium alum	20 gm.
Distilled water	200 cc.
Mercuric oxide	0.5 gm.

[1] Ayre, J. E., and Dakin, E: Cervical Cytology Tests in Cancer Diagnosis. Glycerin Technique for Mailing. Canad. M.A.J. *54:* 489–491, 1946.

[2] Papanicolaou, G. N.: A New Procedure for Staining Vaginal Smears. Science *95:* 438–439, 1942.

[3] From "Pathological Technique," by F. B. Mallory. W. B. Saunders Co., Philadelphia. 1938.

Technic

Dissolve the hematoxylin in the alcohol, the alum in the water by the aid of heat, and mix the two solutions together. Bring the mixture to a boil as rapidly as possible and then add mercuric oxide. The solution at once assumes a dark blue color. As soon as this occurs, remove the vessel containing the solution from the flame and cool by plunging at once into a basin of cold water. As soon as the solution is cooled, it is ready for staining.

We have preferred to use Harris' Hematoxylin *without* the acetic acid, since, in our experience, it gives better nuclear differentiation in this combination of counterstains if the acid is omitted. The length of time of staining in the hematoxylin depends on the strength of the stain and individual preference of density of nuclear stain. We have found, that on the average, staining for three minutes is satisfactory. The slides are washed in running tap water until all excess hematoxylin is removed. The solution of hematoxylin should be filtered after each day's use.

The slides are destained by the five dips in 0.5 per cent HCl. This is made by adding 1 cc. of concentrated HCl to 200 cc. of distilled water. It is important that the slides are not left in the acid longer than for five dips, since, if they are, too much hematoxylin will be removed from the nuclei. After the slides have been in the dilute acid, it is essential that they be washed adequately in running tap water so that all acid is removed. Slides are then placed in the dilute lithium carbonate solution[4] to insure an alkaline *p*H.

The slides are run up through increasing concentrations of ethyl alcohol to the counterstain OG-6, through two rinses of 95 per cent ethyl alcohol, to the counterstain EA-50, followed by three successive rinses in 95 per cent ethyl alcohol, dehydration in absolute alcohol, clearing in xylol and mounted in neutral Canada Balsam.

The prepared counterstains OG-6 and EA-50 may be obtained commercially.[5] We have found these stains very satisfactory. However, early in our work when commercial stains were not available, we prepared our own according to the directions given by Papanicolaou. These are as follows:

	Orange G-6	
Orange G (National Aniline and Chemical Company)	0.5 per cent solution in 95 per cent ethyl alcohol	100 cc.
Acid Phosphotungstic (Merck)		0.015 gm.
	EA-36	
Light Green S.F. yellowish (National Aniline and Chemical Company)	0.5 per cent solution in 95 per cent ethyl alcohol	45 cc.
Bismarck Brown (National Aniline and Chemical Company)	0.5 per cent solution in 95 per cent ethyl alcohol	45 cc.
Eosin yellowish (National Aniline and Chemical Company)	0.5 per cent solution in 95 per cent ethyl alcohol	45 cc.
Acid Phosphotungstic (Merck)		0.200 gm.
Lithium Carbonate	Saturated aqueous solution	1 drop

[4] An aqueous, saturated solution of lithium carbonate is prepared. One cc. of the saturated solution is added to 100 cc. of distilled water.

[5] Ortho Pharmaceutical Corporation, Raritan, N. J.

Since these stains are not as soluble in alcohol as in water, it is best to prepare stock solutions of saturated aqueous solutions and add this to the alcohol, rather than to attempt to dissolve them in the alcohol directly. Both OG-6 and EA-36 are alcoholic solutions, and a certain amount of evaporation takes place, so that it is essential that they be kept in tightly stoppered bottles when not in use.

We have replaced our stains on the average of once every two weeks. This time will vary according to how many slides are stained. Whether the stains should be changed can easily be determined microscopically. If few slides are stained, Coplin jars are satisfactory as staining dishes. For large numbers of slides we have found glass carriers and large staining dishes to work very well. When carriers are used, the technician should be careful that excess fluid is drained off before the carrier is immersed in the next solution. We have used paper towels to blot off excess fluid between dishes.

After the slides are stained, they are mounted in Canada Balsam and covered with a No. 1 coverslip. It is essential in the microscopic interpretation of smears that slides are not wet mounts but completely dry when they are interpreted. Not only because wet preparations are apt to be hard to handle and cannot be thoroughly cleaned, but, more important, because the method depends on the marking of suspicious cells. This is impossible if the coverslip is not set. Ink dots are valueless if the coverslip moves even slightly.

To overcome this difficulty, at the Vincent Laboratory we developed a "cooking" process for rapid drying of the slides. After the smears are mounted, they are placed on a porcelain pan which is on an electric hot plate. The plate is turned on at "medium" heat. Two or three slides are placed on the pan at once. While they are heating, the technician presses gently on the coverslip to squeeze out all excess balsam. As soon as the balsam begins to bubble, the slides are immediately removed, pressure is again applied with the forceps to remove all air bubbles, and the slides allowed to cool. After the slide is cool, the coverslip is immovable, even when the slide is immersed in xylol, so that it may then be thoroughly cleaned. No attempt is made to clean the slides until after the "cooking" process, though they are mounted with as little excess balsam as possible.

We have found this method very satisfactory. Not only do the slides for interpretation come to the microscopist ready to be dotted, but slides may be filed that same say in a "Technicon" drawer without any necessity for separation. This simplifies the filing of large numbers of slides. The "cooking" process does not interfere with the staining properties of cells if the slides are removed as soon as the balsam begins to bubble. All slides used for photomicrographs and drawings in this book have gone through this process.

MICROSCOPIC EXAMINATION OF THE SMEARS

Every smear is covered systematically, field by field, under low power. Any unusual or suspicious cells are examined under high power. If the cell warrants further examination, it is marked by an ink dot. Slides which arc considered to be either positive or doubtful are referred by the technician to the experienced cytologist for final decision. Negative smears are the responsibility of the technician screening thide.

It is extremely helpful to have binocular microseopes for the examination of smears, since it is less tiring in long sessions of microscopic work. We have used 10X oculars, a low power objective of 10X and high power of either 40 or 43X. It is only rarely that the oil immersion objective is used.

TECHNICAL DIFFICULTIES

There is the distinct possibility of contamination. This is a real possibility, and all possible means should be used to avoid it. It is not our impression that the chances of contamination are great in the staining procedure. The slides are washed thoroughly in running tap water so that all free particles should be washed away. However, staining dishes should be examined for any free material and if present, immediately filtered. We believe that the likeliest place for contamination to occur is in the mounting of the slides. If the technician touches the balsam rod to a smear, cells which are not firmly fixed to the slide will stick to the rod, to be placed back in the balsam bottle where they remain perfectly preserved. For this reason, it is most important that the balsam be *dropped* on the slide.

Fortunately it is fairly easy to recognize contamination with malignant cells in most instances. The cells appear distinctly different from other components of the smear and are on a different level. If a smear is suspected to have malignant cells as contaminants, it is called doubtful and immediately repeated. The most common sources of contamination are serous fluids which contain great numbers of malignant cells.

We have found that if the procedures outlined above are carefully followed, good preparations of smears may be obtained. The microscopic interpretation of any type of smear depends to a great degree on the adequacy of the preparations

APPENDIX

Below are the histologic diagnoses on the tumors of the cases used as illustrating exfoliated malignant cells. In a rare instance only a clinical diagnosis was available, and is so indicated.

Chapter IV

SQUAMOUS CELL CARCINOMA OF THE CERVIX

Plate 7	Squamous Cell Carcinoma of the Cervix
Fig. 35	Squamous Cell Carcinoma of the Cervix, Grade III
Fig. 36	Squamous Cell Carcinoma in situ of the Cervix
Fig. 37	Squamous Cell Carcinoma in situ of the Cervix
Fig. 38	Squamous Cell Carcinoma of the Cervix
Fig. 39	Squamous Cell Carcinoma of the Cervix
Fig. 40	Squamous Cell Carcinoma of the Cervix, Grade III
Fig. 41	Squamous Cell Carcinoma in Situ of the Cervix
Fig. 42	Squamous Cell Carcinoma of the Cervix, Grade III
Fig. 43	Squamous Cell Carcinoma of the Cervix, Grade III
Fig. 44	Squamous Cell Carcinoma of the Cervix, Grade III
Plate 8	Squamous Cell Carcinoma of the Cervix, Grade II
Fig. 45	Squamous Cell Carcinoma of the Cervix, Grade III
Fig. 46	Squamous Cell Carcinoma of the Cervix, Grade III
Fig. 47	Squamous Cell Carcinoma of the Cervix, Grade III
Fig. 48	Squamous Cell Carcinoma of the Cervix
Plate 9	Squamous Cell Carcinoma of the Vagina
Fig. 49	Squamous Cell Carcinoma of the Cervix, Grade III
Fig. 50	Squamous Cell Carcinoma of the Cervix, Grade III
Fig. 51	Squamous Cell Carcinoma of the Cervix, Grade III
Fig. 52	Squamous Cell Carcinoma of the Cervix
Fig. 53	Squamous Cell Carcinoma of the Cervix, Grade III
Fig. 54	Squamous Cell Carcinoma of the Cervix, Grade III
Plate 10	Squamous Cell Carcinoma in situ of the Cervix
Fig. 55	Squamous Cell Carcinoma in situ of the Cervix
Fig. 56	Squamous Cell Carcinoma in situ of the Cervix
Fig. 57	Squamous Cell Carcinoma of the Cervix, Grade III
Fig. 58	Squamous Cell Carcinoma of the Cervix
Fig. 59	Squamous Cell Carcinoma in situ of the Cervix
Fig. 60	Squamous Cell Carcinoma in situ of the Cervix

Chapter V

ADENOCARCINOMA OF THE ENDOMETRIUM

Plate 11	Adenocarcinoma of Endometrium
Fig. 62	Adenocarcinoma of Endometrium, Grade II
Fig. 63	Adenocarcinoma of Cervix and Endometrium, Grade III
Fig. 64	Papillary Adenocarcinoma of Endometrium
Fig. 65	Adenocarcinoma of Endometrium
Fig. 66	Papillary Adenocarcinoma of Endometrium and Cervix
Fig. 67	Adenocarcinoma of Endometrium
Fig. 68	Adenocarcinoma of Endometrium
Fig. 69	Adenocarcinoma of Endometrium

Chapter VI

ADENO-ACANTHOMA OF THE UTERUS

Plate 12	Adenocarcinoma of the Endometrium

Chapter VII

OTHER TUMORS OF THE FEMALE GENITAL TRACT

Chapter VIII

RADIATION CHANGES IN NORMAL AND MALIGNANT CELLS

Chapter XII

SQUAMOUS CELL CARCINOMA OF THE LUNG

Chapter XIII

OTHER TYPES OF PULMONARY CARCINOMA

Chapter XV

CARCINOMA OF THE STOMACH

Chapter XVII

MALIGNANT CELLS FROM CARCINOMA OF THE GENITO-URINARY TRACT

Chapter XIX

MALIGNANT CELLS IN SEROUS FLUIDS

BIBLIOGRAPHY

I. General Textbooks

de Allende, I. L. C., Shorr, E., and Hartman, C. G.: A Comparative Study of the Vaginal Smear Cycle of the Rhesus Monkey and the Human. Carnegie Institution of Washington, Publication 557, Contributions to Embryology *31:* 1, 1943.

Gates, O., and Warren, S.: A Handbook for the Diagnosis of Cancer of the Uterus by the Use of Vaginal Smears. Cambridge, Massachusetts, Harvard University Press, 1947.

Papanicolaou, G. N., and Traut, H. F.: Diagnosis of Uterine Cancer by the Vaginal Smear. New York, Commonwealth Fund, 1943.

Papanicolaou, G. N., Traut, H. F., and Marchetti, A. A.: Epithelia of Woman's Reproductive Organs: A Correlative Study of Cyclic Changes. New York, Commonwealth Fund, 1948.

II. Application of Vaginal Smear to Endocrinology

Bonime, R. G.: Application of the Vaginal Smear to the Diagnosis of Pregnancy. Am. J. Obst. & Gynec. *58:* 524, 1949.

Brown, W. E., and Bradbury, J. T.: The Use of the Human Vaginal Smear in the Assay of Estrogens. J. Clin. Endocrinol. *9:* 725, 1949.

Culiner, A., and Gluckman, J.: The Use of the "Phase-Contrast" Microscope in Clinical Gynaecology, A Preliminary Report. J. Obst. & Gynaec. Brit. Emp. *55:* 261, 1948.

Davidson, H. B., Hecht, E. L., and Winston, R. L.: Significance of Abnormal Menopausal Vaginal Smears. Am. J. Obst. & Gynec. *57:* 370, 1949.

Kernodle, J. R., and Cuyler, W. K.: Vaginal Cytology of Post-Menopausal Women, Study I, Cytologic Variations in Vaginal Smears. South. M. J. *41:* 861, 1948.

Kernodle, J. R., and Cuyler, W. K.: Vaginal Cytology of Post-Menopausal Women, Study II, South. M. J. *41:* 869, 1948.

Mack, H. C.: The Glycogen Index in the Menopause. Am. J. Obst. & Gynec. *45:* 402, 1943.

Niesburgs, H. E., and Greenblatt, R. B.: Significance of Specific Estrogenic, Progestogenic and Androgenic Smears in Menstrual Disorders and in Pregnancy. South. M. J. *41:*972, 1948.

Nieburgs, H. E., and Greenblatt, R. B.: Specific Estrogenic and Androgenic Smears in Relation to Fetal Sex during Pregnancy. Am. J. Obst. & Gynec. *57:* 356, 1949.

Papanicolaou, G. N.: The Sexual Cycle in the Human Female as Revealed by Vaginal Smears. Am. J. Anat. (supp.) *52:* 519, 1933.

Papanicolaou, G. N.: Existence of a "Post-Menopause" Sexual Rhythm in Women as Indicated by the Study of Vaginal Smear. Anat. Rec. (supp.) *55:* 71, 1933.

Papanicolaou, G. N.: Diagnosis of Pregnancy by Cytologic Criteria in Catheterized Urine, Proc. Soc. Exp. Biol. & Med., *67:* 247, 1948.

Papanicolaou, G. N., and Shorr, E.: Action of Ovarian Follicular Hormone in Menopause as Indicated by Vaginal Smears. Am. J. Obst. & Gynec. *31:* 806, 1936.

Rakoff, A. E., Feo, L. G., and Goldstein, L.: The Biologic Characteristics of the Normal Vagina. Am. J. Obst. & Gynec. *47:* 467, 1944.

Rubenstein, B. B.: Vaginal Smear: Basal Body Temperature Technic and Its Application to Study of Functional Sterility in Women. Endocrinology *27:* 843, 1940.

Salvatore, C. A.: Cytologic Examination of Uterine Growth during Pregnancy. Endocrinology *43:* 355, 1948.

Shorr, E.: New Technic for Staining Vaginal Smears. Single Differential Stain. Science *94:* 545, 1941.

Shorr, E.: Evaluation of Clinical Applications of Vaginal Smear Method. J. Mt. Sinai Hosp. *12:* 667, 1945.

Schuman, W.: Possible Significance of Vaginal Smears in Diagnosis of Certain Disturbances of Pregnancy. Am. J. Obst. and Gynec. *47:* 808, 1944.

Traut, H. F., Block, P. W., and Kuder, A.: Cyclical Changes in the Human Vaginal Mucosa. Surg. Gynec. & Obst. *63:* 7, 1936.

Traut, H. F., and Papanicolaou, G. N.: Vaginal Smear Changes in Endometrial Hyperplasia and in Cervical Keratosis. Anat. Rec. *82:* 478, 1942.

Ulfelder, H., and Meigs, J. V.: Gynecology: The Vaginal Smear. New England J. Med. *237:* 54, 1947.

Varangot, J., and Labatut, M.: The Use of the Quantitative Vaginal Smear in the Standardization of Estrogens in the Woman. Gynec. et Obst. *47:* 540, 1948.

III. General Characteristics of Malignant Cells

Ayres, W. W.: A Method of Staining Nucleoli of Cells in Fresh Benign and Malignant Tissues. Cancer Research *8:* 352, 1948.

Coman, D. R.: Decreased Mutual Adhesiveness, A Property of Cells from Squamous Cell Carcinomas. Cancer Research *4:* 625, 1944.

Fidler, H. K.: A Comparative Cytological Study of Benign and Malignant Tissues. Am. J. Cancer *25:* 772, 1935.

Guttman, P. H., and Halpern, S.: Nuclear-Nucleolar Volume Ratio in Cancer. Am. J. Cancer *25:* 802, 1935.

Haumeder, M. E.: Cytologic Comparison of Malignant and Non-Malignant Nuclei and Nucleoli. J. Lab. & Clin. Med. *23:* 1046, 1938.

Hauptman, E.: The Cytologic Features of Carcinomas as Studied by Direct Smears. Am. J. Path. *24:* 1199, 1948.

McCutcheon, M., Coman, D. R., and Moore, F. B.: Studies on Invasiveness of Cancer Adhesiveness of Malignant Cells in Various Human Adenocarcinomas. Cancer *1:* 460, 1948.

MacCarty, W. C.: The Malignant Cell. J. Cancer Research *13:* 167, 1929.

MacCarty, W. C.: Has the Cancer Cell any Differential Characteristics, Am. J. Cancer. *20:* 403, 1934.

MacCarty, W. C.: The Value of the Macronucleolus in the Cancer Problem. Am. J. Cancer *26:* 529, 1936.

MacCarty, W. C.: Identification of the Cancer Cell. J. A. M. A. *107:* 844, 1936.

IV. General Application of Exfoliative Cytology in Diagnosis of Malignancy

Adams, G.: Cytologic Method of Cancer Detection. Mil. Surgeon *103*: 341, 1948.

Boyd, W.: Discussion: Symposium on the Cytologic Diagnosis of Cancer. Am. J. Clin. Path. *19:* 341, 1949.

Ferguson, J. H.: Some Limitations of Cytological Diagnosis of Malignant Tumors. Cancer *2:* 845, 1949.

French, W. E., and Golden, A.: Experience with the Examination of Fluid Sediments in the Diagnosis of Neoplasms. Memphis M. J. *23:* 68, 1948.

Fremont-Smith, M., Graham, R. M., and Meigs, J. V.: Early Diagnosis of Cancer by Study of Exfoliated Cells. J. A. M. A. *138:* 469, 1948.

Herbut, P. A., Bohrod, M. G., Freeman, W., Robson, S. M., and Papanicolaou, G. N.: General Discussion: Symposium in the Cytologic Diagnosis of Cancer. Am. J. Clin. Path. *19:* 343, 1949.

Hunter, W. C., and Richardson, H. L.: Cytologic Recognition of Cancer in Exfoliated Material from Various Sources. Surg. Gynec. & Obst. *85:* 275, 1947.

Leslie, Eugenie P., and Chang, Helen: Cytologic Test of Various Body Fluids in Early Diagnosis of Cancer. J. Am. M. Women's A. *3:* 236, 1948.

Marcuse, P. M., and Coulter, W. W.: Examination of Body Fluids for the Diagnosis of Malignancy. Texas State J. Med. *43:* 623, 1948.

Papanicolaou, G. N.: Diagnostic Value of Exfoliated Cells from Cancerous Tissues. J. A. M. A. *131:* 372, 1946.

Papanicolaou, G. N.: The Cell Smear Method of Diagnosing Cancer. Am. J. Pub. Health *38:* 128, 1948.

Quisenberry, W. B.: The Cytologic Diagnosis of Cancer by Smear Technique. I. History, Applications and Future of the Method. Hawaii M. J. *8:* 29, 1948.

Robertson, Edwin M.: Cell Smears in the Diagnosis of Malignancy. Canad. M. A. J. *59:* 148, 1948.

Shushan, Ruth M.: The Value of the Cytologic Smear in the Diagnosis of Cancer. New Orleans M. & S. J. *101:* 108, 1948.

Tilden, I. L.: The Cytologic Diagnosis of Cancer by the Smear Technic. II. Technical Methods. Hawaii Med. J. *8:* 342, 1949.

Ulfelder, H.: Medical Progress: Exfoliative Cytology. New England J. Med. *241:* 236, 1949.
Van Hecke, T. J., and Ziehl, F. L.: Smear Technique for the Diagnosis of Cancer. Marquette M. Rev. *12:* 98, 1947.
Wilensky, Abraham O.: The Identification of Cancer Cells in the Natural Organ Secretions in the Early Diagnosis of Malignancy. M. Rec. *161:* 474, 1948.

V. Carcinoma of Female Genital Tract

Ayre, J. E.: Simple Office Test for Uterine Cancer Diagnosis. Canad. M. A. J. *51:* 17, 1944.
Ayre, J. E.: Vaginal and Cervical Cytology in Uterine Cancer Diagnosis. Am. J. Obst. & Gynec. *51:* 743, 1946.
Ayre, J. E.: Vaginal Cell Examination as Routine in Diagnosis. Study of Vaginal and Cervical Cytology as Related to Abnormal Growths. South. M. J. *39:* 847, 1946.
Ayre, J. E.: Selective Cytology Smear for Diagnosis of Cancer. Am. J. Obst. & Gynec. *53:* 609, 1947.
Ayre, J. E.: Cervical Cancer: A Disordered Growth Response to Inflammation in the Presence of Estrogen Excess and Nutritional Deficiency. Cytological, Clinical, Nutritional, and Pathologic Studies. Am. J. Obst. & Gynec. *54:* 363, 1947.
Ayre, J. E.: Cervical Cytology in Diagnosis of Early Cancer. J. A. M. A. *136:* 513, 1948.
Ayre, J. E.: Preclinical Cancer of the Cervix. J. A. M. A. *138:* 11, 1948.
Ayre, J. E., and Ayre, W. B.: Progression from "Precancer" Stage to Early Carcinoma of Cervix within One Year. Am. J. Clin. Path. *19:* 770, 1949.
Ayre, J. E., and Dakin, E.: Cellular Diagnosis of Uterine Cancer by Centrifuge. Canad. M. A. J. *53:* 63, 1945.
Ayre, J. E., and Dakin, E.: Cervical Cytology Tests in Cancer Diagnosis. Glycerin Technique for Mailing. Canad. M. A. J. *54:* 489, 1946.
Bianchi, Mario: La citologia die neoplasmi dell'apparato genitale femminile, prima e dopo trattamento radium, studiata con il metodo della apposizione. Tumori, Milano *34:*82, 1948.
Bickers, W., Sahyoun, P., and Massie, M.: Vaginal Cervical Smear in the Diagnosis of Uterine Cancer. Virginia Med. Monthly *75:* 568, 1948.
Brown, W. E., Bradbury, J. T., and Kraushaar, O. F.: The Papanicolaou Test in the Cancer Control Program. J. Kansas M. Soc. *50:* 217, 1949.
Brown, W. E., Kraushaar, O. F., and Bradbury, J. T.: Vaginal Smear in Diagnosis of Gynecologic Cancer. J. Iowa M. Soc. *37:* 155, 1947.
Chung, Jos.: Detection of Uterine Carcinoma by the Vaginal Smear Method. Quart. Bull. Northwestern Univ. M. School *23:* 33, 1949.
Cox, J. K.: The Value and Accuracy of the Vaginal Smear in the Diagnosis of Uterine and Cervical Cancer. Texas Rep. Biol. & Med. *6:* 77, 1948.
Cromer, J. K., Platt, Lois I., and Winship, T.: The Colpocytological (Papanicolaou) Method of Diagnosis of Uterine Cancer; A Preliminary Report. M. Ann. District of Columbia *17:* 272, 1948.
Cusmano, L.: The Significance of the Nuclear Structure in the Vaginal Secretion Cells as a Means of Cancer Diagnosis. Am. J. Obst. & Gynec. *57:* 411, 1949.
Diddle, A. W., Ashworth, C. T., Brown, W. W., Jr., and Bronstad, M. T., Jr.: Noninvasive Cervical Carcinoma. Am. J. Obst. & Gynec. *57:* 376, 1949.
Editorial: The Papanicolaou Test for Uterine Cancer. J. A. M. A. *136:* 331, 1948.
Federer, Alice B.: The Technique for Vaginal Smears in Uterine Cancer Determination. J. Am. N. Technologists *8:* 617, 1947.
Foote, F. W., and Li, K.: Smear Diagnosis of in situ Carcinoma of the Cervix. Am. J. Obst. & Gynec. *56:* 335, 1948.
Frech, H. C.: Adenocarcinoma of Ovary Diagnosed by Vaginal Smear. Am. J. Obst. & Gynec. *57:* 802, 1949.
Fremont-Smith, M., et al.: The Vaginal Smear in the Diagnosis of Uterine Cancer. J. Clin. Endocrinol. *5:* 40, 1945.
Fremont-Smith, M., et al.: Cancer of Endometrium and Prolonged Estrogen Therapy. J. A. M. A. *131:* 805, 1946.
Fremont-Smith, M., and Graham, R. M.: Early Diagnosis of Uterine Cancer by Vaginal Smear. S. Clin. North America *27:* 1215, 1947.
Fremont-Smith, M., and Graham, R. M.: The Vaginal Smear: Its Value to the Internist, with Report of Diagnosis of Unsuspected Uterine Cancer in Four Cases. J. A. M. A. *137:* 921, 1948.

Fremont-Smith, M., Graham, R. M., and Meigs, J. V.: The Cytologic Method in the Diagnosis of Cancer. New England J. Med. *238:* 179, 1948.

Fuller, D.: Two New Methods of Staining Vaginal Smears. J. Lab. & Clin. Med. *28:* 1474, 1943.

Gates, O., MacMillan, J. C., and Middleton, M.: The Vaginal Smear as a Means of Investigating Early Carcinoma of the Cervix. Cancer *2:* 838, 1949.

Gates, O., and Warren, S.: Vaginal Smear in Diagnosis of Carcinoma of Uterus. Am. J. Path. *21*: 567, 1945.

Glück, O.: Aspiration of Uterus as Competitive Method to Smear Curettage. Wien. med. Wchnschr. *99:* 173, 1949.

Graham, R. M.: The Effect of Radiation on Vaginal Cells in Cervical Carcinoma. Description of Cellular Changes. Surg., Gynec. & Obst. *84:* 153, 1947.

Graham, R. M., and McGraw, J.: An Investigation of "False Positive" Vaginal Smears. Surg., Gynec. & Obst., in press.

Graham, R. M., Sturgis, S. H., and McGraw, J.: A Comparison of the Accuracy in Diagnosis of the Vaginal Smear and the Biopsy in Carcinoma of the Cervix. Am. J. Obst. & Gynec. *55:* 303, 1948.

Grosjean, W. A.: Cancer Diagnosis by Smears. J. Kansas M. Soc. *48:* 441, 1947.

Guidoux, A.: Diagnostic Précoce du Cancer Utérin par l'Examen Cytologique des Sécrétions Cervicales. Presse Méd. *55:* 358, 1947.

Gusberg, S. B.: Detection of Early Carcinoma of the Cervix. The Coning Biopsy. Am. J. Obst. & Gynec. *57:* 752, 1949.

Herrera, R. C., and Mezzadra, J. M. E.: Valor Actuel del Diagnostico del Cáncer Genital por los Extendidos Vaginales. Prensa Méd. La Paz *35:* 771, 1948.

Hufford, C. E., and Burns, Edward L.: Papanicolaou Test in the Early Diagnosis of Uterine Cancer. Ohio M. J. *44:* 900, 1948.

Ikeda, K.: The Papanicolaou Method of Cancer Diagnosis: An Evaluation. Minnesota Med. *32:* 54, 1949.

Ikeda, K.: Detection of Early Cancer of the Cervix Uteri by the Papanicolaou Method. Minnesota Med. *32:* 488, 1949.

Isbell, N. P., et al: A Correlation Between Vaginal Smear and Tissue Diagnosis in 1,045 Operated Gynecologic Cases. Am. J. Obst. & Gynec., *54:* 576, 1947.

Jones, C. A., Neustardter, T., and Mackenzie, L. L.: Value of Vaginal Smears in Diagnosis of Early Malignancy; Preliminary Report. Am. J. Obst. & Gynec. *49:* 159, 1945.

Kernodle, J. R., Cuyler, W. K., and Thomas, W. L.: The Diagnosis of Genital Malignancy by Vaginal Smears. Am. J. Obst. & Gynec. *56:* 1083, 1948.

Kernodle, J. R., Cuyler, W. K., and Thomas, W. L.: The Use of Vaginal Smears in the Diagnosis of Genital Cancer. North Carolina M. J. *9:* 11, 1948.

Kernodle, J. R., et al: Report of a Fourteen-month Study on the Use of the Vaginal and Cervical Smears in Diagnosis of Genital Malignancy. North Carolina M. J. *9:* 335, 1948.

Kraushaar, O. F., Bradbury, J. T., and Brown, W. E.: The Vaginal Smear in Population Screening for Uterine Carcinoma. Am. J. Obst. & Gynec. *58:* 447, 1949.

Locke, F. R., and Caldwell, J. B.: The Early Diagnosis of Carcinoma of the Cervix with Emphasis on Routine Biopsy. Am. J. Obst. and Gynec. *58:* 1133, 1949.

Lombard, H. L., et al: The Use of the Vaginal Smear as a Screening Test. New England J. Med. *239:* 317, 1948.

MacKenzie, L. L., et al: The Use of the Vaginal Smear in a Gynecologic Service. Am. J. Obst. & Gynec. *55:* 821, 1948.

McClure, G. W.: A Review of the Vaginal Smear Method for Early Diagnosis of Cancer. West Virginia M. J. *43:* 66, 1947.

McClure, G. W., and Cattell, R. B.: Review of Vaginal Smear Method for Early Diagnosis of Cancer. Report of 170 Cases. S. Clin. North America *25:* 550, 1945.

McSweeney, D. J., and McKay, D. G.: Uterine Cancer: Its Early Detection by Simple Screening Methods. New England J. Med. *238:* 867, 1948.

Medina, J., and Silva, A. M.: O Diagnóstico Precoce da Gravidez pelo Exame da Citologia Vaginal. An. Brasil de Ginec. *18:* 342, 1943.

Meigs, J. V., et al.: Value of Vaginal Smear in Diagnosis of Uterine Cancer. Surg., Gynec. & Obst. *77:* 449, 1943.

Meigs, J. V., et al.: Value of Vaginal Smear in Diagnosis of Uterine Cancer: Report of 1015 Cases. Surg., Gynec. & Obst. *81:* 337, 1945.

Meigs, J. V.: The Vaginal Smear: Practical Applications in the Diagnosis of Cancer of the Uterus. J. A. M. A. *133:* 75, 1947.

Merkeley, D. K., and Penner, D. W.: Cytologic Diagnosis of Cancer: A Preliminary Report. Manitoba M. Rev. *28:* 365, 1948.

Nieburgs, H. E., and Pund, E. K.: Specific Malignant Cells Exfoliated from Preinvasive Cancer of the Cervix Uteri. Am. J. Obst. & Gynec. *58:* 532, 1949.

Oxorn, H.: Cervical Cytology: Key to Diagnosis of Early Uterine Cancer. Surg., Gynec. & Obst. *87:* 197, 1948.

Papanicolaou, G. N.: Proc. of the Third Race Betterment Conference, 1928, p. 528.

Papanicolaou, G. N.: A New Procedure for Staining Vaginal Smears. Science *95:* 438, 1942.

Papanicolaou, G. N.: A General Survey of the Vaginal Smear and Its Use in Research and Diagnosis. Am. J. Obst. & Gynec. *51:* 316, 1946.

Papanicolaou, G. N.: Cytologic Diagnosis of Uterine Cancer by Examination of Vaginal and Uterine Secretions. Am. J. Clin. Path. *19:* 301, 1949.

Papanicolaou, G. N., and Marchetti, A. A.: Use of Endocervical and Endometrial Smears in Diagnosis of Cancer and Other Conditions of the Uterus. Am. J. Obst. & Gynec. *46:* 421, 1943.

Papanicolaou, G. N., and Traut, H. F.: Diagnostic Value of Vaginal Smears in Carcinoma of Uterus. Am. J. Obst. & Gynec. *42:* 193, 1941.

Parrett, V., Small, C., and Winn, T.: Vaginal Cytologic Survey in Gynecologic Cancer. Am. J. Obst. & Gynec. *56:* 360, 1948.

Posey, L. C., and Cunningham, J. A.: Impressions of the Vaginal Smear Technic in the Diagnosis of Cervical Cancer. South. M. J. *41:* 221, 1948.

Pund, E. R., and Auerbach, S. H.: Preinvasive Carcinoma of Cervix Uteri. J. A. M. A. *131:* 960, 1946.

Pund, E. R., et al: Preinvasive Carcinoma of Cervix Uteri: Seven Cases in which It Was Detected by Examination of Routine Endocervical Smears. Arch. Path. *44:* 571, 1947.

Rector, Lee T.: Cervical Lesions: Diagnosis of Malignant Disease by Vaginal Smear. J. Florida M. A. *34:* 654, 1948.

Ribeiro, C. S., and da Costa, Gil, Jr.: Citologia vaginal no cancro do utero. Acta. Endocr. Gyn. *1:* 414, 1948.

Rubin, I. C.: Early Cervical Carcinoma in Three Clinically Unsuspected Cases Incidental to Plastic Operations. J. Mt. Sinai Hosp. *12:* 607, 1945.

Scheffey, L. C., and Rakoff, A. E.: The Cytology Test for Uterine Cancer. Phila. Med. *43:* 435, 1947.

Scheffey, L. C., Rakoff, A. E., and Hoffman, J.: An Evaluation of the Vaginal Smear Method for the Diagnosis of Uterine Cancer. Am. J. Obst. & Gynec. *55:* 453, 1948.

Schram, M., and Di Palma, S.: Vaginal and Cervical Smears in Uterine Malignancy. Am. J. Surg. *77:* 191, 1949.

Schtirbur, Issac: Colpocytology in the Diagnosis of Uterine Cancer, Obst. y ginec. latinoam *6:* 201, 1948.

Seibels, Robert E.: Cytology: A Diagnostic Method in Early Carcinoma of the Cervix. South. M. J. *41:* 706, 1948.

Skapier, Joseph: Diagnosis of Cancer by the Papanicolaou Smear Method. J. Am. M. Women's A. *3:* 139, 1948.

Skapier, Joseph: Evaluation of the Cytologic Test in the Early Diagnosis of Cancer. Am. J. Obst. & Gynec. *58:* 366, 1949.

Skapier, Joseph: Diagnosis of Preinvasive Carcinoma of the Cervix. Surg., Gynec., & Obst. *89:* 405, 1949.

Ulfelder, H.: The Use of the Vaginal Smear in the Diagnosis of Cancer. Connecticut M. J. *12:* 513, 1948.

Wilcoxon, G. M., and Falls, Frederick H.: An Experiment with Uterine Cervical Smears in the Diagnosis of Genital Malignancies. Ohio M. J. *44:* 165, 1948.

Yue, H. S., et al.: The Application of a Silver Carbonate Stain for the Diagnosis of Uterine Cancer by the Vaginal Smear Method. Am. J. Obst. & Gynec. *56:* 468, 1948.

VI. Pulmonary Carcinoma

Annotations: Sputum Examination. Lancet, No. 6543, No. IV of Vol. I, Vol. CCLVI, 1949, p. 154.

Appel, M., and Bronk, T. T.: Tumor Cells in Bronchial Secretions. Am. J. Clin. Path. *19:* 320, 1949.

Barrett, N. R.: Examination of the Sputum for Malignant Cells and Particles of Malignant Growth. J. Thoracic Surg. *8:* 169, 1938.

Bibliography

Bergmann, M., Shatz, B. A., and Flance, I. J.: Pulmonary Carcinoma and Tuberculosis. J. A. M. A. *138:* 798, 1948.

Diggs, L. W.: Use of Wright's Stain in Diagnosis of Malignant Cells in Bronchial Aspirations. Am. J. Clin. Path. *18:* 293, 1948.

Dudgeon, L. S.: On the Demonstration of Particles of Malignant Growth in Sputum by Means of the Wet-Film Method. St. Thomas's Hosp. Rep. *1:* 51, 1946.

Dudgeon, L. S., and Patrick, C. V.: New Method for Rapid Microscopic Diagnosis of Tumors with Account of 200 Cases so Examined. Brit. J. Surg. *15:* 250, 1927.

Dudgeon, L. S., and Wrigley, C. H.: On the Demonstration of Particles of Malignant Growth in Sputum by Means of the Wet-Film Method. J. Laryng. & Otol. *50:* 752, 1935.

Farber, S. M., Benioff, M. A., and McGrath, A. K., Jr.: Diagnosis of Bronchogenic Carcinoma by Cytologic Methods, Radiology *52:* 511, 1949.

Farber, Seymour M., Benioff, M. A., and Tobias, Gerd: Primary Carcinoma of the Lung. Diagnosis by Cytological Studies of Sputum and Bronchial Secretions. California Med. *69:* 95, 1948.

Farber, Seymour M., and Tobias, Gerd: Cancer Primario del Pulmon. Rev. Panamer. de Med. y cir del Jorax *1:* 82, 1947.

Farber, Seymour M., et al: Cytologic Studies of Sputum and Bronchial Secretions in Primary Carcinoma of the Lung. Dis. of the Chest *14:* 633, 1948.

Gibbon, John H., Jr., et al.: The Diagnosis and Operability of Bronchogenic Carcinoma. J. Thoracic Surg. *17:* 419, 1948.

Gowar, F. J. S.: Carcinoma of the Lung: The Value of Sputum Examination in Diagnosis. Brit. J. Surg. *30:* 193, 1943.

Gloyne, S. R.: The Cytology of the Sputum. Tubercle *18:* 292, 1936–37.

Herbut, P. A.: Cancer Cells in Bronchial Secretions. Am. J. Path. *23:* 867, 1947.

Herbut, P. A., and Clerf, L. H.: Diagnosis of Bronchogenic Carcinoma by Examination of Bronchial Secretions. Ann. Otol., Rhin. & Laryng. *55:* 646, 1946.

Herbut, P. A., and Clerf, T. H.: Cancer Cells in Bronchial Secretions. M. Clin. North America *30:* 1384, 1946.

Herbut, P. A., and Clerf, L. H.: Bronchogenic Carcinoma: Diagnosis by Cytological Study of Bronchoscopically Removed Secretions. J. A. M. A. *130:* 1006, 1946.

Herbut, P. A., and Clerf, L. H.: Cancer Cells in Bronchial Secretions. Tuberculology *9:* 90, 1947.

Liebow, A. A., Lindskog, G. E., and Bloomer, W. E.: Cytological Studies of Sputum and Bronchial Secretions in the Diagnosis of Cancer of the Lung. Cancer *1:* 223, 1948.

Matthews, W. H.: The Examination of Sputum for Tumor Cells: Canad. M. A. J. *58:* 236, 1948.

McKay, D. G., et al.: The Diagnosis of Bronchogenic Carcinoma by Smears of Bronchoscopic Aspiration. Cancer *1:* 208, 1948.

Papanicolaou, G. N., and Cromwell, H. A.: Diagnosis of Cancer of Lung by Cytologic Method, Dis. of Chest *15:* 412, 1949.

Richardson, H. L., et al: Cytohistologic Study of Bronchial Secretions. Am. J. Clin. Path. *19:* 323, 1949.

Scott, Thornton: Cytologic Studies of Sputum in Bronchogenic Carcinoma. J. Lab. and Clin. Med. *32:* 1543, 1947.

Shatz, B. A., Bergmann, M., and Gray, S. H.: Sputum Cell Study for Pulmonary Carcinoma as a Routine Laboratory Test. J. Lab. & Clin. Med. *33:* 1588, 1948.

Smathers, H. M.: Cytologic Study of Bronchial Secretions in Diagnosis of Bronchogenic Carcinoma, J. Michigan State M. Soc. *47:* 393, 1948.

Wandall, H. H.: A Study of Neoplastic Cells in Sputum as a Contribution to the Diagnosis of Primary Lung Cancer. Acta Chir. Scandinav. (suppl. 93) *91:* 1, 1944.

Watson, W. L., et al.: Cytology of Bronchial Secretions. J. Thoracic Surg. *18:* 113, 1949.

Woolner, L. B., and McDonald, J. R.: Bronchogenic Carcinoma: Diagnosis by Microscopic Examination of Sputum and Bronchial Secretions; Preliminary Report. Proc. Staff Meet., Mayo Clin. *22:* 369, 1947.

Woolner, L. B., and McDonald, J. R.: Diagnosis of Carcinoma of the Lung. J.A.M.A. *139:* 497, 1949.

Woolner, L. B., and McDonald, J. R.: Carcinoma Cells in Sputum and Bronchial Secretions. Surg., Gynec. & Obst. *88:* 273, 1949.

Woolner, L. B., and McDonald, J. R.: Cytologic Diagnosis of Bronchogenic Carcinoma. Am. J. Clin. Path. *19:* 765, 1949.

VII. Gastric Carcinoma

Block, Malcolm, et al.: The Cytology of the Gastric Juice in the Diagnosis of Gastric Carcinoma. Univ. Hosp. Bull., Ann Arbor, *14:* 37, 1948.
Campbell, J. P., et al: Experience with Papanicolaou Stains in the Study of Gastric Contents. Rev. Gastroenterol. *15:* 21, 1948.
Graham, R. M., Ulfelder, H., and Green, T. H.: The Cytologic Method as an Aid in the Diagnosis of Gastric Carcinoma. Surg., Gynec. and Obst. *86:* 257, 1948.
Hess, M., and Hollander, F.: Permanent Metachromatic Staining of Gastric Mucous Smears. J. Lab. and Clin. Med. *29:* 321, 1944.
Hollander, F., et al.: New Technique for Studying the Cytology of Gastric Aspirates in Man. J. Nat. Cancer Inst. *7:* 365, 1947.
Papanicolaou, G. N., and Cooper, W. A.: The Cytology of the Gastric Fluid in the Diagnosis of Carcinoma of the Stomach. J. Nat. Cancer Inst. *7:* 357, 1947.
Pollard, M. H., et al: Diagnosis of Gastric Neoplasms. J. A. M. A. *139:* 71, 1949.
Richardson, H. L., Queen, F. B., and Bishop, F. H.: The Cytohistologic Diagnosis of Cancer, Ulcer and Gastritis in Stomach Washings. Am. J. Clin. Path. *19:* 328, 1949.
Ulfelder, H., Graham, R. M., and Meigs, J. V.: Further Studies on the Cytologic Method in the Problem of Gastric Cancer. Ann. Surg. *128:* 422, 1948.

VIII. Carcinoma of Genito-Urinary Tract

Albers, D. D., McDonald, J. R., and Thompson, G. J.: Carcinoma cells in Prostatic Secretions. J. A. M. A. *139:* 299, 1949.
Chute, R., and Williams, D. W.: Experiences with Stained Smears of Cells Exfoliated in the Urine in the Diagnosis of Cancer in the Genito-Urinary Tract; A Preliminary Report. J. Urol. *59:* 604, 1948.
Daut, R. V., and McDonald, J. R.: Diagnosis of Malignant Lesions of the Urinary Tract by Means of Microscopic Examination of Centrifuged Urinary Sediment. Proc. Staff Meet., Mayo Clin. *22:* 382, 1947.
Fayers, C. M., and Jonsson, C.: Cancer Cells in Prostatic Secretions. Nord. Med. *40:* 2051, 1948.
Foote, N. C., and Papanicolaou, G. N.: Early Renal Carcinoma. J. A. M. A. *139:* 356, 1949.
Haschek, H., and Gutter, W.: Contribution to Early Diagnosis of Prostatic Carcinoma by Demonstration of Tumor Cells in Prostatic Secretion. Krebsarzt *3:* 401, 1948.
Herbut, P. A.: Cytologic Diagnosis of Carcinoma of the Prostate. Am. J. Clin. Path. *19:* 315, 1949.
Herbut, P. A., and Lubin, E. N.: Cancer Cells in Prostatic Secretions. J. Urol. *57:* 542, 1947.
Ludden, T. E., and McDonald, J. R.: Diagnosis of Tumors of the Kidney by Cytologic investigation of Urinary Sediment. Proc. Staff Meet., Mayo Clin. *22:* 386, 1947.
Papanicolaou, G. N.: Cytology of the Urine Sediment in Neoplasms of the Urinary Tract. J. Urol. *57:* 375, 1947.
Papanicolaou, G. N., and Marshall, V. F.: Urine Sediment Smears as a Diagnostic Procedure in Cancers of the Urinary Tract. Science *101:* 519, 1945.
Schmidlapp, C. J., and Marshall, V. F.: Detection of Cancer Cells in Urine: Clinical Appraisal of Papanicolaou Method. J. Urol. *59:* 599, 1948.

IX. Tumor Cells in Serous Fluids

Foote, N. C.: Identification of Tumor Cells in Sediments of Serous Effusions. Am. J. Path. *13:* 1, 1937.
Goldman, A.: Demonstration of Cancer Cells in Pleural Fluids. Dis. of the Chest *6:* 10, 1940.
Graham, G. S.: Cancer Cells of Serous Effusions. Am. J. Path. *9:* 701, 1933.
McDonald, J. R., and Broders, A. C.: Malignant Cells in Serous Effusion. Arch. Path. *27:* 53, 1939.
Merklin, Pr., Waitz, R., and Kabaker, J.: Sur La Cytologie des Épanchements Pleuraux. Presse Méd. *92:* 1828, 1933.
Phillips, S. K., and McDonald, J. R.: An Evaluation of Various Examinations Performed on Serous Fluids. Am. J. M. Sc. *216:* 121, 1948.
Quensel, Ulrik: Ergüsse der Brust und Baudhhöhlen. Acta Med. Scandinav *53:* 765, 1921.
Saphir, O.: Cytologic Diagnosis of Cancer from Pleural and Peritoneal Fluids. Am. J. Clin. Path. *19:* 309, 1949.

Sattenspiel, E.: Cytological Diagnosis of Cancer in Transudates and Exudates. A Comparison of the Papanicolaou Method and the Paraffin Block Technique. Surg., Gynec. & Obst. *89:* 478, 1949.

Schlesinger, Monroe J.: Carcinoma Cells in Thoracic and in Abdominal Fluids. Arch. Path. *28:* 283, 1939.

Zemansky, A. Ph.: The Examination of Fluids for Tumor Cells. Am. J. M. Sc. *175:* 489, 1928.

X. Tumor Cells from Intestinal Tract

Lemon, H. M., and Byrnes, W. W.: Cancer of the Biliary Tract and Pancreas. Diagnosis from Cytology of Duodenal Aspirations. J.A.M.A. *141:* 254, 1949.

Wisseman, C. L., Lemon, H. M., and Lawrence, K. B.: Cytologic Diagnosis of Cancer of the Descending Colon and Rectum. Surg., Gynec. & Obst. *89:* 24, 1949.

INDEX